T0389481

The Uses of Humans in Experiment

Clio Medica

PERSPECTIVES IN MEDICAL HUMANITIES

VOLUME 95

The titles published in this series are listed at *brill.com/clio*

The Uses of Humans in Experiment

Perspectives from the 17th to the 20th Century

Edited by

Erika Dyck
Larry Stewart

BRILL
RODOPI

LEIDEN | BOSTON

Cover illustration: Monitoring electric therapy with Timothy Lane's electrometer. Frontispiece of George Adams's *An Essay on Electricity, Explaining the Theory and Practice of that Useful Science; and the Mode of Applying it to Medical Purposes*, London, 1799.

Library of Congress Cataloging-in-Publication Data

Names: Dyck, Erika. | Stewart, Larry, 1946-
Title: The uses of humans in experiment : perspectives from the 17th to the 20th century / edited by Erika Dyck, Larry Stewart.
Description: Leiden : Brill, 2016. | Series: Clio medica : perspectives in medical humanities, ISSN 0045-7183 ; volume 95 | Includes bibliographical references and index. | Description based on print version record and CIP data provided by publisher; resource not viewed.
Identifiers: LCCN 2015050156 (print) | LCCN 2015049532 (ebook) | ISBN 9789004286719 (E-book) | ISBN 9789004286702 (hardback : acid-free paper) | ISBN 9789004286719 (ebook)
Subjects: LCSH: Human experimentation in medicine--History. | Clinical trials--History. | Human beings--Research--History. | Human experimentation in medicine--Moral and ethical aspects--History. | Clinical trials--Moral and ethical aspects--History. | Human beings--Research--Moral and ethical aspects--History.
Classification: LCC R853.H8 (print) | LCC R853.H8 U85 2016 (ebook) | DDC 174.2/8--dc23
LC record available at http://lccn.loc.gov/2015050156

Typeface for the Latin, Greek, and Cyrillic scripts: "Brill". See and download: brill.com/brill-typeface.

ISSN 0045-7183
ISBN 978-90-04-28670-2 (hardback)
ISBN 978-90-04-28671-9 (e-book)

This book is printed on acid-free paper and produced in a sustainable manner.

Contents

Acknowledgments

We are grateful for the opportunity to work with such talented collaborators whose work is on display in this text. What began in 2008 as a workshop organized around the theme of humans in experiments, evolved into a second consideration in 2009. The fruitful discussions and overlapping themes from those two events brought together early-modern historians of science with modern historians of medicine and prompted us to consider a deeper investigation into the longer history of humans in experiments that crossed the often superficial divide between history of science and social studies of medicine. We therefore thank all the original participants from those early workshops, presenters and participants alike, and give special appreciation to Marc Macdonald and Timothy Nyborg, and the Diefenbaker Centre, which hosted our meetings. We also appreciate the encouragement of the Jo-Anne Dillon, Dean, College of Arts & Science at the University of Saskatchewan. As we began assembling the manuscript, we brought in a number of other authors, and also conscripted the help of talented graduate students who helped to chase up references and wrestle our footnotes into shape; thanks to Marc Macdonald, Adam Montgomery, and Fedir Razumenko for their expert assistance.

Funding for the meetings that inspired this collection formed part of the larger consortium of research that has and continues to stem from the Social Sciences and Humanities Council of Canada's, "Situating Science" network. Led by Gordon McOuat at King's College in Halifax, Nova Scotia, this network of humanist and social science scholars have been probing the intersections of science and society and generating critical scholarship that explores those intersections, and even their policy implications, throughout various places and times. We are proud to introduce this volume as one of the products of this collaboration. Katherine Zwicker, one of our authors here, was also an integral player as a post-doctoral fellow funded through Situating Science and located at the University of Saskatchewan. Finally, we were pleased to work with Michiel Thijssen and the team at Brill/Clio Medica who shepherded this project through to its completion.

List of Figures and Tables

Figures

Tables

Notes on Contributors

Paola Bertucci

is Associate Professor of History at Yale University. She is the author of *Viaggio nel paese delle meraviglie: Scienza e curiosità nell'Italia del Settecento* (2007) and co-editor of *Electric Bodies: Episodes in the history of medical electricity* (2001). She is completing a book on artisanal culture and the sciences in Enlightenment France.

Erika Dyck

is a Professor in the Department of History and a Canada Research Chair in Medical History at the University of Saskatchewan. She is the author of *Psychedelic Psychiatry: LSD from Clinic to Campus* (Johns Hopkins, 2008, which was republished in 2011 by University of Manitoba Press); and *Facing Eugenics: Reproduction, Sterilization and the Politics of Choice* (University of Toronto, 2013), which was shortlisted for the Governor General's award for Canadian non-fiction. Her work focuses on the experimentation, eugenics, psychiatric institutionalization, and the social history of medicine in the 20th century.

Anita Guerrini

is Horning Professor in the Humanities and Professor of History at Oregon State University, where she teaches the history of science and medicine. She has written on the history of animals, medicine, food, and the environment. Her publications include *Experimenting with Humans and Animals: from Galen to Animal Rights* (John Hopkins, 2003) and most recently *The Courtiers' Anatomists: Animals and Humans in Louis XIV's Paris* (Chicago, 2015). Current research projects concern skeletons as scientific and historical objects, early modern urban animals, and the role of history in present-day ecological restoration. She blogs at http://anitaguerrini.com/anatomia-animalia/.

Rob Iliffe

is Professor of Intellectual History and the History of Science in the Department of History at the University of Sussex. He has published a number of articles on early modern history and the history of science, and has written the *Very Short Introduction to Newton* (Oxford University Press 2007). He has edited the *Eighteenth Century Biographies of Newton* (Pickering 2006). He is Editorial director of the online Newton Project, director of the AHRC Newton Theological Papers Project and is also editor of the journal *History of Science*. Prof. Iliffe's main research interests include: the history of science 1550–1800; the role of science and technology in the 'Rise of the West'; techno-scientific and other roots of the current environmental crisis; historical interactions between science and religion; the theological and scientific work of Isaac Newton; and the implications for academic work posed by the increasing digitisation of the scholarly infrastructure.

Paul A. Lombardo
is the Bobby Lee Cook Professor of Law at Georgia State University College of Law. His books include *Three Generations, No Imbeciles: Eugenics, the Supreme Court and Buck v. Bell* (2008) and *A Century of Eugenics in America: From the Indiana Experiment to the Human Genome Era* (ed., 2010). He is a Senior Advisor to the Presidential Commission for the Study of Bioethical Issues, in Washington, D.C., where he recently participated in the investigation of the World War II era Public Health Service STD studies in Guatemala. Professor Lombardo has published extensively on topics in health law, bioethics, and medico-legal history, particularly the history of eugenics. He received both his Ph.D. and J.D. from the University of Virginia, and served on the faculties of the Schools of Law and Medicine there from 1990 until 2006.

Elizabeth Neswald
is Associate Professor in the Department of History at Brock University. Before coming to Brock in 2006, she spent time at the University of Aberdeen, the National University of Ireland at Galway, and the Humboldt University in Berlin. She is the author of *Thermodynamik als kultureller Kampfplatz: Eine Faszinationsgeschichte der Entropie, 1850–1915* (Freiburg im Breisgau: Rombach, 2006) and *Medien-Philosophie. Das Werk Vilém Flussers* (Cologne: Böhlau, 1998). Her current work focuses on the history of metabolism research, nutritional physiology and dietary standards in the nineteenth and early twentieth centuries, nineteenth-century thermodynamics and scientific instruments and the material cultures of experiment.

Joan Steigerwald
is Associate Professor in the Departments of Humanities and Science and Technology Studies, and in the Graduate Programs in STS, Humanities, and Social and Political Thought. She is currently completing a book *Organic Vitality in Germany around 1800: Experimenting at the Boundaries of Life*. She is also editing a volume examining entanglements of instruments and media in investigating organic worlds. Her research interests lie in the history of the life sciences, the German idealism and romanticism, figural representations of nature, and the epistemology of experiment.

Larry Stewart
is Professor of History at the University of Saskatchewan, Canada. He is currently writing a study of experiment during the Enlightenment and first industrial revolution. He has recently edited, with Bernie Lightman and Gordon McOuat, *Circulating Knowledge, East and West* (Brill, 2013), and is currently completing, with Trevor Levere and Hugh Torrens, *The Democratic Vision of Thomas Beddoes; Science, Medicine and Reform.*

Paul Weindling
is Professor of History whose research interests cover the history of eugenics, international health organizations, and Nazi coerced experimentation. His publications include *Health, Race and German Politics between National Unification and Nazism* (1989), *Epidemics and Genocide in Eastern Europe 1890–1945* (2000), *Nazi Medicine and the Nuremberg Trials: From Medical War Crimes to Informed Consent* (2004), *John W. Thompson, Psychiatrist in the Shadow of the Holocaust* (2010), and *Victims and Survivors of Nazi Human Experiments: Science and Suffering in the Holocaust* (2014).

Katherine Zwicker
is the Student Engagement Coordinator for the Faculty of Agricultural, Life & Environmental Sciences, University of Alberta. Formerly a SSHRC Situating Science Postdoctoral Fellow at the University of Saskatchewan, her research in the history of science and medicine focuses on the study and use of radiation in biomedical research, medical practice, and academic discipline-building in the 20th-century United States. She has published on the health consequences of the Manhattan Project.

Introduction

Humans as Subjects and Objects

It is a commonplace that humans have long been at the centre of ethical and practical issues entangled with the history of science and medicine. This is particularly so with humans as experimental subjects. In the 20th century this relationship ultimately gave rise to legal doctrines like a problematic informed consent as part of the broader evolution of human rights and medical therapies. Both rights and remedies have since been confined by statute in strictly narrow legal terms. In our view this also claims the late 20th century as an age of progress in which the status of participants was protected, and even defined, by those with the power and authority over the legitimacy of human experimentation. An historical treatment may challenge this privilege. Examining the rise of the human sciences, more than a generation ago, Michel Foucault's *The Order of Things* posited two great 'discontinuities' in the *episteme* of Western culture.[1] The first, beginning halfway through the 17th century, was characterized as one of order and classification in which the previous search for an ancient and divine discourse in nature was abandoned in favour of a system that gave primacy to tabulation and calculation.[2] To know nature now meant to observe it across a vast (artificial) table encompassing all of its myriad similarities and differences. It also, as Lorraine Daston more recently argued, established the moral authority of the natural world "within a common framework of utility."[3] Even so, to speak of the use of knowledge was not always sufficient reason for the inclusion of man in experiment. We intend to explore the territory of observation – by the experimental operator and by the testimony of the subject which was, immediately, suspect.

Curiously, the early-modern quest for a pure system of classification by which to order nature had actually excluded 'Man' as an object of positive knowledge – partly, perhaps, for moral reasons. If we follow Foucault, the idea of Man in a modern scientific sense did not yet exist. It was only with the second great rupture, beginning in the 18th century, that Man and 'Life,'

1 Michel Foucault, *The Order of Things: An archaeology of the human sciences* (New York: Routledge, 1970).

2 Ibid., 69.

3 Lorraine Daston, "Attention and the Values of Nature in the Enlightenment," in Lorraine Daston and Fernando Vidal, *The Moral Authority of Nature* (Chicago and London: University of Chicago Press, 2004), 100–126, esp. 123, and 9.

 | DOI 10.1163/9789004286719_002

understood not just simply as sentient but as things with organic structure, also gave rise to specialized scientific disciplines of transformation in biology and comparative anatomy. By the 19th century, it was not solely a creature's visible comparative anatomy that dictated its place in nature, but instead the viable, reproductive, dynamic of its organic, reproductive, function. The immense consequence of this shift was that Man was simultaneously inserted into a discourse of nature, while nonetheless remaining arbiter of the very knowledge upon which that very discourse was based. Foucault concluded by warning against any positivist notions that the modern sciences allowed us to escape from the questions raised throughout the early modern and modern periods, stating instead that the *episteme* that began in the 19th century "still forms the immediate space of our reflection."[4]

Foucault's work had challenged scholars to confront both the continuities *and* ruptures intertwined with the rise of scientific knowledge in Western culture. But there is a another question Foucault sought to address in various works – as in *The Birth of the Clinic*, *Madness and Civilization* and in *Discipline and Punish* – which placed the body under an institutional gaze or view. This leads us to the question of the relationship between the experience of investigation and control. In response to a question in 1975, Foucault asserted that, in the everyday, "knowledge constantly induces effects of power."[5] In its incisive challenge of a Whig view of the history of science and the progress of knowledge, *The Order of Things* had highlighted the questions raised in the 17th century and still extant regarding humanity's ambiguous place as both object and subject of scientific and medical knowledge. This commanding yet tenuous position is even further complicated when set against the historically fluid sensitivities of religious, medical, gendered, and racial boundaries.

The current collection may affirm Foucault's chronological and epistemological framework as an important meta-narrative that can nonetheless be complemented by a closer, more nuanced, gaze. Several of these essays make clear that there existed an ethical consciousness in the early-modern world – variously scientific, religious, or metaphysical – that often determined the parameters of experimentation, and refined the acceptance of, or resistance to,

4 Ibid., 419.

5 Foucault, *Power/Knowledge: Selected Interviews & Other Writings, 1972–1977*, ed. Colin Gordon (New York: Pantheon, 1980), 52; Also see, *The Birth of the Clinic: An Archaeology of Medical Perception*, trans. A.M. Sheridan (London: Tavistock, 1976); *Madness and Civilization: A History of Insanity in the Age of Reason*, trans. Richard Howard (New York: Vintage Books, 1973); and *Discipline and Punish: the birth of the prison*, trans. A.M. Sheridan (New York: Vintage Books, 1977).

subjects. Yet, this also implies another issue which these essays illuminate – that is, that the spaces of knowledge created variants and shades of ethical consciousness. These set the parameters to observation generally and to experiment particularly.[6] This point has been made even more complicated when the object of scientific knowledge has also been the subject, as in the case of self-experimentation.[7] Thus, to the modern reader, the sometimes humane, but oft times seemingly callous, manner in which experiment was conducted throughout the early-modern period attests to the existence of fluid discourses concerning 'Man' and of observation. These early moments made 'Man' a topic of discussion within the moral economy of experiment.

Moreover, as the later chapters show, the rise of modern science and ethics after 1800 did not necessarily make answering such questions easier. Nor did this prevent the elision of fluid ethical boundaries, particularly in times of political crisis or war. Perhaps the most enduring element about the late 18th century, and one that several of the collection's essays speak to, is that far from closing the book on questions of the status of the human, the modern scientific *episteme* instead gave a profound instability to defining when a human being was *object* of scientific knowledge and when it was merely *subject* or *subjected to*.

These essays begin with the necessity of human engagement alongside the methods of experiment, and beyond social status, since at least the 17th century. Thus, the roles of the experimenter and of laboratory technicians, the place of the subject, and even the human as a sensuous instrument in its own right ought to be considered much more fully than a field otherwise dominated by our current ethical landscape. For example, debates during the English Restoration had invariably drawn attention to a series of dilemmas. These collapsed the gentlemanly exclusivity to the philosophical investigation of nature and potentially undermined the credibility of the so-called genteel witnesses to experimental effects.[8] Early-modern investigators were influenced, perhaps unduly, by concerns over "discreet and skillful [sic] men, taking care, not to experiment...upon Subjects." They debated whether the possibilities of a cure for a wide, inchoate, variety of medical ailments were sufficient

6 Cf. Daston, *Histories of Scientific Observation*, 82–83.

7 See Stuart Strickland, "The Ideology of Self-Knowledge and the Practice of Self-Experimentation," *Eighteenth-Century Studies* 31 (Summer, 1998), 453–471.

8 See, for example, Steven Shapin, "The House of Experiment in Seventeenth-Century England," reprinted in Shapin, *Never Pure. Historical Studies of Science as if it was Produced by People with Bodies, Situated in Time, Space, Culture, and Society, and Struggling for Credibility and Authority* (Baltimore: Johns Hopkins University Press, 2010), 59–88.

to overwhelm any such sensibilities.[9] There were profound issues at the heart of the human in the early-modern world. Ethical concerns were inevitable. Moreover, it was not the case that the human as an instrument was limited to a quest for cures. Indeed, humans as instruments have long played a role in the testing of chemical or physical natural phenomena. Experiment and therapy often coincided even when avoiding explicit moral or medical matters, and some of the following chapters explore the continuation of this trend into the 20th century.

The convergence of trial and therapy retained some of the logic of the early-modern period when assigning epistemological authority to the experiment, to the experimenter, and to the experimented upon. The definitions and ethical statutes, as Susan Lederer has argued, merely began to codify a longer set of traditions, though not without considerable debate.[10] Amid the pharmacological turn in 20th-century medicine, for instance, and prior to the rigorous enforcement of the randomized control gold standard, researchers enlisted in chemical experimentation continued to debate the nature of authority in relation to informed consent and credible knowledge accumulation. Recent scholars have provided examples of addictions researchers, pharmacologists and psychiatrists respectively who recognized the contested nature of authority when describing drug reactions.[11] Historians in these cases have recently debated the value of an 'objectivity'.[12] Thus, there arises, the academic or impersonal appreciation for a drug's effects, suggesting instead that a more appropriate ethical stance should involve personal experience with a drug's 'objective', measurable, influence or 'affect' measured instrumentally. Hence, in our modern experimental regime, methods and morality may collide. Some scholars have illustrated how these issues played out in terms of soliciting an informed-by-experience kind of consent among incarcerated addicts, while

9 "An Account of More Tryals of Transfusion, Accompanied with Some Considerations Thereon, Chiefly in Reference to Its Circumspect Practise on Man; Together with a Farther Vindication of This Invention from Usurpers," *Philosophical Transactions* 28 (October 21, 1667): 523.

10 Susan Lederer, *Subjected to Science: Human Experimentation in America Before the Second World War* (Baltimore: Johns Hopkins University Press, 1995), p. 73.

11 See for example, Nancy Campbell, *Discovering Addiction: The Science and Politics of Substance Abuse Research* (Ann Arbor: University of Michigan Press, 2007); David Healy, *Mania: A Short History of Bipolar Disorder* (Baltimore: Johns Hopkins University Press, 2010) and Erika Dyck, *Psychedelic Psychiatry: LSD from Clinic to Campus* (Baltimore: Johns Hopkins University Press, 2008).

12 See Lorraine Daston and Peter Galison, *Objectivity* (New York: Zone Books, 2007), esp. 198ff.

others have suggested that questions of subjectivity and objectivity fundamentally framed the experimental enterprise – even prior to clinical engagement.[13] These debates belong instead to a long arc of ethics and experimental cultures, for which historical context helps to explain the convergence of scientific, medical, religious and moral perspectives.

Considerations of an Early-Modern Problem: Bodies and Sites

Experimentation has long been highly contested. In the early-modern world, in an age of emerging scientific societies and new journals, amid a growing philosophical community, lay the territory between commonplace observations, many tests and trials and, literally sometimes, the agonies of experiment. In the late 17th century, among living beings, from the howls of dogs to the anesthetized courage of the human, lay a great uncertainty that raised the spectre of the moral.[14] While soon matters of public debate, trials upon human subjects could nonetheless remain overwhelmingly anonymous and often solitary of which little was ever heard. Even the famous, and apparently crucial, experiment by Isaac Newton on the refraction of rays of light into the spectrum in his college rooms at Trinity, Cambridge, proved difficult to replicate for half a century. His brief descriptions and the lack of any witnesses produced more doubt than accolades.[15] His further claims to have induced colours by putting pressure on his eye by inserting a bodkin between the eye ball and the socket did not appear to have had many takers.[16] But, not unlike much of the century before him, Newton linked the "eye of the body" with the "eye of the mind."[17]

---

13 Healy, *Mania*; and Dyck, *Psychedelic Psychiatry*. For a case study of this debate, see Nicolas Langlitz, *Neuropsychedelia: The Revival of Hallucinogen Research Since the Decade of the Brain* (Berkeley: University of California Press, 2012).

14 Anita Guerrini, "The Ethics of Animal Experimentation in Seventeenth-Century England," *Journal of the History of Ideas* 50 (1989): 391–407, esp.401; cf. Lorraine Daston, "The Empire of Observation, 1600–1800," in Lorraine Daston and Elizabeth Lanbeck, eds., *Histories of Scientific Observation* (Chicago and London: University of Chicago Press, 2011), 82.

15 Simon Schaffer, "Glass Works: Newton's Prisms and the Uses of Experiment," in David Gooding, Trevor Pinch, and Simon Schaffer, eds., *The Uses of Experiment: Studies in the Natural Sciences* (Cambridge and London: Cambridge University Press, 1989), 67–104.

16 Rob Iliffe, "'That puzzleing Problem': Isaac Newton and the Political Physiology of Self," *Medical History* 39 (1995): 433–458, esp. 444.

17 Daston, "The Empire of Observation," 91.

It is notable that instruments like telescopes and microscopes, as those employed by Newton's rival Robert Hooke, evaded the limits of human senses. Yet, most importantly, the body was obviously an instrument in its own right. Self-experiment was, virtually by definition, a solitary vice. Where witnesses were sparse, doubts were plentiful. But even the mathematical Newton was an ardent proponent of the experimental arts advanced by Galileo and his disciples. It is arguable, since the late 17th century, that even when his mathematics remained obscure, experiments secured Newton's reputation.

It is the purpose of this book to explore a series of discrete episodes and particular spaces where humans were engaged in experimental practise. Humans have variously been operators, or assistants and sometimes, as with Hooke, they could be both. They were just as often volunteers as they were unwitting subjects – or sometimes even willingly seduced by payment. Perhaps it was in what Michel Foucault once called the "space of knowledge", or at least those instruments within it, that made such an immense difference to crediblity. Indeed, as Peter Galison has since shown, the "trading zones" of experimental knowledge made a significant difference to how science has been perceived. Thus, radical empiricism reduced nature to observation, but instruments like those devised by Robert Boyle and Robert Hooke made further experiments possible. Like numerous devices, a variety of sites, and encounters of subjects and of experimenters made possible the credibility of scientific knowledge.[18] This is a critical point worth considering. And one of those instruments was the body.

The human role in experimental practise has long been problematic, conflicted over reliable knowledge, and ultimately confused by the concept of consent, especially when often merged with medical promise. Obviously, historical studies of such roles might prove endless. But we need to encourage a closer look. More than a century after Newton undertook his self-experiments, in 1790 the radical democratic doctor Thomas Beddoes was on a mission. He engaged in a bizarre, and many would surely now say offensive, attempt to assess the biological impact of the new chemistry. Beddoes undertook some experiments on what he described as a "distressed negro, to try to whiten part

---

18 Ofer Gal and Raz Chen-Morris, "Empricism Without the Senses: How the Instrument Replaced the Eye," in Charles T. Wolfe and Ofer Gal, eds.,*The Body as Object and Instrument of Knowledge. Embodied Empiricism in Early Modern Science* (London and New York: Springer, 2010), 143–144; Michel Foucault, *The Order of Things. An Archaeology of the Human Science* (London: Tavistock, 1970), xxii; Peter Galison, *Image and Logic. A Material Culture of Microphysics* (Chicago and London: University of Chicago Press, 1997), esp. 798ff.

of his skin with oxygenated marine acid air."[19] Here, within a narrow compass, were the racial, the moral and the sensational which continue to transfix the modern observer. For Beddoes, these were then barely worthy of debate. Curiosity trumped the squeamish, although his victim finally withdrew. Beddoes was the heir of a long period where, at least since the 17th century, experiment had become king. In the 18th century, Ben Franklin's kite in a thunderstorm was but one legend of the dramatic risk.[20] But as with many of Beddoes' contemporaries, the experiment was sufficient justification.

The body became thereby an instrument as much as a microscope or an air pump. Yet, devices and principles might collide. Indeed, even the ideology of the vacuum was highly contested.[21] It was, arguably, its impact on the animal and on the human body that was most telling. There were many such cases, some sensational, some secret, and some that simply remain unnoticed. Yet, as the following essays reveal, such moments can tell us a great deal about how humans have come to play a role in laboratory science even when the intent had little, and occasionally nothing, to do with an overriding search for medical improvement.[22] Nevertheless, the utilitarian and the medical could frequently be conjured up to provide an adequate justification for experimentation and also, not incidentally, to raise questions that were profoundly moral. While there was often anguish over the uses of animals, there was also rationalization by human utility. Take, for example, Newton's contemporary the Reverend Stephen Hales, whose work on airs led to much of the chemical innovation in the 18th century. More directly, in the preface to his famous and distressing

19 [Beddoes], *Considerations on the Medicinal Use, and on the Production of Factitious Airs. Part I. By Thomas Beddoes, M.D. Part II. By James Watt, Engineer. Edition the Third. Corrected and Enlarged* (Bristol: Printed by Bulgin and Rosser; London: For J. Johnson, in St. Paul's Church-Yard, 1796), 45. According to his first biographer, Beddoes made the effort several times. John Edmonds Stock, *Memoirs of Thomas Beddoes, M.D.* (London and Bristol, 1811; reprint Bristol, 2003), 66–67.

20 Cf. Christian Licoppe, "A French Mid-Eighteenth Century Crisis in Experimental Natural Philosophy: Nollet's Electrical Shows vs. the Devious Ways of Franklin's Electrical Atmospheres," in Lorraine Daston and Gianna Pomata, eds. *The Faces of Nature in Enlightenment Europe* (Berlin: Berliner Wissenschafts-Verglag, 2003), 123–135.

21 Steven Shapin and Simon Schaffer, *Leviathan and the Air-Pump. Hobbes, Boyle, and the Experimental Life* (Princeton: Princeton University Press, 1985).

22 See, for example, Gregoire Chamayou, *Les Corps Vils: Experimenter sure les etres humains aux XVIIIe et XIXe siecles* (Paris: La Decouverte, 2008); Jordan Goodman, Anthony McElligott and Lara Marks, eds. *Useful Bodies: Humans in the Service of Medical Science in the Twentieth Century* (Baltimore and London: Johns Hopkins University Press, 2003); cf. Daston, "Empire of Obseravation," 85.

Haemastaticks in 1733, Hales confronted justification, asserting experiments might be, "of great Importance to the Welfare of Mankind." He admitted, even for those who approved of experiments on animals, that the "disagreeableness of the Work did long discourage me from engaging in it..." But, he also asserted the benefit: "...these Experiments do obviously and clearly give an Account of the *Phaenomena*, they may possibly be of Service in the Hands of those who are well skilled in the animal Oeconomy and the History of Diseases..."[23] The argument over the use of animals was more than one of utility, but one that was deeply tied to the moral consequence of the uses of humans. As Sydney Halpern has recently argued, "moral traditions have...powerfully influenced the conduct of clinical research for well over a century."[24] As these essays suggest, a great deal longer than that.

At the very least, the lines between medical trials and philosophical experiments have long been profoundly blurred. Moreover, one serious difficulty that agitates us still relates to the emergence of proper protocols, particularly with the meaning of consent whenever it did arise, and to any attendant legal sanction. Medical ethics have become central to the discourse of our culture and to the definition of the permissible. Yet, moral imperatives, however important, also muddy the historical perspective. As these essays reveal, there was no precise moment of moral discovery, no clear or determined march toward ethical imperatives in the practice of experiment. This is not relativism but contextualism, which influenced evolving definitions of observation, trial, experiment and the ethical. It is the purpose here to open further discussion of the concepts which, over the centuries since the Enlightenment, broadened significantly into the realm of human rights and legal constraints. We have since operated within regimes defined by various authorities who have both policed the boundaries and arguably obstructed the practice of human experiment. Our modern sense of the roles of humans in experiment is thus clearly the result of a long and complex trajectory.

The Early-Modern Subject

The establishment of scientific societies in the 17th century provided a venue for identifying the effects on humans of increasing encounters with the natural,

---

23 Stephen Hales, *Statical Essays: Containing Haemastaticks* (London, 1733) ed. Andre Cournand (New York and London, 1964), xvii–xviii.

24 Sydney Halpern, *Lesser Harms. The Morality of Risk in Medical Research* (Chicago and London, 2004), esp. 2, 18, 31, 39.

philosophical and technical world. The publication of queries in the early *Philosophical Transactions of the Royal Society of London* directed attention to the human consequences of natural knowledge. It is interesting that the very first volume of the *Philosophical Transactions* in 1665 contained one letter reflecting on workers on mirrors who were very badly affected by attacks of palsy. This was only the first of many such observations what would lead to alarm over the occupational health of the many in early-modern industries.[25] By the 17th century, medicine and the compass of scientific observation were already tightly entwined. Indeed, it is perhaps the case that method itself induced moral issues.[26]

It would come as no surprise, immediately after the discovery of the circulation of the blood by Galileo's contemporary William Harvey, that efforts to determine the nature of blood would be the next step in inquiry. Here too the body and the microscope simultaneously functioned as instruments of discovery. But what was the character of the blood, once it was clear that circulation was in order? This, in other words, was not solely an anatomical, physiological, or even a strictly medical question – but one central to the doctrines of natural philosophy best tested experimentally.[27] Experiments on blood were among the first, and certainly most controversial, efforts attempted by fellows of the Royal Society immediately after its founding. While the physician Richard Lower made several trials on the transfusion of blood between animals, it was to the French and particularly to Jean-Baptiste Denys that the honour or, in some minds, the scandal of the first transfusions into humans must perhaps be attached. Interestingly, Denys attempted to justify his ventures on the possibility of medical benefit. And precisely because of this he soon ran afoul of the Parisian Faculty of Medicine, which was deeply opposed to the effort.[28] Dispute with orthodoxy is a recurrent theme.

---

25 See "Extract of a Letter, Written from Venice by the learned Doctor Walter Pope, to the Reverend Dean of Rippon, Doctor John Wilkins Concerning the Mines of Mercury in Fruili,"*Philosophical Transactions* 1 (April, 1665), 21–26, esp. 24.

26 Halpern, *Lesser Harms*, 36ff.

27 See Anita Guerrini, "The Ethics of Animal Experimentation in Seventeenth-Century England," *Journal of the History of Ideas* 50 (July-September, 1989), 391–407; Guerrini, *Experimenting With Humans and Animals: From Galen to Animal Rights* (Baltimore and London: Johns Hopkins University Press, 2003). See also Domenico Bertolini Meli, "The Colour of Blood: Between Sensory Experience and Epistemic Significance," in Daston and Lunbeck, eds., *Histories of Scientific Observation*, 117–134.

28 This particular controversy has surely attracted a great deal of attention of late, especially in Simon Schaffer, "Regeneration: The Body of Natural Philosophers in Restoration England," in Christopher Lawrence and Steven Shapin, eds. *Science Incarnate: Historical*

In an age of imperial conflict, notably between the French and the English, even the efforts to transfuse blood were part of the grand competition. Christopher Wren, soon to be an architect to a charred London following the Great Fire, had been one of those who attempted trials of injecting various fluids into the veins of animals. This led, among the early fellows of the Royal Society, to experiments in transfusing blood from one animal to another during 1665, principally by Richard Lower when, along with the illustrious Robert Boyle, he fled plague to reside in relative Oxford calm.[29] The transfusion trials were then part of the general programme of experiment the Royal Society intended to promote. The limit to these attempts was only that of imagining. But in France it was quickly suggested that the intermingling of the blood of different species of animals might act as "Pyoson and Venome in respect of one another."[30] This conflict of incompatibilities was not determined until the 19th century, but it was reasonably conjectured in the 17th that such a possibility could arise.

Injudicious tampering might produce unwanted consequence. This was especially compelling as such experiments quickly spread throughout Germany, Italy, Holland, England, France and even Sweden – and soon came to involve transfusions into humans.[31] This was not even an early priority dispute. Indeed, repeated moral aversions turned that debate on its head. Further trials, apparently, the English would not countenance being "so tender in hazarding the Life of Man (which they take so much pain, to preserve and relieve) nor so scrupulous to incur the Penalties of Law, which in England, is more strict and nice in cases of this concernment, than those of many other Nations are."[32] This was English one-upsmanship at its best. Even so, any number of difficulties arose which suggest that the business of human experiment was a concern

---

Embodiments of Natural Knowledge (Chicago and London: University of Chicago Press, 1998), 83–120, esp. 89–105; Lisa Jardine, *Ingenious Pursuits: Building the Scientific Revolution* (New York and London: Doubleday, 1999), esp. 118–124; Pete Moore, *Blood and Justice: The seventeenth-century Parisian doctor who made blood transfusion history* (London: John Wiley, 2003); and Holly Tucker, *Blood Work: A Tale of Medicine and Murder in the Scientific Revolution* (New York: Norton, 2011).

29 A.D. Farr, "The First Human Blood Transfusion," *Medical History* 24 (April, 1980): 143–162, 143; Pete Moore, *Blood and Justice*, 51.

30 Farr, "The First Human Blood Transfusion," 150–152.

31 "A Letter Written by an Intelligent and Worthy English Man from Paris, to a Considerable Member of the R. Society of London, Concerning Some Transactions there, Relating to the Experiment of the Transfusion of Blood," *Philosophical Transactions*, no. 54 (December 13, 1669), 1076; "An Account of more Tryals," 519–520.

32 Ibid., 521–522.

well before the 19th century. We have clearly underestimated the distinct awareness of early researchers. Of course, one might argue that these were not really the kind of elevated ethical concerns we now assert. But this view would be very much mistaken. A distinct moral divide between man and animals seemed transgressed. For some, there were obvious theological as well as ethical issues to consider.

A general effect on health was broadly refracted throughout the entire transfusion debate, but this was far from its primary purpose. This should not be surprising. The difficulties of transfusion, of the use of vein or artery, crural or carotid, in dogs, lambs and a calf had produced enough anguish and death of the subjects that it was not immediately obvious that the potential improvement for humans was worth the risk – even in the absence of sophisticated protocols and legislative pontificating.[33] Human subjects proved highly problematic. As Simon Schaffer has pointed out, various moral failings among humans could be construed as madness or, if effectively addressed by new blood, as a material basis for all manner of behaviour and illness that might be deemed a proper focus of experiment. This was the essential debate in the Restoration. But it also drew as much attention to the subject as to the process. As the critic, Henry Stubbe, wittily remarked in 1670, "To argue from the cures of Madmen, or from what they suffer without hurt, is not for a Physician, but for one that deserves to be sent to Bedlam."[34] This was both a prescient and recurrent theme. A bedlamite character could just as readily be attached to the experimenter as much as to the unfortunate subject.

It has to be said, however, that the early transfusion trials were full of difficulties, some quite gruesome, and others quite hilarious. Thus, the noted diarist and virtuoso Samuel Pepys once remarked, just as London shook off the ashes of the Great Fire,

> ...there was a pretty experiment of the blood of one Dogg let out (till he died) into the body of another on one side, while all his own run out on the other side. The first died upon the place, and the other very well, and likely to do well. This did give occasion to many pretty wishes, as of the blood of a Quaker to be let into an Archbishop and such like; but...may, if it takes be of might use to man's health, for the amending of bad blood by borrowing from a better body.[35]

---

33 Moore, *Blood and Justice*, 97.

34 Quoted in Schaffer, "Regeneration," 105.

35 Quoted in Moore, *Blood and Justice*, 84.

Selecting the Subject

The uses of humans in any form of experiment inevitably encountered the serendipity of selection. The menagerie had provided no escape from the experimenter's probe. Humans, on the other hand, might still have choice. The availability and the willingness of human engagement were never predictable. The most common way around this was the payment of volunteers assuming, of course, the meaning of volunteer was not entirely obscure. Certainly, in the case of human transfusion in Paris, one subject was a 34 year old man, possibly suffering from some form of dementia or psychosis the symptoms of which resemble hyperthyroidism. Denys claimed that the obvious "phrensy", exhibited as long as 7 to 8 years before these trials, was alleviated by the transfusions from a calf. The fact remains that the subject soon died, possibly of arsenic poisoning administered by an irate and disappointed wife. But even with the initial signs of relief, the fits of the volunteer were "periodical". Initially, his wife had attempted to have him confined, especially when he had the habit of running naked through the streets and threatening the neighbours with burning down their houses. Even trying to calm him with the help of a porter who had himself previously undergone such a trial, and even upon the wife insisting on further applications, his episodes at the full moon continued. Denys then effected the transfusions before "persons worthy of Credit," and in the presence of several well-known physicians and nobility.[36] In any case, it was finally clear that death or any discomfort would be laid at the door of the experimenters and this made the selection of subject even more crucial – especially as the Academy of Medicine was watching carefully.

The tale of payments here is also a sorry one, including the perjury of witnesses in a trial following accusations by the Paris physicians against Denys. Even the assistance of a labourer of 45, who had previously experienced the trials, was not aided with the poor patient adjourning to the tavern to drink the proceeds of his courage.[37] According to one account, the subject was "a Brittan by birth, and the Original of his Madness, Love." Despite the numerous reports of success that had been received from England, Germany, Italy and Holland,

36 J. Denis, "An Extract of a Letter, Written by J. Denis, Doctor of Physick, and Professor of Philosophy and Mathematicks at Paris, Touching a Late Cure of an Inveterate Phrensy by the Transfusion of Blood," *Philosophical Transactions of the Royal Society* 2 (1666–1667): 617–623; cf. Steven Shapin, *A Social History of Truth: Civility and Science in Seventeenth-Century England* (Chicago and London: University of Chicago Press, 1994).

37 Harcourt Brown, "Jean Denys and Transfusion of Blood, Paris, 1667–1668," *Isis* 39 (May, 1948): 15–29, esp. 23; and Farr, "The First Human Blood Transfusion," 153.

Denys' subject might have survived "had it not been for the debauches in Wine and Brandy that he fell to…"[38] For Denys, this problem of selection had been a matter of serious consideration. While animal to animal transfusion was thought possible because they were controllable and expendable, this was hardly the case with humans. According to Robert Boyle, when early attempts were made in London, a subject initially proposed by the French ambassador, was "an inferior Domestick of his that deserv'd to be hanged."[39] But, in Paris when first he began considering the possibilities, Denys refused to "beg some condemned Criminal, on whom to make the first Essay" essentially because the circumstances were such that the subject might exhibit "faintings and other accidents, which would undoubtedly be ascribed to the Experiment by such as decry it."[40] Such niceties continually complicated trials.

Certainly the Royal Society took the entire matter seriously and, momentarily at least, surely thought about the priority claim that might link transfusions with cures. This was reflected in initial reaction to reports from Paris. Indeed, at a meeting in 1667 the Society instructed some of its members to consult with the resident physician Thomas Allen, about the possibility of securing "some mad person in the hospital of Bedlam" as a suitable subject. Apparently, Allen refused to permit it.[41] But it is remarkable that the choice of subject followed upon the supposition that madness might be mended by new blood – of the lamb or otherwise. Indeed, the notion that blood disorders were at the root of many a disease continued to attract the attention of many an experimentalist. But, once again a volunteer before the Royal Society, the Cambridge graduate Arthur Coga, a "poor and debauched man," was either mad, drunk, or both. According to the Society's secretary, Henry Oldenburg, while Coga seemed physically sound, "the wildness of his mind remains unchanged – or perhaps we ought rather to consider that the improvement of his brain is prevented by the incurable intemperance of which he is guilty."[42] Such subjects, trapped between disdain and desperation, could never hope

38 "A Letter Written by an Intelligent and Worthy English Man from Paris," 1076.

39 Quoted in Schaffer, "Regeneration," 96.

40 Quoted in Farr, "The First Human Blood Transfusion," 152.

41 Thomas Birch, *A History of the Royal Society of London, vol.* 2 (London: A. Millar, 1746), 202; Anita Guerrini, "The Ethics of Animal Experimentation in Seventeenth-Century England," *Journal of the History of Ideas* 50 (July-September, 1989): 391–407, esp. 403. Allen was the author of a note on Anna Wilde, a hermaphrodite, published in the same number of the *Philosophical Transactions* as Denis' transfusion report from Paris in 1667. Thomas Allen, "An Exact Narrative of an Hermaphrodite Now in London," *Philosophical Transactions* 2, no. 32 (February, 1668): 624–625.

42 Quoted in Schaffer, "Regeneration," 100, 103.

to gain any kind of moral or spiritual relief when there were uncertainties about the sanctity of species or experimental intervention in the lives of men.

Desperation and Therapy

It is transparent, even in these early-modern ventures, that trials upon the ill-informed, the desperate, and the unreliable raised clear and serious alarm in the minds of the experimenters. Here were early examples of precisely the problems that much human experimentation has since repeatedly encountered. Even with the best intentions, there remained a population of those able to be coerced, or who did not have a great deal of choice, or even of the unsuspecting who, precisely because they were unaware, were often the best subjects.[43] Inevitably, where medical improvement was the professed motivation, this added yet another dimension to the experimenter's dilemma: Was a potential cure worth the possible risk? And how could risk be identified without the trial, in any case? Early experimenters recognized and debated these issues.

This is especially obvious in the disputes surrounding inoculation for smallpox in the early 18th century. Once again, the pre-eminence of the Royal Society proved central to the matter. The discovery by the English of the Turkish method of inoculation involved physicians with connections to Constantinople. During Newton's presidency, and around the time of the Hanoverian succession in 1714, certain Whig elements in the Society were actively promoting the utility of inoculation. To this end the Society's *Philosophical Transactions* published letters from Dr. Emanuel Timoni, physician to the English ambassador to Turkey, and from Dr. Jacob Pylarini, Venetian consul to Smyrna who had encountered inoculation in Constantinople as early as 1701.[44] By 1716, consequently, there was great interest among the English, famously promoted by Lady Mary Wortley Montagu, wife of the ambassador. This is a well-known tale, but it is particularly revealing for our purposes because it tells us much about the application of a method and the selection of subjects. In Constantinople, Lady Mary had her young son inoculated and, after returning to England, under the shadow of a possible smallpox epidemic she set out to promote the

43 Anita Guerrini, *Experimenting with Humans and Animals*, 1.

44 Emanuel Timonius, "An Account, or History, of the Procuring the Small Pox by Incision, or Incoculation; As it Has for Some time Been Practised at Constantinople," *Philosophical Transactions* 29, no. 339 (1714), 72–82; and Jacobus Pylarinus, "Noca & Tuta Variolas Excitandi per Tansplantionem Methodus, Nuper Inventa & in Usum Tracta," *Philosophical Transactions* 29, no. 347 (1716), 393–399.

method amongst the fashionable. As it happens, the very same articles published in the *Philosophical Transactions* had attracted the attention of Cotton Mather in Massachusetts, and by 1721, amid yet another epidemic, physicians' sons and black servants were being inoculated in the colonies. Of course, neither children nor slaves were necessarily given much choice in the matter. More to the point, in London, within days of the trials in the colonies, "some physicians" had approached the King to attempt the procedure on condemned felons at Newgate prison.[45]

In these circumstances, it is not unreasonable to wonder what constitutes consent. In the remarkable tale of inoculation, there are many intersections between the incarcerated and the innovative, which have since echoed over virtually three centuries. Morality in experiment is not a recent discovery. Indeed, it could be argued that condemned felons were perfect subjects for trials otherwise too risky to perform on common volunteers. This is one reading of the proposition already put, in 1676 in Paris, by the physician and botanist Denis Dodart that the condemned should be used to test antidotes to poisons.[46] The point is not that this was a matter of a drastic morality, but rather that the issue was then fairly acknowledged and well considered. In the case of inoculation in London, with the blessing of George I in August 1721, six condemned were chosen, with regard to "Ages, Sexes, and different Temperaments" all of whose names we now know: Anne Tompson, 25; Elizabeth Harrison, 19; Mary North, 36; John Cawthery, 25; Richard Evans, 19; and John Alcock, 20. By September all had survived the experiment and all were pardoned.[47] By all accounts, this should have been regarded as an immense success but not very surprisingly there was plenty of alarm and not all of it came from those who believed that such an intervention was immoral or ungodly. Here was a significant dispute that could not simply be resolved by medical men. Ministers must have their pulpit. Thus, in the words of one clergyman:

---

45 Genieve Miller, "Putting Lady Mary in Her Place: A Discussion of Historical Causation," *Bulletin of the History of Medicine* 39 (Spring, 1981), 2–16, esp. 3, 6–7; Chamayou, *Les Corps Vils*, 100ff.

46 Denis Dodart, *Memoires pour servir a l'histoire des plantes* (Paris, 1676), 10. See Londa Scheibinger, "Human Experimentation in the Eighteenth Century: Natural Boundaries and Valid Testing," in Lorraine Daston and Fernando Vidal, eds. *The Moral Authority of Nature* (Chicago and London: University of Chicago Press, 2004), 384–408, esp. 394.

47 See Hans Sloane, "An Account of Inoculation by Sir Hans Sloane, Bart. Given to Mr Ranby, to be Published, Anno. 1736, Communicated by Thomas Birch, D.D.," *Philosophical Transactions* 49 (1756), 516–520; and Scheibinger, "Human Experimentation," 397. Cf. Halpern, *Lesser Harms*, 19–23.

> The successful Use of Inoculation in other Parts of the World, was a just Inducement to make the Experiment among us; especially when Criminals, upon their own Choice, were the first who underwent it... these Wretches thankfully preferr'd it to Hanging; and I believe any Man in the like Circumstances...would approve their Choice; So severe were the Agonies they underwent![48]

The logic of the so-called "Antiprovidential Project", allegedly contrary to Scripture, seemed to deny any protection to Christian believers, but to make it available to Infidels.[49] The hijacking of morality in human trials by the studiously or theologically ethical resolved little. This could not to be left simply to priests, lawyers, legislation, or committee. Yet, the difficulty could be explored seriously, and there were significant issues indeed. There were profound philosophical problems to be addressed in the epistemology of experiment. But morality had already muddied the waters.

Approximately 25 physicians and surgeons had witnessed the trials on the Newgate felons. This thereby made of Newgate a laboratory subject to the same rules that had then evolved amongst the gentlemen of the early Royal Society to ensure the veracity of any experimental claims. In other words, no private, solitary, philosopher or physician could credibly present results without collaboration of appropriately sanctioned witnesses.[50] In this instance, ultimately, the proof lay in the pustule. It did not help that wags in the press declared that "any Person that expects to be hang'd may make Use of it."[51] Critically, however, the Princess of Wales was sufficiently impressed that other subjects were soon inoculated at the wish of the Crown, and orphans in the relatively posh St. James's Parish provided evidence of survival. On the other hand, this did not necessarily prove immunity of any duration. One of the original subjects, nineteen-year old Elizabeth Harrison, was required to nurse a young boy ill with smallpox, even to lie in his bed at night for a period of six weeks. She survived with no ill effect. While the promise of immunity appeared to have been answered, this did not allay all alarms. In Boston, Mass., Dr. William Douglas criticized the local effort as a "rash and thoughtless Procedure in a

48 *A Letter to the Reverend Mr. Massey, Occasioned by his Late Wonderful Sermon Against Inoculation* (London, 1722), 7.

49 Ibid., 9, 16.

50 Miller, "Putting Lady Mary in Her Place," 8. On witnessing, see Shapin, *A Social History of Truth*.

51 *Applebee's Journal*, quoted in Guerrini, *Experimenting with Humans and Animals*, 53.

Medical Experiment of Consequence."[52] The fear was that inoculation might actually spread the disease by giving it to people who were otherwise free of the infection.

Inoculation as experiment was not always acceptable as prophylactic. In the 18th century, the process was clearly counter-intuitive. Indeed, this suggests one of the fundamental difficulties in dealing with humans in open-ended biological experiments. One could not be sure of the consequences, even in the case of well-controlled populations. It is fair to say that a few high-profile deaths in England amongst adults curbed early enthusiasm for inoculation. But this was not necessarily the case in circumstances with less cumbersome obstacles, as in plantations in Jamaica amongst a slave population. Even after the London College of Physicians gave their approval in 1755, there was no apparent rush to employ the practise in England. Despite the risks, some were not inordinately squeamish. In Jamaica, Dr. John Quier undertook extensive trials amongst slaves who presumably had little opportunity to object. Quier examined a wide range of issues, including inoculation of those suffering from another disease, or whether it was safe for women during menstruation or pregnancy. The apparent tendency to miscarry after inoculation was one alarm. Much of this, not surprisingly, resulted in comparisons made between slave populations and European women. As Londa Scheibinger has noted, "Quier's work raised a new and potentially explosive question: whether medical experiments done on blacks were valid for whites" – especially while some suggested that much more delicate and refined European women should not be subject to the treatment.[53] It is especially revealing that Quier's numerous trials drew attention to distinctions of effectiveness or danger in populations of different racial origins and over comparatively large numbers, another trend that continued into the 20th century.

Dr. James Jurin, Secretary of the Royal Society, addressed the issues of population and mortality from smallpox in the 1720s. Between 1723 and 1727 Jurin collected accounts and statistical data regarding the method of inoculation, days of illness, the pustules and the outcome. But Jurin was intent on more than collecting anecdotal evidence. He trolled the published Bills of Mortality, from at least the turn of the century, for information relating to smallpox. This provided one index of the difference in mortality amongst those who could not have been inoculated and those who were.[54] Evidence in the aggregate

52 William Douglas, *The abuses and scandals of some late pamphlets in favour of inoculation of the Smallpox...* (Boston: J. Franklin, 1722), Introduction.

53 Scheibinger, "Human Experimentation," 402–403.

54 Miller, "Putting Lady Mary in Her Place," 10; Larry Stewart, "The Edge of Utility: Slaves and Smallpox in the Early Eighteenth Century," *Medical History* 29 (1985), 54–70, esp. 59

overwhelmed the purely anecdotal. This was increasingly an issue where reporting was so various and so idiosyncratic as in physician's reports and private correspondence. Between the experiment and the publication there were many stages, involving the inoculator as well as the inoculated.

In early-modern experimental reports, neither a procedural template nor uniform conventions were consistently adopted. In these circumstances, the human subject was essential for a reading of the effect of some natural event or experimental procedure. Even in the case of auto-experimentation, however self-deceived, the sentient being could reveal what the inanimate and the animal could not.[55] Morality and method were especially conflicted when there was little ability to predict results. These were matters of continuing difficulty. But they emerged primarily within a version of the experimenter's regress – that is, groping in the dark, in trials with no clear, defined, end in sight.[56] As the promotion of experimental method had become the most essential characteristic of early-modern science, the body became another instrument of revelation as much as a telescope or microscope. Clearly, the role of the human in experimentation proved epistemologically complex.

Consent

The human is not only an observer, but he/she is also potentially subject, and at the very least interpreter of natural phenomena. The material culture of the laboratory cannot utterly overwhelm the human. This is one of the lessons of encounters in the laboratory. Localities only limit what experimenters can

---

Halpern, *Lesser Harms*, 21–22. See also Andrea Rusnock ed. *The Correspondence of James Jurin (1684–1750): Physician and Secretary to the Royal Society. Clio Medica*, 39 (1996) and Rusnock, "Biopolitics: Political Arithmetic in the Enlightenment," in William Clark, Jan Golinski, and Simon Schaffer, eds. *The Sciences in Enlightened Europe* (Chicago and London: University of Chicago Press, 1999), 49–68, esp. 65.

55 Cf. Scheibinger, "Human Experimentation," 389. On the debate on the self, see Jan Goldstein, "Mutations of the Self in Old Regime and Postrevolutionary France," in Lorrain Daston, ed. *Biographies of Scientific Objects* (Chicago and London: University of Chicago Press, 2000), 86–116.

56 See the discussion in H.M. Collins, "A Strong Confirmation of the Experimenter's Regress", *Studies in the History and Philosophy of Science* 25 (1994), 493–503, esp. 502; and Benoit Godin and Yves Gingras, "The experimenter's regress: from skepticism to argumentation," *Studies in the History and Philosophy of Science* 33 (2002), 137–152; Halpern, *Lesser Harms*, 117ff.

understand and communicate.[57] The human, thus, is always instrumental and functional even in the most austere and apparently objective laboratory settings. This is especially revealed in cases where human subjects of a quite deliberate trial have very little knowledge or awareness of the natural or biological process in which they have been engaged. This becomes especially obvious, of course, in the case of children or of those without any previous access to the laboratory world. Importantly, many of these situations were not commonly medical in any obvious way.

Take, for example, the research into the newly discovered world of electrical phenomena in the early eighteenth century. Electrical displays became a constant in the philosophical lectures of the period, notably amongst the disciples of Newton who would give demonstrations of static electricity in a partially evacuated jar in a darkened room, where the discharge provided a strange blue glow for which there was then no suitable explanation. While the Royal Society was exploring the benefits of inoculation, they were also promoting an experimental approach to electricity by way of static generators. The experimentalist Stephen Grey attempted various trials of electrical effects, using ivory, metals, even soap bubbles, an umbrella and, as subject as much as witness, a 47-pound child, likely an orphan at the Charterhouse charity in London. What is fascinating about the use of the child is the uniting of compliance (as consent has no obvious meaning here) with experimental innovation. At least the child could describe the immediate sensations of electrical charges.[58] At this stage the voltages and subsequent discharge were so low as to be otherwise measurable simply by observation of the attraction of scraps of paper.

One puzzle, as the century wore on, was the distance through which grounded electricity might travel. This was the cause of experiments done by Fellows of the Royal Society in the middle of the century who ran a wire across the New Westminster Bridge to a distance of some 400 yards. The gentlemen of the Society then tested the force, or sensation, of the discharge on the side farthest from the electrical generator. While the electrical "commotion" could be felt, the repetition of the experiment was interrupted several times by the "great Concourse of People." But the experiment was regarded to be such a success, even over distances of up to two miles further along the Thames, that it was repeated several times despite the interruptions occasioned by crowds and the occasional horse. At one attempt, near Shooter's Hill (near Greenwich),

57 Cf. Peter Galison, *Image and Logic: A Material Culture of Microphysics* (Chicago and London: University of Chicago Press, 1997), esp. Chapter 9.

58 J.L. Heilbron, *Electricity in the 17th and 18th Centuries: A Study of Early Modern Physics* (Berkeley and Los Angeles, University of California Press, 1979), 247.

the gentlemen of the Society, after the discomfort caused by increasingly large shocks, convinced two country folk to join them in a chain where they all joined hands. "Upon the Explosion they were all so strongly shocked in their Arms and Ankles, that the Countrymen could by no means be prevailed upon to try the Experiment again."[59] This was only one of a long series of experiments of electrical forces undertaken in the middle of the eighteenth century measured not mechanically or instrumentally, but by way of human sensation.

This led to some fairly dramatic demonstrations that verged on parlour games. Public performance unlocked philosophers' worlds. Human actors enhanced the entertainment and spectacle. Blurring any distinction between trial and demonstration was particularly evident where there were human subjects, either in displays of great strength in city markets or of the forces of nature people sought to tame. Perhaps the most well-known of such efforts were those by the famous Abbé Nollet in Paris who made a distinguished career as a scientific lecturer and electrical experimenter in the last half of the eighteenth century. He made and sold instruments, provided dramatic entertainments at the royal palace at Versailles, and made himself generally amusing to the French aristocracy. In the presence of the King, Nollet sent a bolt of electricity through 180 gendarmes, and then through over 200 Carthusian monks in Paris.[60] Beyond the obvious drama with both symbols of authority, neither monks nor monarch seemingly recorded any ethical consideration.

Given the uncertainties and even the demonstrable power associated with electrical charge, there were some who wondered if there might not be a medical use to be extracted from the phenomenon. If a bolt from the clouds could kill, could electricity revive? This was even more the case when the Leyden Jar enabled electrical charges to be retained for some time and even transported into new settings, from laboratory to drawing room and clinic. Therapeutic promise provided a profound inducement. Certainly Nollet became well known in Italy precisely because of European debates over the benefits of electricity in cases of paralytic disorders. While unsuccessful, the prolonged experiments on injured guards, approved by the King, turned Nollet into an arbiter of the efficacy of

59 William Watson, "A Collection of the Electrical Experiments Communicated to the Royal Society by Wm. Watson, F.R.S. Read at Several Meetings between October 29, 1747, and Jan. 21. Following," *Philosophical Transactions*, 45 (1748): 49–120, esp. 80; Simon Schaffer, "The consuming flame: electrical showmen and Tory mystics in the world of goods," in John Brewer and Roy Porter, eds. *Consumption and the World of Goods* (London and New York, Routledge, 1993), 489–526, esp. 496.

60 Heilbron, *Electricity in the 17th and 18th Centuries*, 280–289, 318.

electrical treatments during a visit to Italy.[61] The potential to harness the power of electricity in medical ailments was irresistible, we submit here, precisely because it was so much an unknown. Even those unfortunates who decided to repeat Benjamin Franklin's foolishness with a kite in a thunderstorm provided ample evidence of lightning upon the human form. And even when there was some debate as to the identity of lightning and the kind of electricity generated in the laboratory, it was sufficiently obvious that this was something physicians might usefully explore. They only needed willing subjects. Among the sick there were also the sufficiently desperate. The point needs to be emphasized once again – it was not simply medicine that first promoted experiment on humans but precisely the programme of experiment that suggested human trials were the next logical step in achieving a deeper understanding – despite the consequences for those involved, whether experimenter or experimented upon.

The Uses of Humans in Experiment

For some time historians have been exploring these tensions and have been particularly attuned to inconsistencies that emerged in the experimental landscape. One such area of interest emerges in the study of attempts by medical and political communities to establish ethical codes of conduct to guide these decisions and standardize approaches by scientific and medical players. These codes attempt to balance the appetite for research and knowledge while safeguarding the rights of human subjects. But as historians have shown us, this process in itself has a long history, emerging at least from debates in the French Revolution, and was no less complicated by the introduction of international ethical codes. For example, Susan Lederer reminded modern historians that in spite of landmark legal codes emerging in the 20th century, that drew global political and medical attention to unethical uses of humans in experiments, physicians had long held to ethical guidelines when it came to distinguishing right from wrong. She also showed how anti-vivisectionists and animal rights activists significantly shaped bio-medical research with a specific eye to moral and ethical considerations.[62] Her work reminds us of the significant ways

61 See Paola Bertucci, "Sparking Controversy: Jean Antoine Nollet and medical electricity south of the Alps," *Nuncius*, 20 (2005): 153–187; also "Revealing Sparks: John Wesley and the religious unity of electrical healing," *British Journal for the History of Science*, 39 (September, 2006): 341–362.

62 Susan E. Lederer, *Subjected to Science: Human Experimentation in America Before the Second World War* (Baltimore: Johns Hopkins University Press, 1995).

social and political contexts have influenced conceptualizations of morality, whether regarding animal experimentation, military contexts, or medical approaches to new treatments.

The Nuremburg Code of 1947 followed the atrocities of the Second World War and the gruesome revelation that under the veil of science humans had been tortured, exterminated and studied literally to death in an effort to test the limits of human physiology.[63] The World Medical Association later followed suit by issuing the Declaration of Helsinki, perhaps as a reminder that in spite of the rigid terms laid out by the Nuremberg trials, few nations or societies had designed mechanisms to police experimental activities in the latter half of the century, many believing that their work was simply better than that of Nazi war criminals.[64] American scientists faced international scrutiny when the Tuskegee experiments were exposed in the 1970s, indicating that black subjects had been knowingly given syphilis and left untreated in an effort to better understand the pathology of the disease, particularly as it reached its tertiary stages.[65] Ethical breaches have continued in spite of international codes and tribunals revisiting the policies and standards of practice, drawing attention to the multiple interpretations of such codes, an inability to implement them, and the challenges of imposing legal definitions onto a longer tradition of medical and scientific encounters. Legal measures thus only scratched the surface of the complex relationship between humans and experiments. The editors of *Dark Medicine* pointed to the folly of seizing upon international tribunals, codes of declarations as markers of bioethical successes and instead cautioned that historical context and cultural circumstances create more fluid rationales for research that is at one moment judged necessary, while at another unethical. The contexts of war, reconstruction, new technologies,

---

63 Several historians have looked into this history. See for example: Robert Proctor, *Racial Hygiene: Medicine Under the Nazis* (Cambridge, MA: Harvard University Press, 1988); Paul Weindling, *Nazi Medicine and the Nuremberg Trials: From Medical War Crimes to Informed Consent* (Houndmills, Eng.: Palgrave Macmillan, 2006); and Michael H. Kater, *Doctors Under Hitler* (Chapel Hill, NC: University of North Carolina Press, 1989).

64 One particularly good example comes from the Tuskegee experiments in the United States, which are explored in this volume. See also, Jenny Hazelgrove, "The Old Faith and the New Science: The Nuremburg Code and Human Experimentation in Britain, 1946–73," *Social History of Medicine* 15, no. 1 (2002): 109–135.

65 See Allan Brandt, "Racism and Research: The Case of the Tuskegee Syphilis Study," in *Sickness and Health in America: Readings in the History of Medicine and Public Health*, 3rd ed., eds. Judith Walzer Leavitt and Ronald L. Numbers (Madison, WI: University of Wisconsin Press, 1997), 392–404.

and both early and late life-stage illnesses created conditions for rationalizing experiments even in the context of practices that otherwise appeared out of step with international medical guidelines.[66] These studies help to demonstrate the historical elasticity that has come to define modern ethics, while at the same time draw concerted attention to the need to historically contextualize these moments in an effort to extract meaning from apparent ethical transgressions and to track a longer, less codified, history of ethics regarding experiments with humans.

Historians of science and medicine have likewise questioned the modern preoccupation with bio-ethics and reinserted such issues into a more contextualized history. Harry Marks did so by exploring how the 'progress' of experiments were increasingly defined by profit margins, which readily dictated what was fit for experimentation and when such an object ceased to be an experiment and transformed into a therapeutic [marketable] option. His study of market forces on pharmaceuticals revealed the pernicious influence of a multi-billion-dollar exercise that perverted the pursuit of curiosity-based inquiry on individualized trials into a massive industry aimed to minimize risk and maximize profits. The resultant perversion of concern for patients into a profit-oriented language of evaluation fundamentally altered the parameters by which one assessed efficacy and risk.[67]

The alarm continues to be raised where morality and money meet.[68] Scholars have enthusiastically seized the opportunity to criticize the pharmaceutical industry for setting its own rules on risk, disclosure, and transparency in what appears to be a losing battle between curiosity and social, value-driven, research and a commercially-driven industry. David Healy's trenchant critiques in this vein have exposed many of the ways in which profit-seeking corporations acquired significant political power to change the context of experimentation and the distribution of information to consumers. Beyond measuring progress in dollars, as Marks suggests, Healy goes further to argue that the very meaning of illness or that which requires medication is constructed first by corporate desires, then the same actors design solutions to the problem they defined. The intractable feedback loop created by this pharmaceutical industry absorbs biomedical research and reorients modern science to suit a different set of

66 William R. Lafleur, Gernot Böhme, and Susumu Shimazono, eds. *Dark Medicine: Rationalizing Unethical Medical Research* (Bloomington, IN: Indiana University Press, 2007).

67 Harry Marks, *The Progress of Experiment: Scientific and Therapeutic Reform in the United States, 1900–1990* (New York, NY: Cambridge University Press, 1997).

68 See, for example, Ben Goldacre, *Bad Pharma* (London: Fourth Estate, 2012).

ideals: that is, goals that are neither therapeutic nor knowledge-based, but which conform to contemporary ethical guidelines.[69] It is hardly surprising given the funding and the objectives that even peer-review has been deemed pernicious.[70]

Beyond the doomsday, albeit convincingly bleak, image created by some scholars and practitioners, they effectively point to ways in which humans are being used in experiments in a manner no longer clearly within the reach of acceptable medical codes. In the case of the pharmaceutical industry, these authors contend that consumers emerge as the new research subjects. They are complicit and willing subjects by virtue of their participation, while the context of information and thus informed consent, reaches new degrees of uncertainty. Here too, though, there remains a longer more complicated history of consumers as subjects. Nancy Campbell has recently turned that idea around somewhat, by suggesting that while jailed drug addicts in Lexington had on the one hand been captive subjects of drug experimentation in the 1950s, they also readily emerged as knowledgeable advisors in drug research. By virtue of their experience, as with heroin use, convicted drug users became a critical source of expertise in on-going research into the effects of both heroin use and withdrawal. Under the circumstances, users or consumers provided researchers with greater insights than they might have gleaned on their own, although the trials were conducted under conditions that later revealed unorthodox and unethical approaches, even by contemporary standards.[71] Moving beyond confined criminals to middle-class suburbs, historians of prescribed drugs have routinely shown how middle-class consumers have increasingly assumed the role of subject without the protection of medical oversight or ethical guidelines. On and off-label use of pharmaceuticals coupled with unprecedented advertising budgets now place pills in the ultimate human trial – in the marketplace. This context moves well beyond the medico-scientific arena both in terms of knowledge production, but also as concerns legal liability and risk. The rules of engagement in this terrain are guided by market forces and as such succumb to a different culture of ethics that introduces another layer of complications.

---

69 See for example, Healy, *Mania*; and Richard DeGrandpre's *The Cult of Pharmacology: How America Became the World's Most Troubled Drug Culture* (Durham: Duke University Press, 2006).

70 See, for example, "How science goes wrong," *The Economist* (October 19, 2013).

71 Nancy D. Campbell, *Discovering Addiction: The Science and Politics of Substance Abuse Research* (Ann Arbor, Michigan: University of Michigan Press, 2007).

Essays

These essays set out to explore a series of places and practices associated with the use of humans in the expanding realm of observation and experiment since the 18th century. Many of these papers developed out of a series of workshops hosted by the editors and funded by the Situating Science network of scholars in the philosophy and history of science. By bringing together historians and science with historians of medicine this collection aims to bring such scholars into a deeper conversation about the uses of humans in experiment from the 17th century onward.

They begin by exploring bodies as sites of experimentation, meaning bodies as instruments, gauges and interpreters of phenomena that contribute to the growth of systematic knowledge. Essays by Guerrini, Steigerwald and Iliffe point out that re-defining the human body as a site of experiment and of spectacle served to reinforce the notion that humans provided the kind of opportunities otherwise unavailable, even in animals. There was, for example, much to be gained by expanding trials of newly-discovered airs with candles and mice to human subjects. Testimony therefore came not simply from witnesses, however credible, but directly from human reactions. Reflecting upon an era which sometimes redefined the uses of the body, Anita Guerrini's chapter takes a closer look at the burgeoning culture of human experimentation and explores the 'monstrous though wondrous' bodies of hermaphrodites as sites that displayed contestations over gender and human anatomy and physiology in both popular and learned cultures. Hermaphrodites' dual physical nature provided both public spectacle but also medical and scientific curiosity that drew sustained attention to these particular bodies as sites of investigation. Rob Iliffe and Joan Steigerwald continue this exploration of human physiology as they explore Galvanic experiments from different angles. Steigerwald examines the intersections between human, animal and instrumental development that gave rise to new knowledge and new approaches relying on bodies as instruments, in this case, of electrical currents. Rob Iliffe moves this subject in a slightly different direction by exploring what Galvanic theories explained about animal magnetism and how these explorations influenced the development of scientific medicine in the 19th century.

Paola Bertucci's essay begins to challenge that view of experimentation by providing an electrifying look at the junction between parlour displays of electricity and electrotherapeutics. Her work explores gender in the experimental context by illustrating how fashionable electric displays incorporated common gendered tropes about women's and men's bodies and their inner unruly desires. Unlike Bertucci's willing performers, Larry Stewart examines this relationship through pneumatic chemistry and the radical Thomas Beddoes. Stewart's portrayal of experimental therapies reveals how struggles over desperation and risk entered the experimental domain in growing tensions between experimentalist, therapist, and subject.

Returning to 19th century dilemmas, Elizabeth Neswald offers a careful examination of the physiological experiments conducted throughout the 19th century in an effort to examine the science of nutrition. Her essay underscores the complexity of agency as it affects subjects, subjectivities and experimenters themselves (both as subjects and observers) as they engage in experimental research that moves between animal and human metabolic studies. Neswald's historical analysis of nutrition experiments pinpoints the challenges associated with choosing the appropriate subjects and the implications these choices have for elements of participation and discipline.

Into the late 19th and early 20th centuries, Katherine Zwicker traces the career paths of physicists and clinicians whose fates become intertwined as cancer treatments embraced the highly technical and ultimately risky but effective new therapies alongside the emergence of laboratory-informed, hospital-based practices. The professional collaboration, however, creates new subjects as physicists increasingly play a role in the clinical context and move decidedly from the laboratory into the clinic and face a new set of subjects. Zwicker's chapter further underscores the longer tradition of experiments within the context of scientific knowledge production, and not narrowly as therapeutic application.

Moving solidly into the 20th century Paul Lombardo examines how contemporary 1920s ideology reified racial and gendered characteristics in early American eugenics programs, and did so in an effort to justify experimentation on men who did not fit an ideal profile. Lombardo shows, however, that assembling an ideal profile had been riddled with challenges, many of which were explained away through adherence to an ideological belief in racial degeneration. His focus on a pre-Tuskegee case in the southern United States is a careful reminder of the racial and eugenic context that overlaid the experimental enterprise when it came to evaluating black bodies.

Paul Weindling shifts our attention into the Second World War in Europe to examine the politics of experimentation that unfolded under Nazi rule, also within eugenics contexts, and he demonstrates how that event has profoundly shaped our understanding of subjects, victims, and consent. His quantitative and detailed study of subjects in this context encourages us to rethink how ethics and ideology interacted to offer temporary justifications for widespread experimentation.

Erika Dyck explores an institutional environment in western Canada that became home to the country's most aggressive eugenics program, a program that also supported a series of experiments aimed at defining a link between degeneration and psychotic disorders among males. The institutional context provided something of an incubator, both for containing subjects, but also for diffusing public criticism that may have otherwise been levelled at post-World War II experiments conducted largely on children. These experiments continued into the late 1960s, well past the international condemnation of eugenics research in Europe and the United States, which again draws attention to the localized cultures of ethics and how investigators justified their continued work.

Together this collection reminds us that the uses of humans in experiments have a long history and a complicated future. These essays speak to sites and discourses – scientific, political, gendered, racial, and religious, to name some – which have a continuing trajectory that is far from linear. Despite the consequences of modern science and medicine, many of the ethical questions that perplexed early-modern natural philosophers continue to resonate. These force us to contend with a history that is sometimes "strange, undefined, [and] ineffaceable."[72] And yet, if there is one thing that this history proclaims: even the strongest philosophical and political challenges have not prevented science from a making sustained effort to use the human body as both *subject* of experiments and *object* of the knowledge that such experiments produce. These are the connections of the human *as* instrument. So long as there are unanswered questions about exactly how 'life' functions, we can be certain that such experiments will continue – in agreement with, and sometimes in spite of – political, economic, and religious interests. After all, does this quest to comprehensively understand the body not bring us one step closer to a deeper grasp of what makes us distinctly human? However one chooses to ponder that question, perhaps these essays point to the debate ahead.

---

72 Foucault, *The Order of Things*, 400, 405.

CHAPTER 1

The Hermaphrodite of Charing Cross

Anita Guerrini[1]

Hermaphrodites were much discussed in the early scientific societies of the late seventeenth and early eighteenth centuries. Long a topic of speculation and wonder, hermaphrodites became one subset of a large category of "monsters" that early modern natural philosophers yearned to debunk. Unlike other monsters, hermaphrodites, although wondrous, did not portend remarkable events, and by the end of the seventeenth century they had become mundane although no less compelling. From at least the sixteenth century onward, the determination of the sexual status of a hermaphrodite was a medical question, although it impinged on issues of legal status. Ambiguity was not an option, and medical examination determined if the relevant parts were male or female. Multiple accounts of hermaphrodites dotted the medical literature.[2] The *Philosophical Transactions* of the Royal Society of London and, from 1699, the *Mémoires* of the Paris Academy of Sciences included several accounts of hermaphrodites, sometimes written in Latin to discourage the prurient curiosity of the wider public.

These accounts aimed at a natural history of their subjects and sought a thorough physical description that did not aim to demonstrate causes. William Harvey had pioneered this historical method in his work on the circulation of the blood.[3] The natural history of animals (which included humans) employed a number of observational techniques, among them dissection. Harvey further claimed that his observational method, particularly the dissection of live animals, was also experimental. It was not merely descriptive but intervened in the ordinary course of nature and therefore led to the creation of new knowledge.[4] This knowledge was not knowledge of causes but of the operations of nature which had hitherto been hidden.

---

1 I am grateful to the editors and to the anonymous reviewers for their comments. My research for this project was funded by the Horning Endowment in the Humanities at Oregon State University. Thanks also to Michael A. Osborne.

2 See Lorraine Daston and Katharine Park, "Hermaphrodites in Renaissance France," *Critical Matrix* 1:5 (October 1985); idem, "The Hermaphrodite and the Orders of Nature: Sexual Ambiguity in Early Modern France," *GLQ* 1(1995), 419–438.

3 On Harvey's method see Anita Guerrini, *The Courtiers' Anatomists*, Chapter 2 (University of Chicago Press, 2015).

4 On the definition of early modern experiment see Peter Dear, *Discipline and Experience* (Chicago: University of Chicago Press, 1995).

 | DOI 10.1163/9789004286719_003

At the same time, the physicians who penned these descriptions worked within the boundaries of the medical case history. Pioneered by the ancient Greek physician Hippocrates, and revived first by Renaissance humanists and more recently by the English physician Thomas Sydenham, the case history aimed to describe a pathological condition, again without assigning causes but giving a possible prognosis. While the detailed examination of hermaphrodites did not resemble any contemporary medical examination of other humans, it partook of both of these genres, natural history and the case history. It described a "preternatural" condition outside the ordinary course of nature that may or may not also have been pathological.

The uneasy relationship between the anatomists who performed these examinations and these particular subjects served as a lens to magnify the uneasy relationship between anatomists and their publics: potential patients, curious spectators, and readers of broadsides and newspapers.[5] To most potential patients, a good bedside manner was more important than anatomical skill. Anatomists were suspected, with reason, of desecrating graves, and there were many stories of corpses that revived while under the anatomist's knife. The physician James Parsons, who figures in our story, demanded that he not be buried until his corpse began to deteriorate, to avoid both of these circumstances.[6] Indeed, earlier anatomists had declared their disappointment that human vivisection was not legally or morally permissible.[7] In addition, nearly all of the hermaphrodites thus examined were on public display, and in this way came to the attention of learned men. When anatomists sought out the monstrous to examine them, were they any different from the gawking public, or did they seek a deeper knowledge than was possible with normal humans? It was this interplay of learned and popular observation, of different kinds of natural histories joined to case histories with a wealth of circumstantial details, that led to the learned recognition of the truth about monstrous births.

My argument in this essay is that anatomists' examination of hermaphrodites in this period constituted a form of human experimentation, an extension of Harvey's historical method of research, which he had defined both as

5 On the relationship between science and its publics, see most recently Jonathan R. Topham et al., "Focus: Historicizing 'Popular Science,'" *Isis*, 100 (2009) 310–268; see also James A. Secord, "Knowledge in Transit," *Isis*, 95 (2004), 654–672.

6 *Oxford Dictionary of National Biography*, s.v. Parsons, James (1705–1770). Henceforth cited as *ODNB*.

7 In the sixteenth century, both Vesalius and Barengario da Carpi had expressed this, and their critics commonly claimed that they practiced it. See, for example, Jacopo Barengario da Carpi, *A Short Introduction to Anatomy*, trans. L.R. Lind (Chicago: University of Chicago Press, 1959), 10.

experimental and as leading to true knowledge. Although these examinations were not surgical – and therefore do not qualify as vivisection – the particular techniques that anatomists used to examine these subjects were invasive and at times painful, attempting to reach as far as possible into the living human body. I argue in addition that the monstrous identity of hermaphrodites, and their status as objects of display, placed them outside normal bounds of morality. This ambiguous status gave anatomists license to perform physical examinations and create illustrations that, by the standards of the medical practice of the day, were often regarded as indecent and unseemly.

The learned appropriation of hermaphrodites also reveals the slippery and porous boundaries between observation and intervention, between private science and public display, between anatomy and pornography, and between the learned and the popular. I will not look at the discourses surrounding the causes of hermaphroditism, but instead look at what anatomists did, and the identities of their experimental subjects, concluding with some comments on the mutually dependent relationship between natural philosophers and public display.[8]

•••

Late in 1714, advertisements appeared in two consecutive issues of the London newspaper *Post Man* to announce the display of "the greatest Wonder in Nature, being an Hermaphrodite, compleat Male and Female." It (as the advertisement referred to the person) "has given a general Satisfaction to the most Ingenious Physicians, who have inspected and confirmed the same." The second advertisement gave more detail, noting the hermaphrodite's age as 18, and naming the price – 2s 6d – for private viewings. The viewings took place from 9 in the morning until 9 at night at the Red Lyon Inn at the end of Pall Mall, near Charing Cross. Such advertisements appeared alongside a growing number touting scientific lectures, exhibits, instruments, and books.[9] A broadside

8 On the question of causes and medical explanations, see most recently Palmira Fontes da Costa, "Mediating sexual difference: The medical understanding of human hermaphrodites in eighteenth-century England," in *Cultural approaches to the history of medicine: Mediating medicine in early modern and modern Europe*, ed. Willem de Blécourt and Cornelie Usborne (Basingstoke: Palgrave-Macmillan, 2003), 127–47; idem, "Anatomical Expertise and the Hermaphroditic Body," *Spontaneous Generations*, 1 (2007), 78–84.

9 *Post Man and the Historical Account* (London), November 30-December 2, 1714; December 4-December 7, 1714. On advertising, see Jeffrey R. Wigelsworth, *Selling Science in the Age of Newton* (Burlington, VT: Ashgate, 2010).

employed similar language, with a few differences in the details, announcing a public display from 1 in the afternoon until 9 at night, with mornings reserved for private viewings. The venue had changed, to the nearby King's Head coffee-house, and the price was now one shilling. A paper lantern over the door of the King's Head declared, "*The Hermaphrodite is to be seen here without a Moments loss of Time.*"[10] The change of venue was not unusual, for such exhibits regularly traveled through London, from fairs to coffee-houses and back, another aspect of the circulation of knowledge in the metropolis.[11]

Although the advertisements declared the authority of physicians and surgeons to ensure the authenticity of the spectacle, the only documented examination of this particular hermaphrodite occurred early in the next year. On 16 February 1715, the physician and anatomist James Douglas delivered a paper to the Royal Society of London. The manuscript, now in the library of the University of Glasgow, begins "some account of a young woman said to be an hermaphrodite, to be seen every day at Charingcross, with the description and figure of the externall appearance of the parts that serve for generation." Douglas's description, which included illustrations, was deemed to be "very fit to be printed in the Transactions," but in fact it never appeared in the *Philosophical Transactions* of the Royal Society, although Douglas published many other essays there. Two of Douglas's illustrations, with a short description, appeared in the 1722 edition of William Cheselden's *Anatomy of the Humane Body*.[12]

---

10 "By His Majesties Permission," broadside, in "Collection of Advertisements," shelfmark NTAB 2026/25, British Library. All quotations from broadsides in the text come from this collection, which was made by Hans Sloane. Dating of this broadside is uncertain; reference to "His Majesty" and "Vivat Rex" place it sometime after Queen Anne's death in August 1714, but it is impossible to know whether it precedes or succeeds the newspaper advertisements.

11 For details, see Anita Guerrini, "Advertising Monstrosity: Broadsides and Human Exhibition in Early Eighteenth-Century Britain," in *British Ballads and Broadsides, 1500–1800*, ed. Patricia Fumerton and Anita Guerrini with Kris McAbee (Burlington, VT: Ashgate, 2010), 109–127.

12 James Douglas, "Drawings of the Pudendum Muliebre with an Account of a Hermaphrodite," MS Hunter DF 60 (1–7), Glasgow University Library. While it is not absolutely certain that this is the same hermaphrodite, it seems too much of a coincidence for it not to be, and the details of age, date and location roughly coincide. The title of the manuscript was later added by William Hunter. *Journal Book*, vol. 12 (June 1714-April 1720), f. 44 (17 February 1714/15), Royal Society, London. William Cheselden, *The Anatomy of the Humane Body*, second edition (London: S. Collins for James Knapton et al., 1722), Tab. XXV, pp. 318–319; this illustration appeared in subsequent editions until 1740, when it was supplanted by the Angolan hermaphrodite (see below). It should be noted that Douglas read

Douglas's treatise is of particular interest because it gives a detailed account of the life of the subject as well as an anatomical description. Douglas, born in Scotland in 1675, was a well-known physician, surgeon, man-midwife, and anatomist. After study in Paris and Utrecht, he obtained a medical degree from Rheims (a well-known diploma mill for foreign students) and moved to London around 1699. He began his obstetrical career as an assistant to Paul Chamberlen of the famous family of man-midwives. The Chamberlens were adept at the use of obstetrical forceps, which they held as a secret for over a century. Douglas had apparently adopted these in a case in 1702, but there is no evidence he used them subsequently, and Alexander Pope referred to Douglas's "soft obstetric hand" in his 1728 poem *The Dunciad*.[13] Paul Chamberlen was better known for his "anodyne necklace" which supposedly relieved the pain of childbirth and which he advertised heavily in newspapers.[14] Douglas recorded numerous case histories during his time with Chamberlen, from 1699 to 1711.

During the same period, Douglas developed his skills as an anatomist. He had attended the lectures of Joseph-Guichard Duverney at the Jardin du Roi in Paris and those of George Rolfe in London, and advertised his own private anatomy course in 1706. He did anatomical demonstrations before the Royal Society, which elected him a fellow in 1706, and was later an anatomical lecturer to the College of Surgeons. Douglas pioneered the peculiar juxtaposition of life and death, obstetrics and anatomy, which he passed on to his pupils James Parsons and William Hunter.[15]

Douglas did not explain in his essay why he examined the hermaphrodite and wrote his account. Like many of his contemporaries, he was fascinated by the "monstrous" and in the *Philosophical Transactions* he had described monstrous births he had encountered in his practice. In addition, his knowledge of the female body, gained through dissection and his obstetrical practice, made him the perfect observer. His curiosity in this matter was certainly not unusual among his peers.

---

over fifty papers to the Royal Society that were never published (ODNB, s.v. Douglas, James (1675–1742)). All translation from Latin and French are mine.

13 Adrian Wilson, *The Making of Man-midwifery: Childbirth in England, 1660–1770* (Cambridge, MA: Harvard University Press, 1995), 68, 86.

14 ODNB, s.v. Chamberlen family.

15 For biographical detail see Anita Guerrini, "Anatomists and Entrepreneurs in Early Eighteenth-Century London," *Journal of the History of Medicine and Allied Sciences*, 59 (2004): 219–239; ODNB, s.v. Douglas, James (1675–1742).

"By way of preface," wrote Douglas, "give me leave to introduce my selfe," quoting a Latin passage from the seventeenth-century Dutch anatomist Theodor Kerckring. Translated, it reads:

> And so, let licentious wantonness have no part in this discussion; be far, far from here, profane and lecherous ears. We write for physicians, whom necessity permits to touch, from time to time, those things that nature, the watchful preserver of modesty [*natura pudicitiœ conservatrix*], will take care should be covered. Thus, for these people it will not be a burden to hear something about these parts even beyond the usual, whenever that very all-producing womb [*mater*] labors.[16]

At the outset, Douglas established that as a physician he had special privileges not only to examine this particular subject, but also to talk about it to a suitable audience. By quoting Kerckring in Latin he confirmed that particular audience – the ordinary rabble would not know Latin – and echoed earlier accounts of hermaphrodites such as Thomas Allen's, published in Latin in the *Philosophical Transactions* in February 1668 "for the view of the Learned," or the even earlier one of Realdo Colombo, both of which he quoted in his treatise.[17] In addition, he established that even though his subject matter was prurient, his audience was not to take "profane and lecherous" pleasure from his account, unlike, presumably, others who attended the coffeehouse display. Yet the nature of the hermaphrodite as a sexual monster, as well as the nature of the examination Douglas performed, make it difficult to ignore a level of sexual as well as anatomical violation of his subject.

Like Allen, Douglas began with a biographical sketch of this hermaphrodite, but he gave more details beyond what had become the standard narrative arc, which consisted of the subject being identified with one sex at birth, and then exhibiting signs of the other sex later in life, often at puberty. The kind of circumstantial details Allen had offered of the name – Anna Wilde – and birthplace served to anchor his narrative in allegedly concrete facts. But Douglas went farther and seems to have engaged in conversation with his subject,

16 Douglas, "Drawings." The manuscript is not paginated, so I will not footnote subsequent quotations from this text. Many thanks to Jole Shackelford for help with translation of this passage.

17 [Thomas Allen], "An exact Narrative of an Hermaphrodite now in London." *Philosophical Transactions*, Vol. 2, no. 32 (1668): 624–625, at 624; Realdo Colombo, *De re anatomica* (Venice: Nicolo Bevilacqua, 1559), 268–269.

eliciting a life story, and making his subsequent examination of her even more problematic to our eyes.

Douglas refers to his subject throughout as she, not the "it" of the advertisements, in keeping with his overall theory that hermaphrodites were in fact deformed women, a theory which had gained credibility over the course of the seventeenth century. She was born in Deptford in 1697, making her 17 or 18 at the time of his examination. Her name was "Constant or Constantia" Boon, and she was the daughter of the "lieutenant of a ship," that is, a ship's mate. She had been put out as an apprentice to a quilter a year earlier. Thus far she was resolutely average: according to Peter Earle, most London girls started work between the ages of 15 and 17, and the needle trades were viewed as the most "respectable," although the industry was so crowded as a result that it was also noted for its low wages. Needlewomen were accordingly often suspected of prostitution.[18]

The case of Constantia Boon well illustrates the irony of this situation. If she was sent to learn quilting as a respectable profession, she was soon sadly disillusioned. In Douglas's words, Constantia

> was cloathed and brought up like a girle and no body ever doubted of her sex till about a year agoe she was put out apprentice to a Quilter when her mistris observing something extraordinary about her lower parts had her examined by some old women who immediately pronounced her an hermaphrodite as partaking of both sexes being a mixture of male and female upon which her mistris designing to make a penny of her exposed her as a wonderfull prodigy of nature first in Smithfield about half a year agoe.

Douglas did not explain how her mistress obtained a view of her "lower parts."

By apprenticing her to a quilter, Constantia's parents presumably had thought to avoid for her a fate as a domestic or some other lower-class employment. Unfortunately we know nothing of her parents' feelings about her situation. Once Constantia was apprenticed, she was beyond parental control. Apprentices were considered to be part of the master's family and as such were under the complete control of the master. Although the application of this model varied greatly, legally the apprentice was bound to follow the master's wishes.[19] The possibilities for abuse were obvious, and apprentices sometimes

---

18 Peter Earle, "The Female Labour Market in London in the Late Seventeenth and Early Eighteenth Centuries," *Economic History Review*, N.S. 42 (1989), 328–353, at 344–345.

19 Steven R. Smith, "The Ideal and Reality: Apprentice-Master Relationships in Seventeenth Century London," *History of Education Quarterly*, 21 (1981), 449–459; see also Smith, "The

sued their masters because of their mistreatment: in one case, a female apprentice had been hung by her thumbs and whipped.[20] A further irony in the case of Constantia is that the terms of apprenticeship included the provision of moral and religious training, which it is unlikely she received while exposing herself in a booth for twelve hours a day.

Rather than learning how to quilt, then, Constantia was out on display. Douglas described her as "of a feminine air and voice," "fatt and well set," with long thick hair she had recently cut to make a wig which she wore. One of a throng of humans on display in London in its fairs, taverns, and coffeehouses, these unfortunates also tenanted booths at popular squares and crossroads, including Charing Cross, Temple Bar and Lincoln's Inn Fields, the latter also the site of anatomy lectures and later of Surgeons' Hall. These sites exhibited the range of human anomaly, including giants, dwarfs, conjoined twins and unusual medical conditions and were widely advertised in broadsides and newspapers. Only a few, however, were hermaphrodites. The noted physician and future president of the Royal Society, Hans Sloane, was fascinated with such anomalies and assembled a large collection of broadsides dating from about 1680–1740. But this collection includes only one other hermaphrodite apart from Constantia Boon, and newspaper advertisements from the same period reveal possibly one or two more.[21] This rarity, along with the combination of medical and sexual curiosity that hermaphrodites aroused, made Constantia Boon a prized commodity. Her prices of up to 2s 6d put her on the high end of exhibits, some of which charged as little as a penny. Her travels from place to place also indicated a certain demand. Douglas notes that she was first exhibited at Smithfield, a lower-priced venue (particularly during Bartholomew Fair), before moving to various coffeehouses around Charing Cross.

Taverns and coffeehouses played many functions in this era, including hosting public lectures on a wide variety of topics, from history and languages to anatomy and chemistry. In this context, the public exhibition of anomalous humans could take on a certain scientific aura, however spurious, and physicians and natural philosophers could visit under the guise of legitimate curiosity rather than prurient voyeurism. The course of anatomy lectures that James

London Apprentices as Seventeenth-Century Adolescents," *Past & Present*, 61 (Nov. 1973), 149–161; Deborah Simonton, "Earning and Learning: Girlhood in Pre-industrial Europe," *Women's History Review* 13 (2004), 363–386.

20 Smith, "London Apprentices," 152.

21 On Sloane's collection and newspaper advertising see Guerrini, "Advertising Monstrosity"; the Burney Collection of eighteenth-century British newspapers at the British Library contains many examples of such advertising.

Douglas had advertised in 1706 was to take place at a tavern known as the Blew Boar in Fetter Lane, near the Old Bailey and its hanging grounds.[22]

We can assume that Douglas came to the King's Head coffeehouse sometime early in 1715 and paid his shilling. He was conducted to a room at the back and allowed not only to view the private parts of Miss Boon, but thoroughly to examine and manipulate them. He was undoubtedly granted this privilege because his validation that she was a genuine hermaphrodite could be used as a selling point. At least one human exhibition, of the "bristly boy" in the early 1730s, even quoted an article from the *Philosophical Transactions* in its broadsides to validate the boy's condition as genuine.[23]

What did Douglas do and see? After interviewing the subject and taking note of her stature and bearing, he, like all who visited the back room, bade her lift her skirts, and he looked at her private parts. James Paris du Plessis, who compiled an album of his observations of street exhibits which he later sold to Hans Sloane, depicted this act in the style of the "flap anatomies" of the sixteenth and seventeenth centuries, in which layers of paper could be removed to reveal anatomical details, a kind of paper dissection. These "fugitive sheets" were widely distributed, unlike large and expensive illustrated anatomy texts. Du Plessis made a flap skirt which could be lifted to show the hermaphrodite's private parts, perhaps also echoing pornographic illustrations in this style.[24]

Unlike other visitors, Douglas did not simply look. After questioning her closely about her menstrual cycle and her sexual preferences, he pulled open Constantia Boon's labia to see the clitoris, which was of the shape and size of a penis, and caused it to become erect (he did not say how). Viewing a "great tuft" of pubic hair above the labia – looking more male than female – he noted the similarity in size and shape of the oddly hairless labia to the male scrotum

22 Douglas advertised this course in the back matter of his *Myographiae comparatae specimen* (London: for George Strahan, 1707), but it is unknown if he ever offered the course, although Helen Brock asserts that he did in *ODNB*. I have not found a newspaper advertisement for it, although the book was advertised, e.g. *Daily Courant*, issue 1407, 17 October 1706.

23 Guerrini, "Advertising Monstrosity," 124–127.

24 On James Paris du Plessis, see Dudley Wilson, *Signs and Portents: Monstrous births from the Middle Ages to the Enlightenment* (London: Routledge, 1993), 90–95; his drawing of a hermaphrodite is reproduced on p. 93. On flap anatomies, see Andrea Carlino, *Paper Bodies: A Catalogue of Anatomical Fugitive Sheets 1538–1687*, trans. Noga Arikha, *Medical History*, suppl. 19 (London: Wellcome Institute for the History of Medicine, 1999); on pornography, see Paula Findlen, "Humanism, Politics and Pornography in Renaissance Italy," in *The Invention of Pornography: obscenity and the origins of modernity, 1500–1800*, ed. Lynn Hunt (New York: Zone, 1993), 49–108; there is an example of flap pornography on p. 61.

and palpated them to perceive a testicle-like hardness inside, while noting that the texture of the skin was not like that of the scrotum.

Douglas continued his examination, seeking the opening of the urethra, which was not in the penis-like clitoris but "opens more within the vagina than in women, and cannot without putting her to some pain be seen." However, he did feel the "protuberating papilla" at the orifice of the vagina, but concluded that proof that this was indeed the opening of the urethra "only may be probed by autopsy itselfe." He peered into the vagina, noting its color and texture, "very red and with an abundance of blood vessels" and the absence of a hymen. But the opening of the vagina "soon contracts and grows so very nervous…that one cannot pass the end of one finger but a very little way without putting her to a great deal of pain." However, he managed to pass a catheter "severall inches up" until he hit "something hard," probably the cervix.

Douglas sketched pictures of the relevant parts at the home of a surgeon, Mr Rowsat, who was treating Constantia for a "serous ulcer," probably a boil, in her armpit. Rowsat suggested that the clitoris could be pierced to see if the "seminal matter" might be discharged through it. Douglas noted that Allen had made much the same suggestion about Anna Wilde, quoting the 1668 *Philosophical Transactions* article in its original Latin. "I should be glad," said Douglas, "to sie [sic] so curious and nice an operation performed by so dextrous [sic] a hand as that of Mr Rowsat." If Rowsat did indeed perform such an operation Douglas did not report it. We may note, however, that in the context of receiving medical treatment for something quite unrelated to her sexual parts, Constantia nonetheless had those parts again examined.

Douglas's examination of Constantia differed considerably from the usual medical examination. In this era, most physicians only touched the patient to take the pulse. A doctor examined a patient by taking a narrative of symptoms and by visually noting signs of illness. He might also examine urine, blood, or other discharges. If he touched a female patient it was in a highly mediated way. For example, when the physician George Cheyne examined Catherine Walpole in the early 1720s for a cluster of symptoms that included amenorrhea and a persistent throbbing pain in her side, he did not look at her private parts, and felt the throbbing in her side not only through her clothes but through her iron stays.[25] Even if we take into account the class differences between Catherine

25 On touching see Barbara Duden, *The Woman Beneath the Skin*, trans. Thomas Dunlap (Cambridge, MA: Harvard University Press, 1991), 81, 84–86; Anita Guerrini, *Obesity and Depression in the Enlightenment: The Life and Times of George Cheyne* (Norman: University of Oklahoma Press, 2000), 110.

Walpole (daughter of the Prime Minister) and Constantia Boon, Douglas's examination far exceeded the usual boundaries of medical examination.

While Douglas took an oral history of Constantia, his subsequent poking and prying was unheard of even in obstetrical cases. At this time, male midwives were called in mainly to deal with intractable cases, and even then they touched the female body only as necessary to release the child. Early in the seventeenth century, the French midwife Louise Bourgeois had advised drawing the bed-curtains in such a way that the male midwife and his patient could not see each other.[26] While the female midwife used her hands, only man-midwives used forceps, which became increasingly popular after the Chamberlens' monopoly ceased in 1733. Female midwives could touch parts of the female body (such as the "protuberating papilla") which their male counterparts, in normal circumstances, could not.[27]

The 1726 case of Mary Toft, the "Rabbit-Woman of Godalming", displayed the consequences of this customary reticence of man-midwives. Numerous experts examined Toft's claim that she had given birth to rabbits and the parts of other animals, but none could definitively prove she was a fraud. Douglas himself, called in as an expert witness, did not examine her cervix for signs that she had given birth and was among those who later defended themselves against charges of gullibility.[28] Displays upheld this division of the sexes: when Samuel Pepys and his wife saw the bearded woman Barbara Urslerin in 1668, "They offered [to] show my wife further satisfaction if she desired it, refusing it to men that desired it there."[29] The "further satisfaction" apparently consisted of pulling on her beard.

Douglas's examination of Constantia Boon was, indeed, closer to a dissection than to a conventional medical examination. Twice in his account he mentioned that he could not probe farther because his subject was (inconveniently?) still alive, but the language of opening, prying, and inserting instruments were the invasive terms of dissection, with distinctively sexual overtones. Douglas had advertised the availability of "fresh bodies" as one of the selling points for his anatomy course, and certainly this body was as fresh

26 Lianne McTavish, *Childbirth and the Display of Authority in Early Modern France* (Burlington, VT: Ashgate, 2005), 57–59.

27 Wilson, *Making of Man-midwifery*, Chapter 3, *passim*.

28 Ibid., 108. On the Toft case, see Dennis Todd, *Imagining Monsters* (Chicago: University of Chicago Press, 1995).

29 *The Diary of Samuel Pepys*, ed. Robert Latham and William Matthews, 11 vols. (Berkeley: University of California Press, 2000), 9:398 (21 December 1668).

as it could be.[30] Yet, as anatomists well knew, truth could only be found in the fully opened body, and no amount of probing and poking, stretching and pulling apart, could reveal the entire truth of this body. Unlike many other monsters, which were stillborn or which died shortly after birth, hermaphrodites commonly lived to adulthood.[31] The only well-known post-mortem of a hermaphrodite was of an infant who had died shortly after birth.

The publications that Douglas cited all described living subjects, not autopsies, and were much less detailed than his own. Allen's description consisted first of a narrative of the gradual revelation over time of Anna Wilde's sexual ambiguity: the emergence of testicles at the age of six, of a penis at 13, of a beard at sixteen, while she was raised as a girl and continued to display feminine characteristics such as menstruation. Like Constantia, she was attracted to women rather than to men, her attraction measured by the erection or flaccidity of her "penis." But Allen's examination of Anna Wilde was no more thorough than that of any other visitor to her display.[32]

Douglas compared Constantia Boon to two other accounts of hermaphrodites besides Allen's: that by the Dutch physician and anatomist Isbrand van Diemerbroeck in 1668, and by Realdo Colombo from 1559. Diemerbroeck described a hermaphrodite he saw in Utrecht who was almost certainly Anna Wilde. His account, like Allen's, was superficial and owed most to the narration of her "Governour." Diemerbroeck perceived her "yard" to be "half a Finger long" but the Governour assured him that "this Yard would upon venereal and lascivious Thoughts erect itself a Fingers length."[33]

A century and a half before Douglas, the renowned Italian anatomist Realdo Colombo had described a hermaphrodite in the final section on rare anatomical occurrences of his *De re anatomica* (1559). This "Ethiopian" woman had a "penis" the size of a little finger. Like Douglas, Colombo inserted the tip of his finger into the woman's vagina but found he could insert it no farther. However, he did not then insert a probe.[34]

---

30 Douglas, *Myographia Comparatae Specimen*, "An Account of what Dr. Douglas obliges himself to perform in a Course of *Human* and *Comparative* ANATOMY," pp. 209–216.

31 I have found no accounts of an autopsy of a hermaphrodite other than that of De Graaf, cited below.

32 Allen, "Exact Narrative"; Ruth Gilbert, *Early Modern Hermaphrodites: Sex and other Stories* (Basingstoke: Palgrave, 2002), 144–146.

33 Isbrand van Diemerbroeck, *The Anatomy of Human Bodies*, trans. William Salmon (London: for Edward Brewster, 1689), 183. Originally published 1669. Cf. Gilbert, *Early Modern Hermaphrodites*, 146–147.

34 Colombo, *De re anatomica*, 268–269. Colombo identified male and female hermaphrodites.

The Dutch physician and anatomist Regnier de Graaf managed to dissect a hermaphrodite. In his work on female anatomy, he described an infant girl he had seen in Delft with such a large clitoris that the midwife and others attending the birth had mistaken the child for a boy, and had given it a boy's name. Born in June 1670, the child lived only a month. The family's physician persuaded the parents to allow the body to be dissected "for the satisfaction of the curious." Only in the dead body could the full truth be discovered.[35] Although Douglas did not cite this case, he almost certainly must have known of it.

Douglas also did not cite the only other account of a hermaphrodite to appear in the *Philosophical Transactions* before 1715. She may be depicted in this engraving from ca. 1690 (Figure 1.1). In 1686, Monsieur Veay, a physician of Toulouse, described the case of Marguerite Malause or Malaure. This young maidservant had been raised as a girl and had all the characteristics one would attribute to a young woman, including an attractive face and neck, and "breasts as well made as one could desire in a girl." Like others, Veay did not hesitate to insert his fingers into her private orifices. Although her "parties honteuses" appeared sufficiently female, Veay could not penetrate her vagina beyond the depth of two fingertips ("deux petits travers de doigts"). And in the midst of this appeared a "membre viril" that, when erect, was some eight inches long. Perhaps most extraordinarily, semen, urine, and menstrual blood all flowed from this "membre." Veay admitted that the latter especially was hard to believe, but he had observed this with his own eyes, and had examined the member in question "fort exactement" at the time of the monthly courses over a period of several months. Veay referred to Marguerite Malause initially as a hermaphrodite, claiming that she possessed both female and male parts. But after consulting several of his fellow physicians, Veay declared her to be a male. Henceforth he would be called Arnaud and would dress as a man. The reasoning, said Veay, was that he could function as a man but not as a woman.[36]

---

35 Regnier de Graaf, *De mulierum organis generationi inservientibus Tractatus novus*, in *Opera omnia* (Leiden: Oficina Hackiana, 1677), 147–427, at 175–176 and 388–394.

36 "An Extract of a Letter Written by Mr. Veay Physician at Tholouse to Mr. de St. Ussans, concerning a Very Extraordinary Hermaphrodite in That City. Communicated by Dr. Aglionby. Reg. Soc. S.," *Philosophical Transactions*, vol. 16, no. 186 (1687), 282–283. The article was in French. Ruth Gilbert quotes the *Philosophical Transactions* to the effect that the article was published in French, "it being judged improper to appear in English," but I have not found this statement on the cited pages: Gilbert, *Early Modern Hermaphrodites*, 147; on this case, see also Daston and Park, "Hermaphrodites in Renaissance France." It was not uncommon for articles in French to appear in the *Philosophical Transactions*. On the legal status of hermaphrodites, see Fontes da Costa, "Anatomical Expertise" (mainly focused on the eighteenth century); Daston and Park, "The hermaphrodite and the orders of nature."

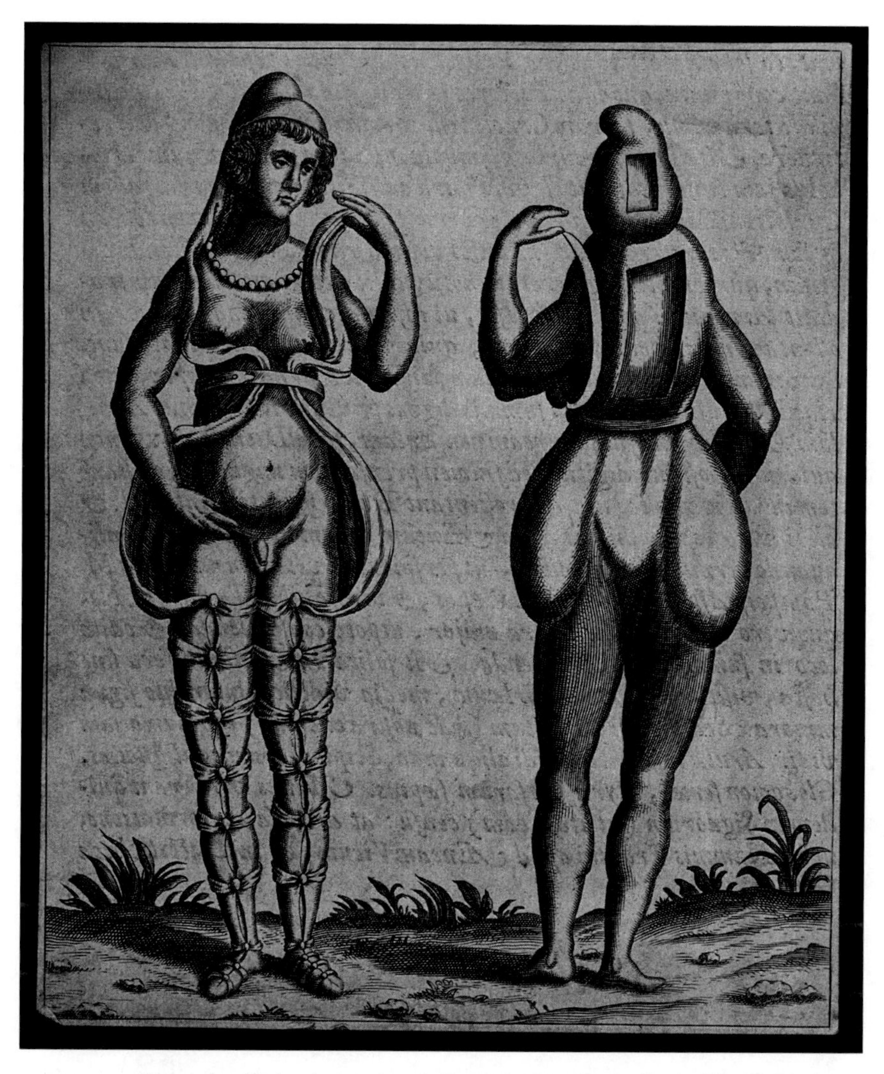

FIGURE 1.1 *Hermaphrodite in a loose costume. Engraving, ca. 1690. Wellcome Library, London.*

However, physicians and surgeons continued to examine Marguerite/Arnaud, who, it turned out, really was female. Without skills and unable to find employment as a man, she turned to public display of herself to make a living, moving from town to town. In 1693, the Paris surgeon and man-midwife Barthélémy Saviard examined her and then forced her to urinate in front of an audience at the Hôtel-Dieu Hospital in Paris. In order to prove that urine only came through her "penis", he held the lips of her vulva apart to show that no urine issued from elsewhere. Saviard, however, ultimately concluded that Marguerite was not a hermaphrodite but a woman who suffered from a severely prolapsed uterus, a condition he had demonstrated by the dissection of a number of female corpses.[37]

Saviard's commentary on the case of Marguerite Malause was first published in his collected works, *Nouveau recueil d'observations chirurgicales* (1702). Saviard strongly implied that because she allowed herself to be exhibited for money, his own examination of her, however invasive, was perfectly justified. Because he intended to reveal the truth, unlike other physicians and surgeons who simply accepted her claims about herself, he alone was capable of seeing through her subterfuges. At first he entirely refused to see her, but when he finally agreed (having already decided what her problem really was) he willingly paid the penny she demanded from each of the witnesses at the Hôtel Dieu.[38] Her status as a "monster" placed her outside the ordinary course of nature; but in making herself a commodity she placed herself outside the common bounds of acceptable human social behavior.

Would James Douglas have ventured to examine Constantia Boon had she not been on display? His description situated Constantia in the natural philosophical and medical discourse on hermaphrodites, which in turn fed into heated contemporary debates on generation and inheritance. This discourse relied heavily on popular representations. With the exception of the infant dissected by de Graaf, all of the learned descriptions of hermaphrodites I have found for this period were of individuals who displayed themselves for money. Douglas concluded that Constantia was not a true hermaphrodite, in that she did not have functioning sexual organs of both sexes. She was, he said, a woman with some irregular parts:

> Now from the external proportions of her body, from her mean, nature and complexion, and especially from the structure of her genital parts I cannot but conclude that this creature (reckoned amongst that kind of monster they call hermaphrodite or Androgyni) is a real woman with the addition only of something extraordinary in her labia pudendi but what that is and with what parts it dos [sic] communicate can never be thoroughly discovered till after death. however [sic], I humbly submitt my opinion to the gentlemen here present.

---

37 Barthélémy Saviard, *Nouveau recueil d'observations chirurgicales* (Paris: Jacques Colombat, 1702), 70–78; previous cases are described on pp. 49–66. Cf. the account in Gilbert, *Early Modern Hermaphrodites*, 147–149.

38 Saviard, *Nouveau recueil*, 73.

Constantia Boon appears to have been the only human hermaphrodite of note in London for over two decades. By the late 1720s, the anatomist and physician Frank Nicholls could report that "at present, the Existence of praeternatural Hermaphrodites seems universally denied," at least the human variety. His statement came in an article with four full-page figures describing a hermaphrodite lobster.[39] But the best-known hermaphrodite of the century appeared only a decade later.

In October 1740, the *London Daily Post* noted "the Dispute lately arisen between several Physicians and Surgeons concerning the African Hermaphrodite."[40] This was the "Hermaphrodite (Lately brought over from Angola)" advertised on a broadside with text in English and Latin. The English text gave a general external description, with no reference to sexual parts, while the Latin was an explicit and detailed account of those parts. As in Allen's account in the *Philosophical Transactions* over 70 years earlier, Latin served to hide the explicit nature of the description from all but the learned. "The arrival of the *Angolan* woman in town," said James Parsons, encouraged him to write his *Mechanical and Critical Enquiry into the Nature of Hermaphrodites*, published in 1741.[41] Parsons (1705–1770), a protégé of James Douglas, like him combined anatomical studies and medical practice as a man-midwife. Also like Douglas, Parsons was fascinated by the monstrous and had described several different monsters in the pages of the *Philosophical Transactions*.

Parsons discreetly summarized his book in the *Philosophical Transactions*. There were, he concluded, no true hermaphrodites; "those so reputed," he wrote, "were either perfect Men or Women, having only some Deformity or Disease in the Parts of Generation."[42] The previous published account of hermaphroditism, Giles Jacob's 1718 *Tractatus de hermaphroditis, or, A Treatise of Hermaphrodites*, had come to much the same conclusion. Jacob's Latin title gave his work a spurious learned authority, and he began his treatise, "The Secrets of Nature have in all Ages been particularly examin'd by Anatomists

---

39 Frank Nicholls, "An Account of the Hermaphrodite Lobster Presented to the Royal Society on Thursday May the 7th, by Mr. Fisher of Newgate-Market, Examined and Dissected, Pursuant to an Order of the Society," *Philosophical Transactions*, Vol. 36, no. 413 (1730), 290–294, at 290. See also *London Evening Post*, no. 376 (7–9 May 1730).

40 *London Daily Post*, no. 1860, 9 October 1740.

41 James Parsons, *A Mechanical and Critical Enquiry into the Nature of Hermaphrodites*. (London: J. Walthoe, 1741), liv.

42 James Parsons, " A Letter from James Parsons, M.D.F.R.S. to the Royal Society, Giving a Short Account of His Book Intituled, A Mechanical Critical Inquiry into the Nature of Hermaphrodites. London,1741. in 8vo," *Philosophical Transactions* Vol. 41 no. 459 (1741), 650–652, at 652.

and others," and claimed he followed the "Methods of Anatomical Writings." Jacob's work contained little that Douglas and others had not already mentioned, with the exception of the five-part classification of hermaphrodites of the French physician Nicolas Venette. But Jacob was not a physician, and his work was appended to a notorious pornographic work, *A Treatise of the Use of Flogging in Venereal Affairs*.[43]

Jacob's work was not illustrated, but neither was the *Philosophical Transactions*, which did not include illustrations in its articles on human hermaphrodites. Douglas's images of Constantia Boon had appeared, however, in several editions of William Cheselden's anatomy textbook, in the context of the anatomy of reproduction. Cheselden, a surgeon, was a popular public lecturer and also lectured at Surgeons' Hall. His anatomy text, written in English, was directed at an audience of interested gentlemen as well as medical and surgical students.[44] A century earlier, most anatomy texts were not illustrated, reflecting learned skepticism about visual evidence and the verisimilitude of reproductions. By the early eighteenth century, however, illustrations played an increasingly important role in such texts, as memory aids and as support for complex arguments.

Douglas ventured out again in 1740, this time to the Golden Cross coffeehouse near Charing Cross, to examine the "Female, Male, Moor, monster miracle of the world" and to draw her private parts.[45] His drawing was then engraved and appeared in a large fold-out illustration in Parsons' book, accompanied by smaller versions of his earlier drawings of Constantia Boon (Figure 1.2).

Unlike Douglas's 1715 paper, there is no extended account of his examination of the Angolan. Douglas briefly detailed her background: she had been taken as a slave from Africa to America, and was subsequently brought to Bristol. He did not state how she then came to be exhibited in London. He described her as 26 years old and quite feminine in appearance, unlike the broadside, which

43 [Giles Jacob], *Tractatus de hermaphroditis; or, A Treatise of Hermaphrodites*, in John Henry Meibomius, *A Treatise of the Use of Flogging in Venereal Affairs...to which is added, A Treatise of Hermaphrodites* (London: E. Curll, 1718), separately paginated, 1–2. Jacob is identified as the author in ODNB, s.v. Jacob, Giles (bap. 1686, d. 1744). Jacob was a lawyer and legal writer, and he claimed that hermaphrodites were not legally monsters. Nicolas Venette (1633–98) wrote the much-reprinted *Tableau de l'amour conjugal;* see Roy Porter, "Spreading Carnal Knowledge or Selling Dirt Cheap? Nicolas Venette's *Tableau de l'amour conjugal* in Eighteenth-Century England," *Journal of European Studies*, 14 (1984), 233–255.

44 Guerrini, "Anatomists and Entrepreneurs."

45 The broadside advertising the Angolan is headed "Faemina, Mas, Maurus, Mundi mirabile monstrum." Parsons credits Douglas with the main illustration of the Angolan in *Mechanical and Critical Inquiry*, 133–134.

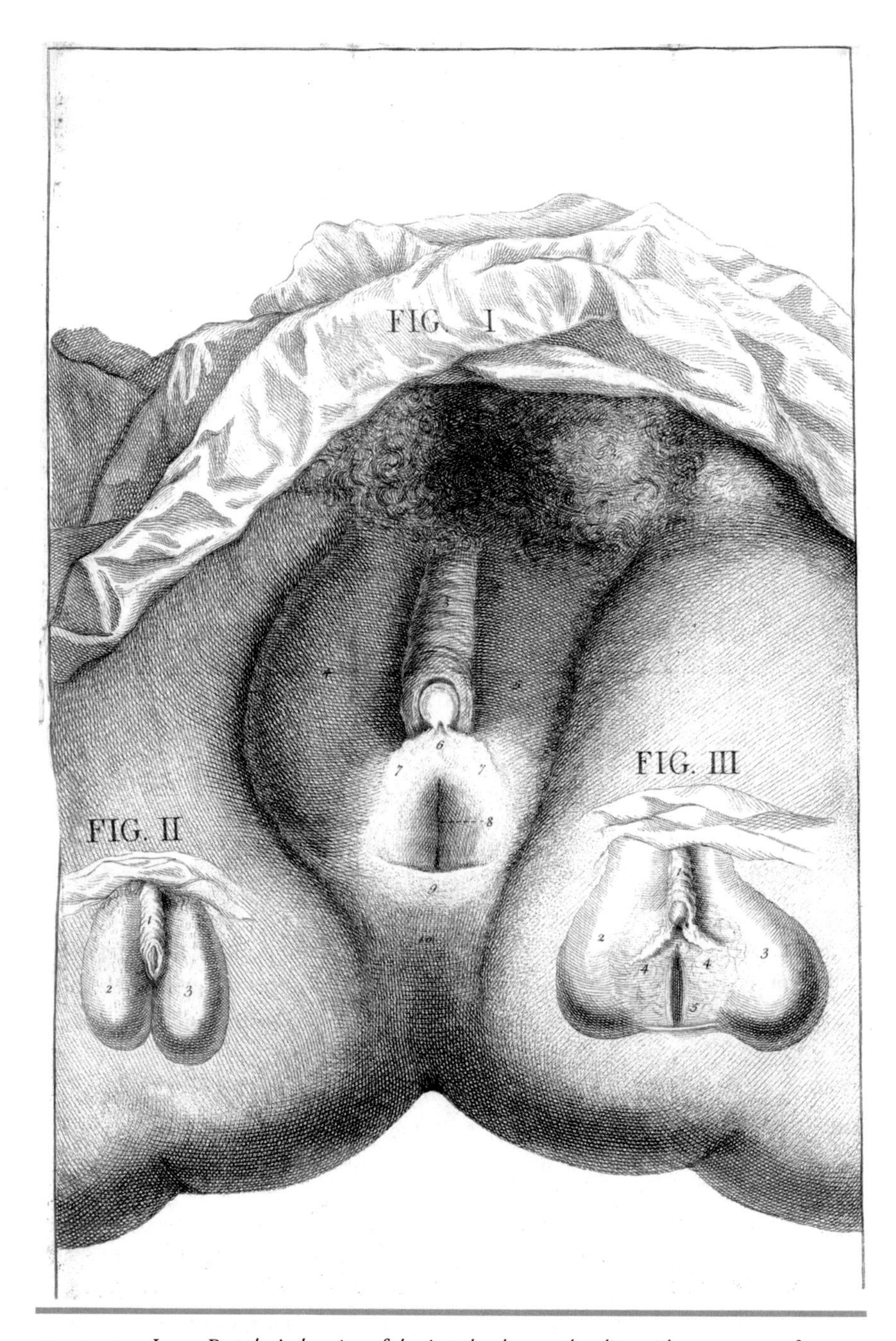

FIGURE 1.2 *James Douglas's drawing of the Angolan hermaphrodite, with two images of Constantia Boon superimposed on either thigh. James Parsons,* A Mechanical and Critical Inquiry into the Nature of Hermaphrodites, *1741. Wellcome Library, London.*

emphasized her masculine-looking muscular build. Douglas's commentary on his illustration, however, gives evidence of an examination as least as invasive as that of Constantia Boon. Although he described his drawing as "A view of the external Parts of Generation in the *African* Woman," as with Constantia, he endeavored to look as far inside her body as he could. He folded back her labia and palpated what was thought to be a testicle but that he concluded was a herniated ovary, referring not only to Diemerbroeck for authority, but also to his own collection of preparations "of the morbid uterine Parts." But a final determination could only be made "by ocular Inspection" of the internal parts – that is, after death. Parsons claimed, however, that Douglas's inspection had revealed parts of the body, including the interior of the vagina, that even de Graaf had failed to inspect during his 1670 autopsy of the hermaphrodite infant. Nonetheless, noting that there were other opinions on the particular case of the Angolan – including those of Cheselden and Hans Sloane – Parsons concluded that "as none of these Opinions can be ascertained without a fair Dissection of such a Subject" the question must remain unresolved.[46] We might compare the exhibition of the Angolan woman to that half a century later of the "Hottentot Venus," Sara or Saartje Baartman, noted for her prominent buttocks, who was exhibited in London beginning in 1810. She died in 1815 and was dissected by none other than Georges Cuvier; a cast of her body was on display in Paris until the 1970s.[47]

Ruth Gilbert recently called Douglas's drawing of the Angolan a "graphic illustration of the eroticization of the colonial gaze," although the images of Constantia Boon's parts engraved on either thigh make it less erotic than clinical. Like the examination itself, it revealed what was normally concealed, not, like most anatomical illustrations, in the context of the dissection of a dead body, but in the examination of someone who was very much alive and, to a degree, a participant in the depiction. The details of dark skin and nappy pubic hair make it clear that this is a particular individual. The image makes explicit what Ludmilla Jordanova has called the "sexual-cum-intellectual penetration" of the dissecting room, but it is not the product of a dissection. While the image had affinities with the brutally explicit obstetrical images of William Smellie and later, of William Hunter, its purpose was not for obstetrical instruction.[48]

---

46 Parsons, *Mechanical and Critical Inquiry*, 138–143 and Fig. I–III.

47 A good survey of the voluminous literature on Baartman is Sadiah Qureshi, "Displaying Sara Baartman, the 'Hottentot Venus'," *History of Science*, 42 (2004), 233–257.

48 Gilbert, *Early Modern Hermaphrodites*, 164; Ludmilla Jordanova, "Gender, Generation and Science: William Hunter's Obstetrical Atlas," in *William Hunter and the Eighteenth-Century Medical World*, ed. W.F. Bynum and Roy Porter (Cambridge: Cambridge University Press, 1985), 385–412, at 401.

Apart from Douglas's drawing, there were at least two other illustrations of the Angolan. Parsons included a second drawing that showed the labia closed where Douglas had shown them open. The reason for a second view, said Parsons, was so that "the Reader may the better understand, how easily the ignorant or Superstitious might be deceived at the Sight of such Parts."[49] In other words, the superficial impression gained by the casual viewer did not give the whole truth; only the anatomist, whose gaze extended beneath the surface, could find and convey this truth. Parsons relied heavily on his anatomical authority to promote his explanation of the hermaphrodite as a deformed woman, but not all agreed.

Yet another view appeared in the fifth edition of Cheselden's *Anatomy*, published in 1740. Previous editions, from the second in 1722 to the fourth in 1730, had used Douglas's images of Constantia Boon. In 1740, however, Cheselden offered instead "The Parts of an hermaphrodite negro, which was neither sex perfect, but a wonderful mixture of both. This person was twenty-six years of age, and in shape perfectly male." (Figure 1.3) This can be no other than the Angolan, which Cheselden readily accepted at face value as a true hermaphrodite. His credulity mirrored that of his book's (and lectures') broad and unspecialized audiences and also reflected the disagreements, even among experts, that the *London Daily Post* had mentioned. His drawing is a "closed" view and emphasizes the penis-like qualities of the clitoris; neither the drawing or the accompanying text betray any further exploration of the subject other than this external view.

Accompanying this drawing is a re-engraved version, without attribution, of one of Douglas's drawings of Constantia Boon, again with only superficial explanation of the parts. Cheselden's title page emphasized the verisimilitude of his illustrations by showing an artist – presumably himself – using a camera obscura to view an anatomical model suspended in front of him on a tripod.[50] Since other evidence shows that the Angolan began to be exhibited in London in the autumn of 1740, it is apparent that Cheselden, like Parsons, acted quickly to capitalize on her fame, although with rather differing motives. Parsons aimed to debunk the entire notion of a hermaphrodite, while Cheselden's text, accepting the popular valuation of the subject, added the Angolan as an update to the earlier image of Constantia Boon. His text, with this illustration, had a

49 Parsons, *Mechanical and Critical Inquiry*, 155 and Tab. I. Douglas's previous images of Constantia Boon had indeed given both views. Parsons was noted as an illustrator and it is likely he drew this view.

50 Cheselden, *The Anatomy of the Human Body*, fifth edition (London: William Bowyer, 1740), 314 and Table XXXIII. My thanks to Sarah Cohen for help in interpreting the title page.

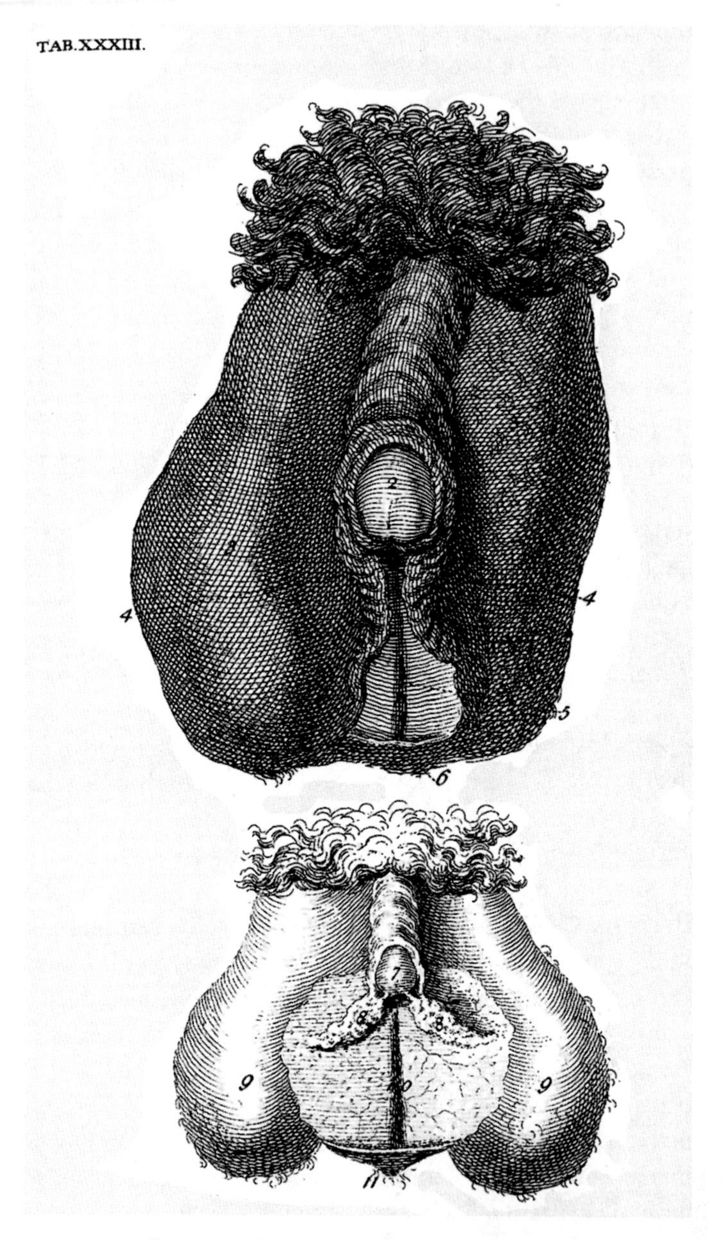

FIGURE 1.3 *Illustration of the Angolan (above) and Constantia Boon (below). William Cheselden,* The Anatomy of the Humane Body, *seventh edition, 1750 (images first published 1740). Wellcome Library, London.*

long life, reaching a thirteenth edition in 1792. It may be the only illustration in the text based on a living rather than a dead person.

The intermingling of learned and popular, of anatomy and display, was not confined to hermaphrodites. I have shown elsewhere how print, either in broadsides or in learned articles, functioned to control the narrative of display of monsters more generally, and how different individuals and groups attempted to appropriate that narrative.[51] In conclusion, let me examine more closely the intimate relationship between anatomists and the public display of hermaphrodites. The culture of curiosity has been much examined, and most of this historiography posits a widening split between science and popular culture by the eighteenth century. Lorraine Daston and Katharine Park, in particular, argue that wonder among the learned is tamed and domesticated by the naturalistic explanations of science, leaving the wondrous to what Parsons called "the ignorant or Superstitious."[52] But such an assessment takes the natural philosophers' evaluations of themselves at face value and overlooks the continued interaction between learned and vulgar.

The obsession of the Royal Society – and Hans Sloane in particular – with monsters of every sort can be seen as a trivial and lingering relic of the past. Satirist William King twice took Sloane (editor of the *Philosophical Transactions* from 1693 to 1713) to task for his attention to the trivial in *The Transactioneer* (1700) and in *Useful Transactions in Philosophy* (1709). But in King's work, it is the virtuoso who expresses wonder and credulity while the "gentleman" debunks what he considers to be common-sensical and even mundane occurrences.[53]

Popular display and natural philosophy quite often were about the same things, and this is particularly true when we enter the realm of defining the human. One reason that anatomists continued to pay attention to human and animal anomalies and their public display was because the preternatural, as Francis Bacon had long ago suggested, could help to define the natural.[54]

---

51 Guerrini, "Advertising Monstrosity." The concept of appropriation derives from Roger Chartier, *The Cultural Uses of Print in Early Modern France*, trans. Lydia Cochrane (Princeton: Princeton University Press, 1987), 6–8.

52 Lorraine Daston and Katharine Park, *Wonders and the Order of Nature* (New York: Zone Books, 1998), 305 and Chapter 9; Parsons, *Mechanical and Critical Inquiry*, 155.

53 [William King], *The Transactioneer* (London, 1700); [idem], *Useful Transactions in Philosophy* (London: Bernard Lintott, 1709). The latter appeared briefly as a periodical. There is much more to be said about King's critiques; see Roger D. Lund, "'More strange than true': Sir Hans Sloane, King's 'Transactioneer', and the deformation of English prose," *Studies in Eighteenth-Century Culture*, 14 (1985), 213–230.

54 Francis Bacon, *The New Organon*, ed. Lisa Jardine and Michael Silverthorne (Cambridge: Cambridge University Press, 2000), 148, discussing "deviant instances."

After all, the eighteenth century was the great era of classification, of organizing nature's productions into a logical order that would tell us something about the purposes and order of nature. In this context, anomalies had to be accounted for or otherwise explained away.

As King suggested, the virtuoso inserted himself into the narrative of a popular display in order to control its message, to make the vulgar rational. So James Douglas classified Constantia Boon as a human woman, not a monster. She was not outside the ordinary realm of nature, but rather showed the variety and fecundity of that nature. By publishing his account of the Angolan, Parsons went farther, asserting that the category of "hermaphrodite" was always invalid. He repeated this assertion when another hermaphrodite made "some noise in town" in 1750. Parsons hastened to Ludgate to "undeceive such as are mistaken about it," subjecting the young French person to the kind of invasive examination we have already seen. He noted that if the skin that overgrew the perineum (which he poked his finger under to feel a vagina) was "snipp'd with scissars [sic]" this would be obvious to all and not just to his expert eyes.[55]

But Parsons was mistaken in thinking that he could control the narrative of this display, or of any hermaphrodite, simply by snipping some superfluous skin. As the case of the Angolan showed, multiple narratives, popular and learned, existed side by side, with none gaining hegemony. Parsons' attempt to appropriate the narrative in 1741 did not succeed, since the Angolan returned to display as a hermaphrodite in 1743. An advertisement from that year in the *London Daily Post* proudly proclaimed:

> Just Arriv'd from ABROAD.
> To be SEEN,
> At the *Rummer-Tavern, Charing-Cross*
> From Ten in the Morning to Ten at Night
> The Famous AFRICAN HERMAPHRODITE, mention'd by *William Cheselden*, Esq., in Page 314 of his Anatomy, and by Dr. *Parsons* in his Treatise of Hermaphrodites, though erroneously describ'd by this latter Author; being a Native of Angola, about 27 Years old.

The advertisement continued with a "particular Description of the Parts" in Latin, and noted that "An Engrav'd Figure of the Parts, lower'd to 6d. each

55 James Parsons, "A Letter to the President, concerning the Hermaphrodite Shewn in London: By James Parsons M.D.F.R.S.," *Philosophical Transactions* vol. 47 (1751–52), 142–145.

Figure" was also available.[56] I do not know if this engraving was one of the three published in 1740–41.

Neither Douglas nor Parsons appear to have considered that if the women they examined were not in fact hermaphrodites, then their examinations far exceeded the bounds of ordinary medical, or moral, behavior. In order for them to examine these subjects, the anatomists had also to accept the public's valuation of their status. This valuation bridged the space between observation and intervention and made these investigations possible. Unlike the usual medical examination, the investigations of these hermaphrodites were indeed experimental.

56 *London Daily Post*, issue 3780, 20 September 1743. The Latin differs from the description on the 1740 broadside.

CHAPTER 2

Galvanic Humans

Rob Iliffe

Two revolutionary discoveries announced in the last decade of the eighteenth century transformed what was known about the nature of electricity and the role it played in animal life. The advent of galvanism and then within a decade, of the Voltaic pile, accelerated major changes that were already taking place in the medical and physical sciences. These events were transformative both because they had immense theoretical implications for understanding the basis of life and matter, but also because they provided extraordinary research tools with which topics in these fields could be investigated. Galvanism served to further destabilise existing relations between medical and scientific disciplines that were already being rearranged. Accordingly, it changed the traditional roles of anatomists, natural philosophers and others, and enabled – or demanded – cooperation between practitioners in a wide range of fields. Within medicine, galvanic experiments played a key role in the foundation of experimental physiology, which investigated the body by means of new techniques and theories drawn from physics and chemistry.[1]

The revelation of residual electrical activity in amputated limbs and recently expired cadavers brought into question any strict demarcation between life and death. Galvanism offered the potential for determining the time and indeed the nature of absolute, irrevocable death, and it promised to shed light on the peculiar liminal states of being that were akin to death – such as drowning and asphyxiation (of which hanging was an example) – from which some humans had recovered. Some even argued that various states of suspended animation could be engineered by means of galvanism in order to permit certain surgical procedures to be carried out, after which the patient could be revived by the application of another galvanic current. As for the *post mortem* vitality revealed by galvanism, it was initially unclear whether it was the same phenomenon as the 'irritability' in various organs that had been demonstrated by Albrecht von Haller and his followers. Similarly, it remained to be decided whether galvanic electricity was the same as 'ordinary' electricity, or whether it was to be

1 For a general account of historical links between electricity and medicine, see M. Rowbottom and C. Susskind, *Electricity and Medicine: A History of their Interaction*, (San Francisco: Macmillan, 1984).

 | DOI 10.1163/9789004286719_004

identified with the 'nervous fluid' that apparently effected intentional action as it coursed from the brain through the nervous system.[2]

In this paper I look at the main features of European research in the first phase of galvanic enquiries. I trace the development of galvanism from the pioneering researches on frogs carried out by Galvani, up to the use of the voltaic battery on the remains of people who had been killed by means of the guillotine. I discuss the role of animals in galvanic research, and show how galvanic practices demanded a seamless transition from experiments on animals to those on humans. Although electrical experiments had been tried out for some time on humans for medical reasons, the application of a steady electrical current promised new ways of providing and standardising medical treatments for both physical and mental ailments. I examine the various techniques adopted by self-experimenters, notably the Scottish physician and philosopher Richard Fowler, the German natural philosopher Alexander Humboldt, and his one-time assistant Johann Ritter. I finish by briefly looking at the rage for using a Voltaic pile to perform experiments on guillotined victims that lasted from 1802 to 1804, and discuss the social and political contexts in which these new trials took place.

Galvanism

In 1791 Luigi Galvani, the Bolognese Professor of Anatomy and Obstetrics at the University of Bologna announced in an obscure publication entitled *De Viribus electricitatis in motu musculari Commentarius* that he had definitively proved the existence of 'animal electricity' within living creatures. Based on a discovery made five years earlier, Galvani described an extremely simple experimental arrangement according to which the exposed thigh muscle of a recently sacrificed frog could be made to twitch by joining its attached nerve to two different metals. Galvani noted that muscular contractions were more powerful when nerves were coated in metal foil, and they only conducted electricity when they were moist. Galvani had been working on bioelectricity for over a decade before his initial observation. In the early 1770s he had already become enamoured of the notion that animals had a special kind of intrinsic electricity, and that muscular contractions could be effected by electrical stimulation. More generally, electricity played a key role in his general conception of how the mind and body operated together. He argued that an electrical force was

2 See H. Steinke, *Irritating Experiments. Haller's Concept and the European Controversy on Irritability and Sensibility, 1750–1790*, (Amsterdam: Rodopi BV, 2005).

triggered by the mind when it willed an act, and the appropriately stimulated nerves then gave rise to activity in 'voluntary' muscles. With the new discovery, Galvani claimed that the muscle fibre and attendant nerves functioned like Leyden Jars (i.e. capacitors, devised in the middle of the 1740s). These Jars were able to store electrical charge and then discharge it when the internal and external coatings (or 'armatures') were connected by means of a metal wire or hook. According to Galvani and his followers, what took place in the nerve-muscle combination was akin to the operation of the Jar. Electricity inside and on the outside of individual muscle fibres was in a state of 'disequilibrium' that was disturbed when a nerve allowed electricity to flow from the inside to the outside of the muscle. This discharge produced contractions in the muscle.[3]

Galvanic effects immediately provoked questions about the nature of galvanic fluid, such as whether it was the same as frictional or atmospheric electricity. The existence of the galvanic circuit promised to shed new light on standard problems such as whether the electrical fluid was the same as the nervous fluid, and whether animal electricity resided only in those muscles that were susceptible to the will. Overshadowing these issues was the question of the relationship between animal electricity and 'irritability', which was prominent in debates in medicine and natural philosophy in the second half of the eighteenth century. In the middle of the century. Haller had attacked the older theory according to which the brain controlled muscle movements by generating animal spirits that travelled down the nerves as 'nervous fluid'. Instead, on the basis of thousands of experiments performed on both living and dead animal tissue, he argued that organs had an intrinsic capacity to contract (when touched by an external object) independent of the action of the brain and nerves. Moreover, a number of animals seemed to show no sign of possessing a nervous system. While critics such as Robert Whytt continued to maintain the view that the brain was the dominant organ in the body, Hallerians saw the heart as the main source of life and energy. By showing that the brain

3 Galvani's publication actually appeared at the start of 1792. See M. Pera, *The Ambiguous Frog: The Galvani-Volta Controversy on Animal Electricity* (Princeton, 1992); N. Kipnis, "Luigi Galvani and the debate on animal electricity, 1791–1800," *Annals of Science*, 44 (1987), 107–42, esp. 114–6; C. Blondel, "Animal electricity in Paris: from initial support to its discredit and eventual rehabilitation," in M. Bresadola and G. Pancaldi, eds *Luigi Galvani. Workshop Proceedings*, (Bologna 1999), 187–204; P. Bertucci and G Pancaldi, eds, *Electric Bodies: Episodes in the History of Medical Electricity*, (Bologna, 2001). Comprehensive early histories of galvanism can be found in P. Sue, *Histoire du Galvanisme; et Analyse des différens Ouvrages publiés sur cette Découverte, depuis son origine jusqu'à ce Jour*, 4 pts (Paris, 1802–5); J.B. Trommsdorff, *Geschichte des Galvanismus, oder der galvanischen Elektricität*, (Erfurt, 1803), and C.H. Wilkinson, *Elements of Galvanism, in Theory and Practice, with a comprehensive View of its History*, 2 vols (London, 1804).

exercised control over muscles by means of its capacity to mobilise animal electricity, Galvani's work reinstated the mental organ as the primary source of life. Nevertheless, the unresolved question concerning the role of the heart meant that its susceptibility to galvanism remained a primary experimental target for researchers.[4]

The second great electrical discovery of the epoch was revealed by Alessandro Volta in 1800. Volta, who had repeated Galvani's results by the end of March 1792, was a professor of experimental physics at the University of Pavia and had already gained a major reputation both as a theoretician and as a brilliant manipulator of electrical instruments. Having been interested for some time in so-called 'weak' electricity, he thought that a galvanised frog's leg could be a very sensitive electrometer. Although at this stage he had not ruled out the existence of animal electricity, his investigations led him to deny that any sort of 'vital' element within the animal body was *essential* for producing Galvani's electricity. By the time he wrote two letters to the natural philosopher Tiberius Cavallo in London towards the end of 1792, he had decided that the metals were not merely conductors but were electromotors that provided *extrinsic* electric force to the circuit. On this showing the animal tissue was merely a passive receptacle of the current produced by the metals. In response, Galvani argued in 1794 that experimental evidence indicated that there was a continuously circulating current in the animal tissue that was disrupted whenever the circuit was created or broken. Crucially, he reported that he had been able to excite contractions in a frog's leg with no metallic element playing any role in the arrangement.[5]

Volta then sought to exclude everything organic from his experimental set-up, and he concentrated on the role played by moist bodies ('second class' conductors in his terminology) in the various galvanic phenomena that he and Galvani had produced. In 1797 he announced that he had been able to measure with extremely fine instruments the weak electricity produced when two metallic parts of the circuit came into contact, proving that it was this contact

4 See W. Bernardi, "The controversy on animal electricity in eighteenth-century Italy: Galvani, Volta and others," *Revue d'histoire des Sciences*, 54 (2001), 53–70; Kipnis, 'Animal electricity', 131–4 and more generally Bernardi, *I fluidi della vita: Alle origini della controversia sull'electricittà animale*, (Firenze, 1992); Haller, 'Irritating experiments'; M. Sirol, *Galvani et le Galvanisme. L'Électricité Animale*, (Paris, 1939).

5 Volta, "Account of some discoveries made by M. Galvani, of Bologna," *Philosophical Transactions*, 83 (1793), 10–44. See Kipnis, 'Animal electricity', 119–121; M. Bresadola, "Animal electricity at the end of the eighteenth century: The many facets of a great scientific controversy," *J. Hist. Neurosciences*, 17 (2008), 8–32, esp. 9–19; Pera, 'Ambiguous Frog'; and especially G. Pancaldi, *Volta: Science and Culture in the Age of Enlightenment*, (Princeton, 2005).

that occasioned the current. Although Galvani showed in the following year that animal contractions could be generated by a wholly organic circuit, (composed entirely of nerves), his death soon afterwards ended his part in one of the great controversies in the history of science. As the culminating act in this exchange, at the end of 1799 Volta created a 'pile' or battery that made use of alternating silver (or copper) and zinc plates, each of which was separated by a brine-soaked piece of cardboard (or cloth) that acted as a humid conductor (or electrolyte). Connecting each end with a wire created a continuous current that lasted a long time and which could be more easily controlled than with other electrical instruments. The battery became a marvellous research tool in medicine, physics and chemistry. Not only did it facilitate major new discoveries in these fields, but it also promised to shed new light on the role of electricity in life. A bonus was the fact that its scaled up effects made it a much more suitable demonstration device than the standard galvanic circuit.[6]

Working with Animals

Following a division standard in comparative anatomy, early galvanic researchers routinely divided up their results between cold- and warm-blooded animals. The former appeared to be much better research objects than their counterparts on account of their capacity to hold their 'vitality' much longer. Like many other researchers who were interested in irritability, Galvani commonly used frogs to conduct experiments on living and dead organic tissue. The University of Bologna had a long-standing research programme aimed at elucidating the possible electrical nature of the nervous fluid, and frogs were the most commonly used organism in the institution. In particular, Galvani's practice was based on the procedures used by the Bolognese Hallerians Felice Fontana and Leopoldo Caldani, who perfected techniques for triggering a response in a properly prepared dead nerve-muscle specimen by means of an electric conductor. In the 1780s Galvani modified the core techniques used by his colleagues, which involved bisecting a frog so as to leave only the lower

---

6 Bresadola, 'Animal electricity', 13–15; Pancaldi, 'Volta'; and Volta, "On the Electricity excited by the mere contact of conducting Substances of different kinds," *Philosophical Transactions*, 90 (1800), 403–31. See also J. Golinski, "From Calcutta to London: James Dinwiddie's Galvanic Circuits," in Bernard Lightman, Gordon McOuat and Larry Stewart, eds., *The Circulation of Knowledge Between Britain, India and China: The Early-Modern World to the Twentieth Century* (Brill, 2013).

limbs with crural nerves and spinal cord exposed. As Marco Bresadola has pointed out, thanks to the training he had received at Bologna in both anatomy and surgery, Galvani had developed practical skills that would be essential for his work with frogs and other animals. Indeed, in his *de Viribus* Galvani attributed the success of his experiments, which confirmed that nerves were electrical conductors, to the careful way in which the frogs had been prepared.[7]

Frogs had long been the basic sacrificial object in anatomy and surgery, and they fared even worse after Galvani made his fateful discovery. Virtually all writers on galvanism commended frogs for their availability, their ease of preparation (the thigh muscle could easily be dissected along with the nerve) and the way in which they conserved activity after bisection and maintained 'vitality' after death. Aside from the use of frogs *per se*, the mode of preparation used by Galvani became the dominant technique for reproducing the basic phenomenon. This was passed on in many descriptions of the galvanic effect, or it was learned from personal contact with individuals who had performed the experiment. In any case, the basic set up of the preparation, i.e. the use of frog thigh muscles with an exposed crural nerve, was soon well established in anatomical and surgical establishments across Europe. In the decade following the announcement of galvanism, untold numbers of frogs were used in experiments although for reasons of comparison and also for generating a riveting spectacle, researchers resorted to other sorts of animals. Very quickly, as we shall see, they turned to humans as experimental subjects.[8]

Other animals became highly relevant to galvanic research. Aside from the frog, the most significant experimental object was the torpedo fish, a creature whose ability to discharge a stunning force had been known since the times of the Romans. A fish that could deliver an even more powerful shock, the South American *gymnotus electricus* had also been recently described. With the

---

7 See A. Cunningham, *The Anatomist Anatomis'd: An Experimental Discipline in Enlightenment Europe*, (Farnham, 2010), 162–3, 223–31; J. Heilbron, "The contributions of Bologna to Galvanism," *Historical Studies in the Physical and Biological Sciences* 22 (1991), 57–85; M. Bresadola, "Medicine and science in the life of Luigi Galvani, (1737–1798)," *Brain Research Bulletin* 46 (1998), 367–80, esp. 368–74 and Bresadola, "At play with nature: Luigi Galvani's experimental approach to muscular physiology," in F.L. Holmes, J. Renn and H-J. Rheinberger, eds, *Reworking the Bench: Research Notebooks in the History of Science*, (Kluwer 2003), 67–92, esp. 70–6.

8 See Wilkinson, 'Elements', 1: 199–200 and 292; F.L. Holmes, "The old martyr of science: The frog in experimental physiology," *J. Hist. Biol.* 26 (1993), 311–28, esp. 317–19; T.H. Levere, "Dr. Thomas Beddoes (1750–1808): Science and medicine in politics and society," *British Journal for the History of Science* 17 (1984), 197, and A-H Maehle and U. Tröhler, "Animal experimentation from Antiquity to the end of the eighteenth-century" in N.A. Rupke, ed., *Vivisection in Historical Perspective*, (London, 1990), 14–47.

advent of the Leyden jar researchers conjectured that the discharge from the torpedo was of a similar electrical nature, and in the early 1770s a number of experiments and dissections were performed on specimens of both species by John Walsh and John Hunter. These investigations seemed to throw light on the mechanism by which the torpedo could extrude an electrical shock by an act of will, though it left other issues unresolved. For example, the discharges given off by electric fish differed considerably in their effects – for example, no electrical sparks could be drawn from the torpedo – from those produced by the Leyden jar, or by normal (static or atmospheric) electricity.[9]

The Torpedo fish became a central research tool for both Galvani and Volta. From one perspective it was an ideal object for studying the existence of electricity in animal tissue and, in particular, Galvani believed that the fish's ability to voluntarily discharge electricity demonstrated the general way in which the brain disseminated its influence around the body. Galvani's use of the torpedo was in part a response to the criticism of his work made by Volta, who had accused him of only experimenting with dead animals. In turn, Galvani's work motivated Volta to work intensively on electric fish, and it was the latter's encounter with the peculiar arrangement of the electric organs of the torpedo and the electric eel that prompted him to develop the battery in the way he did. In the letter to the Royal Society in which he revealed the existence of the pile, he explicitly stated that his design was based on the internal anatomical structure of the torpedo. Indeed, although he strove to create an entirely material circuit, he was fascinated by the ways in which the galvanic circuit might shed light on the working of the animal kingdom, and was concerned with the medical benefits that might arise from the use of the battery.[10]

Human Remains

Both before and after the invention of the pile, galvanism practically demanded that human tissue be presented for examination. Humans had long been the subjects of analytical dissections in medicine, but the way in which human bodies were examined, and findings were interpreted, was undergoing a dramatic transformation, particularly in Paris. Given the paucity of proper bodies for dissection, comparative anatomy was the best means of seeing how parts of a body

9 W. Cameron Walker, "Animal electricity before Galvani," *Annals of Science*, 2 (1937), 84–113; Kipnis, 'Animal electricity', 112–3 and in particular M. Piccolino, *The Taming of the Ray: Electric fish Research in the Enlightenment, from John Walsh to Alessandro Volta*, (Florence, 2003).

10 Piccolino, 'Taming of the Ray'; Bresadola, 'Animal electricity', 19–21.

cohered, and the manipulation and dissection of body parts was an essential part of surgical training. However, its predominance in medicine was challenged by the new disciplines of experimental physiology and pathology. As the number of anatomical schools increased throughout Europe, so the acquisition of sufficient numbers of suitable cadavers became a pressing concern. In Italy anatomists had access to bodies of people who had recently died in hospitals but in many European cities criminal executions provided the only ready supply of human subjects. After the 'Murder Act' of 1752, London's College of Surgeons were supposed to receive all bodies executed for murder. However, except where the state intervened proactively on behalf of medics, it remained notoriously difficult to procure sufficient bodies for the burgeoning numbers of students attending 'new' styles of medical anatomy. Grave-robbing remained a prevalent problem in many cities and gave rise to serious friction between relatives of the dead and the 'agents' of anatomists, known colloquially as 'resurrection men'.[11]

Apart from dissecting intact humans, physicians and natural philosophers had for some time been performing electrical trials on amputated limbs. Both Galvani, who envisaged a number of medical benefits from his discovery, and the Pisan physician Eusebio Valli used amputated limbs in their early demonstrations of galvanism. Given the difficulty in obtaining a copy of Galvani's initial publication, Valli's European tour was central to the initial transmission of galvanic theory and practice. In March 1792 he spent some time at Pavia with Volta, and performed a number of galvanic experiments using frogs; however, he broke with the physicist as Volta's disillusionment with the notion of animal electricity became evident. In the summer, having already published some preliminary results, Valli travelled to France as an 'ambassador' of Galvani. At a time of much political tension, there was nonetheless substantial interest from French physicians and natural philosophers in galvanic phenomena. Valli initially worked with a commission set up by the Académie des Sciences, and soon afterwards he repeated his experiments at the Société Royale de Médicine, where he worked in the laboratory of Jean-Claude de Lamétherie. Many of these trials involved finding different ways of sacrificing various species of animals, and occasionally he was successful in 'reviving' chickens. Over the following year he published 9 'Lettres' on galvanism in the journal *Observations et Mémoires sur la Physique*, edited by de Lamétherie.[12]

---

11 See Cunningham, 'Anatomy' and W.F. Bynum, "Reflections on the history of human experimentation," in S. Spicker, I. Alon, A. de Vries and H. Engelhardt eds, *The Uses of Human Beings in Research*, (Dordrecht 1988), 29–46.

12 See Valli *Experiments on Animal Electricity*, (London, 1793), 91–110; Wilkinson, 'Elements', 1: 70–2; Blondel, 'Animal electricity', 188–90. Valli had undertaken inoculation experiments

Valli travelled to London in the Autumn of 1792, attending the November meeting of the Royal Society where Tiberius Cavallo gave the second part of a Voltian (i.e. bimetallic) account of the galvanic phenomena. However, Valli's ports of call were mainly hospitals. He did galvanic tests on a horse with the remarkable veterinarian and Himalayan explorer William Moorcroft at his equine hospital in Oxford Street, and then did experiments on amputated limbs with William Blizard at the London Hospital. Valli and Blizard, who believed that electricity could be used effectively to alleviate certain kinds of deafness, succeeded in eliciting contractions when they used a silver conductor to touch a nerve coated with tinfoil, though the limb lost its 'vitality' after half an hour. Galvanic experiments on amputated limbs were repeated throughout Europe, for example, by Baron Dominique-Jean Larrey (later Surgeon-General to the French Army in Egypt), Leopold Vacca Berlingheri and Carl Caspar Creve in Germany.[13]

Because they could help determine the nature and extent of consciousness, very recently dead cadavers were perceived to be of much greater potential service than amputated limbs for pursuing electrical experiments. The advent of the guillotine provided researchers with an extraordinary device for conducting research on this topic. The early experiments on its victims were performed in the context of vigorous medico-political debates over the ethics of its use in state-sponsored execution, and in the light of discussions regarding the role of the brain in mind-body interaction. Defenders of the revolutionary death machine argued that it was a humane killing technique that offered an immediate despatch, with no trace of lingering pain. However all researchers agreed that this question could best be decided by experimental enquiry. Although it scarcely featured in discussions, religious convictions underpinned the views of many of those who found the guillotine obnoxious. Defenders of traditional beliefs tended to believe that it was at least conceivable that the soul might outlive its severance from the body, while materialist proponents of an organised systemic view of the body had no need of the hypothesis in the first place.[14]

---

both on himself and others in the late 1780s and would die in Cuba in 1816 as a result of self-experiments involving Yellow Fever.

13 Valli, 'Experiments', 15–16, 70–2, 161–5, 195–6, 316–7; Blizard, "Observations on the uses of electricity in deafness, communicated in a letter to Dr. Simmons" *London Medical Journal*, 11 (1790), Article IV.

14 R. Borgards, "'Kopf ab': Die Zeichen und die Zeit des Schmerzes in einer medizinschen Debatte um 1800 um Brentanos Kasperl und Annerl," in G. Brandsetter, G. Neumann eds, *Romantische Wissenpoetik: die Künste und die Wissenschaften um 1800*, (Wurzburg 2004), 123–50; D. Arasse, *The Guillotine and the Terror*, (Harmondsworth 1989), 32–47; D. Outram,

Early experimental procedures carried out by Samuel Thomas Soemmerring (Professor of Anatomy at the University of Mainz), involved touching various parts of the severed head with an electric probe, and were performed to determine whether there was any *post mortem* consciousness or any cranial muscular activity at all in beheaded victims. In 1795 Soemmerring wrote a letter to the radical Prussian journalist Konrad Engelbert Oelsner arguing that the fact that amputated limbs exhibited galvanic activity *post mortem*, and also that heads displayed signs of consciousness *post guillotinem*, indicated that the brain was the seat of volition (he located the soul in the cerebro-spinal fluid) and the chief organising centre of general human activity. Along with the physician Jean-Joseph Sue he concluded that the guillotine was undignified and inhumane – unlike hanging – because the capacity for awareness survived beheading. Independently of the feeling pain, the individual might also experience the ultimate terror of being aware of his/her own death.[15]

In response to Soemmerring and Sue, a number of defenders of the guillotine claimed that any signs of movement after decapitation were signs of automatic and non-conscious irritability exhibited naturally by all parts of the body. Urban myths about severed heads talking or even blushing (as allegedly happened in the case of Charlotte Corday, the mistress and murderer of Marat), were either the feverish fantasies of the vulgar, or were stories put about by those who were unsympathetic to the new regime. Pierre Jean Georges Cabanis claimed that the systemic operation of the holistic organism ceased to function at the moment the guillotine cut through the victim's spinal cord. He equated this conception of the body, which lacked a specific location for an organising principle, to the hierarchical order of systemic functions within a state governed by the General Will. Jean Sédillot (le jeune) offered a related rationale for supporting the guillotine, although his argument more overtly compared the fate of criminals to those of animals. Accordingly, he claimed that its effects were no different from those achieved by countrymen, butchers or cooks, who killed animals immediately by striking them in the neck.[16]

The Body in the French Revolution, (New Haven 1989), and L. Jordanova, "Medical meditations: mind, body and the guillotine," *History Workshop Journal*, (1989), 39–52.

15 S.T. Soemmering, "Lettre de M. Soemmerring à M. Oelsner," *Magasin Encyclopédique*, 3 (1795), 468–77; J-J Sue, "Opinion de J-J Sue sur la douleur qui survit à la decollation," *Magasin Encyclopédique* 4 (1795), 170–89. See also M. Hagner, "The soul and the brain between anatomy and *Naturphilosophie* in the early nineteenth century," *Medical History* 36 (1992), 1–33.

16 Cabanis, "Note addressée aux auteurs du Magasin Encyclopédique, sur l'opinion de Messieurs Oelsner et Soemmering et du Citoyen Sue, touchant le supplice de la guillotine, par le citoyen Cabanis," *Magasin Encyclopédique* (1795), 155–74; Sédillot, *Réflexions*

In alliance with this new scientific instrument, galvanism seemed to offer an exact experimental means of deciding whether consciousness survived decapitation, and thus whether guillotining was truly ethical. Galvanic experiments on the heads and torsos of decapitated animals performed by Sue indicated that there was substantial 'vitality' in the body, particularly the heart, for many hours after the execution. Here was evidence that there was sufficient residual *post mortem* activity for feeling and thus suffering to occur. The most extensive experiments on guillotined bodies using galvanic stimulation were performed by Xavier Bichat at the end of 1798. As part of the transformation of clinical medicine in Paris, Bichat pioneered new techniques in tissue pathology in which the analysis of organ tissues became the primary means of determining how different diseases had taken their course. Bichat's experiments with the guillotine were undertaken as part of a series of trials on warm-blooded animals. The deceased were brought to him about 20 minutes after execution but the quality of his results varied considerably between individual subjects. He was able to stimulate the heart when "mechanical exciters" were applied to its fleshy fibres, but was not able to activate the heart by creating a galvanic circuit out of the spinal marrow, brain and heart. These results were corroborated by trials on pigs and dogs. He concluded from his experiments that organic functions associated with involuntary muscles were not connected to the immediate influence of the brain, and that therefore the actions of the heart were completely independent of it.[17]

Galvanic Benefits

Galvanism promised even more practical benefits in the case of living beings, and its use as a possible cure for various ailments came nearly half a century after the Leyden Jar had been used for the same reasons. In his *History of Electricity* of 1767, Joseph Priestley attributed the first attempt to apply electricity to medicine to Christian Gottlieb Kratzenstein, who attempted to alleviate

---

Historiques et Physiologiques sur le Supplice de la Guillotine, (Paris, 1795), 14–22. See Outram, 'Body', 106–23 and Jordanova, 'Medical meditations', 41–2 and 44–7.

17 J.-J. Sue, "Recherches physiologiques et expérimentales sur la vitalité," *J. de Physique* 46 (1798), 226–35 and Bichat, *Physiological Researches on Life and Death*, trans. T. Watkins (Philadelphia 1809), 269–78. See also F. Duchesneau, *La physiologie des lumières. Empiricisme, modèles et théories*, (Boston and London, 1982) and J. Lesch, *Science and Medicine in France: The Emergence of Experimental Physiology, 1790–1855* (Cambridge, Mass. 1984).

arthritis in a patient in 1744. However, the advent of the Jar the following year made it much easier to control the shocks applied to the sufferer. Initially, experimenters reported great successes with the new equipment, especially in treatment of paralysis and palsies. However, in the wake of research by Nollet, Franklin and others, early claims for the efficacy of the process proved to be greatly overdone. Franklin became the most authoritative witness for the belief that electricity had only a very narrow potential for positively affecting diseases. He used the Leyden Jar on many patients in the late 1740s and early 1750s, initially believing that electricity might be beneficial in common palsies such as strokes. However, the effects he achieved were never more than temporary, and he told the Royal Society in 1757 that he had little hope that electricity would help in the majority of afflictions. Although his initial success with the instrument was confined to treating a patient suffering from hysteria, in the 1774 (second) edition of his *Experiments* he claimed that a series of substantial shocks from the Jar might alleviate madness. In 1783 Jan Ingenhousz confirmed that strategically placed 'cranial' shocks could cure melancholy.[18]

Medical electricity remained immensely popular. In Britain, James Graham was a prominent and popular proponent of medical cures, although by the 1770s his dubious practices, and those of many others, were being denounced as forms of quackery. A plethora of new electrical instruments such as Lane's electrometer appeared in the 1770s and 80s, expanding the range of techniques used in the procedures. Electrical practitioners like the physician Erasmus Darwin and Tiberius Cavallo, who published *An Essay on Medical Electricity* in 1780, felt increasingly compelled to distinguish their work from those who had been condemned in Britain and France as charlatans. Over a number of decades Darwin used many different sorts of electrical source to effect cures, and when the Voltaic battery became available at the end of his medical career he was quick to adopt it for medical purposes. Cavallo was particularly critical of the claims made by the Abbé Pierre Bertholon, arguably the most famous proponent of electrical cures in Europe. Even more so than in England, medical electricity was viewed with suspicion by French natural philosophers and physicians, having been linked to the discredited doctrines of mesmerism and animal magnetism in the 1770s and early 1780s. Nevertheless, although there had been many fraudulent claims made on its behalf, Cavallo believed that electricity could produce real benefits and was quick to test the efficacy of the

18 Priestley, *History and Present State of Electricity*, 2nd ed. (London 1775), 472–3; S. Finger, "Benjamin Franklin on the electrical cure for disorders of the nervous system," in H.A. Whitaker, C. Upham, M. Smith and S. Finger, *Brain, Mind and Medicine: Essays in Eighteenth Century Neuroscience*, (New York 2007), 245–56, esp. 48–52.

galvanism. His most important sources of information about such effects were the surgeon Myles Partington and James Lind, physician to the chronically ill George III. Lind offered testimony about real cures, which Cavallo included in his book, and the pair corresponded over a number of experiments reported in the earliest papers on galvanism by Galvani and Volta.[19]

Although it had entirely novel characteristics that were connected to the issue of control of voluntary muscles and also to the nature of vital functions, galvanism offered new means of addressing practical questions that had been raised earlier with other sorts of electricity. For example, it promised new means of determining true death, and for possibly reviving 'drowned' or 'asphyxiated' victims. Ascertaining absolutely irrevocable death was deemed to be crucial at a time when there were reportedly numerous examples of hasty burials, or precipitous declarations of death on surgeons' operating tables. Moreover, there were widespread rumours of individuals being resurrected in coffins or in anatomical theatres. Whether in the hands of an ordinary citizen or a skilled natural philosopher, in principle one could decide once and for all whether specific (involuntary) muscles could be stimulated with a powerful galvanic current, and thus define death in galvanic terms. Debate over the utility of galvanism for this purpose was vigorous. Creve performed a number of experiments on recently dead bodies to determine the putative moment of 'real' death, and concluded that such an analysis was possible. Others like Cristoph Pfaff decided that galvanism was insufficiently sensitive to perform the task. To a much greater extent than Hallerian irritability, Galvanism called into question the entire understanding of death, whether from the liminal conditions of existence that were the concern of the humanitarian societies across Europe, or from the galvanic sensitivity that briefly characterised bodies and parts of bodies *post mortem*.[20]

---

19 Bertholon, *De l'Électricité du Corps humain dans l'État de Santé et de Maladie* (Paris, 1786) (the first edition of which had appeared in 1780); P. Bertucci, "Medical and animal electricity in the work of Tiberius Cavallo, 1780–1795," in M. Bresadola and G. Pancaldi, eds, *Luigi Galvani. Workshop Proceedings*, (Bologna 1999), 147–65, esp. 151–2; G. Sutton, "Electric medicine and mesmerism," *Isis* 72 (1981), 375–92; Blondel, 'Animal electricity', 199–200 and 203 and M. Bresadola, "Early galvanism as technique and medical practice," in Bresadola and Pancaldi, 'Galvani', 157–79, esp. 164–9.

20 Wilkinson, 'Elements', 1: 408–16 and 2: 463–5; C. Creve, *Vom Metallreize, einem neuendeckten untrüglichen Prüfungsmittel des wahren Todes* (Leipzig, 1796) and more generally C. Milanesi, *Mort apparente, Mort imparfaite. Médicine et Mentalités au XVIIIe Siècle*, (Payot 1991) and C. Sleigh, "Life, death and galvanism," *Studies in the History and Philosophy of the Biological Sciences*, 29 (1988), 219–48.

Self-galvanism

In the event that frogs, electrical fish or the sick were unavailable, the experimenters' own bodies were particularly appropriate for galvanic trials. In his second letter to Cavallo (of 25 October 1792), Volta had mentioned that he had tried numerous experiments on many different animals, finding that the weak electric current produced by the galvanic circuit produced 'irritation and movement' only in voluntary muscles, and 'not at all in those muscles on which the will has no direct power.' The latter class of organs included the brain, the heart and the intestines. Moving on to the amputated limbs of humans, Volta found substantial galvanic effects on voluntary muscles provided they were properly prepared, that is, with the 'coating' of relevant metals applied to bare muscle. So much for experiments on the tissue of 'dead' bodies, but what if the experiment were tried on living, exposed human tissue? Volta decided to use the tongue, ideally suited to the purpose since it was susceptible to the will – and what better than his own organ?[21]

Volta's decision to use the tongue unwittingly rehearsed the procedure recorded in 1767 by Johann Georg Sulzer in his *Nouvelle Théorie des Plaisirs*. While performing experiments on various aspects of 'taste', Sulzer had mentioned that two different metals applied to different parts of the tongue and then brought together would engender an acrid taste. Volta told Cavallo that he had placed a piece of tinfoil on the upper surface of his tongue and had then brought it into contact with a silver spoon that was in contact with the lower surface. Looking at himself in a mirror to get a better view, he witnessed no contraction of the muscle but experienced the same taste – produced continuously as long as the metals were in contact – reported by Sulzer. This, he deduced, was because the nerves dealing with taste were in the area affected by the muscle, whereas those that controlled movement were further back towards the 'root' of the tongue. This was confirmed by applying a galvanic circuit to the severed root and upper surface of the tongue of a recently killed lamb, and the experiment succeeded on a host of other animals. The results effectively refuted Galvani's idea that the circuit made use of an electrical source that was inherent to the muscles, since Volta was concerned only with the nerves and the metallic components of the circuit.[22]

21 Volta, 'Account', 34–5 and 41–2. For the earlier tradition of self–experiment see S. Schaffer, "Self evidence," *Critical Inquiry* 18 (1992), 327–62.

22 Volta, 'Account', 42–3. In the 'Memoria secunda' he described using lead and gold as alternatives to tin and silver.

In Scotland the primary setting for reproducing the experiments reported by Volta and Valli was the Edinburgh Medical School. Although members of the School were in contact with James Lind, who had been a student there, the primary conduits of information to Scotland were Joseph Black (who had himself been informed of Galvani's work in August 1792 by Lavoisier's assistant Armand Séguin), and a report in the 1792 volume of *Medical Facts and Observations*. The redoubtable anatomist and surgeon Alexander Monro *secundus* reproduced the basic phenomenon described by Galvani on 3 November 1792, and the following week tried variations on the same on a large number of frogs and rabbits. Richard Fowler, a student of Monro and a member of both the Royal Medical and Speculative societies of Edinburgh, performed the most influential series of galvanic experiments. At the time Fowler undertook his experiments he had not yet completed his MD, which was awarded in the September of 1793 on the topic of inflammation. He had heard lectures on galvanism by Monro in Edinburgh, and earlier by Matthew Baillie, in London (in October 1792), and had begun to perform experiments with friends at the RMS in November. The Royal Medical Society encouraged both self-experiment and animal vivisections and these were prominent in Fowler's reports.[23]

Fowler's experiments were undertaken to determine whether galvanic effects could be attributed to known laws of electricity, and whether the same effects could bring about positive influences on the sick. He recorded a series of experiments on both warm and cold-blooded animals, and demonstrated from trials with decapitated frogs that Valli was incorrect in assuming that the brain was the 'common receptacle' for electrical fluid. On the other hand, the fact that the torpedo fish was able to control its electrical discharge at will, seemed to suggest that it was a different species of force from galvanism. Fowler concluded that the galvanic influence was largely owing to the activity of the external metals but having ascertained that 'healthy' frogs could mitigate the effects of the influence to some degree, he argued that the final effects arising from the application of galvanism were not the result of the metals alone. Otherwise, he managed – contrary to the results of previous investigators – to stimulate the hearts of a frog and a cow by 'irritating' the nerves with zinc and silver. This seemed to support a Hallerian position though Fowler argued that

23 A. Monro (Secundus), "Experiments relating to Animal Electricity," *Transactions of the Royal Society of Edinburgh*, (1792) 3–7. Accounts of Galvani's *De Viribus* and on Valli's letters in the *Journal de Physique* appeared in *Medical Facts and Opinions*, 3 (1792), 180–9 and 190–211. For the Edinburgh medical context see Chris Lawrence, "The Edinburgh Medical School," *History of Universities*, (1988), 259–86 and G. Risse, *New Medical Challenges during the Scottish Enlightenment*, (Amsterdam, 2005), 74–5 and 78–84.

further experiments were required to determine whether and to what extent the brain could affect the galvanic fluid in different organs.[24]

Fowler's work was of most interest to researchers because of the innovative experiments he did on his own body, in conjunction with self-experiments performed by a number of his fellow members of the RMS. Volta's tongue experiment was easy to reproduce and many researchers quickly replicated it in various social settings. For example, the natural philosopher and recent member of the RMS, Thomas Beddoes, tried the phenomenon, as did a number of ladies who were present. Fowler performed the experiment under a number of different conditions, varying the temperature of the tongue, trying it on inflamed parts of the body, and also comparing the sensations produced by galvanism with that engendered by normal electricity. Completing a new circuit by inserting different metals into his ears resulted in a "disagreeable jirk" of his head, and overnight haemorrhaging of blood. Unwilling to repeat this performance, he now tested the galvanic current on his eyes, creating an eyeball-tongue-head circuit. This produced a 'pale flash of light' over the whole of his eye, and the effect was greatly magnified if zinc and gold were used instead of tinfoil and silver. A further elaboration of this involved putting a silver rod up his nose and connecting it with some zinc placed on his tongue. This likewise produced a flash, arguably more powerful than one resulting from a circuit applied directly to the eye, and Fowler believed he had shown that the iris could be a reliable detector of the galvanic influence. Finally Fowler reported that his friend, George Hunter from York, who had assisted him in nearly every experiment, had tried placing different metals between the lip and gum of the upper and lower jaws. This had resulted in another 'flash' that was different in kind from the one experienced by Fowler, and which appeared to be located over the whole face. These experiments would be repeated many times in the next decade, and further variations on them were developed by researchers who believed that they shed a powerful light on the arrangements of the facial nerves.[25]

---

24 Fowler, *Experiments and Observations relative to the Influence lately discovered by M. Galvani, and commonly called Animal Electricity*, (Edinburgh 1793), ii, 1, 7, 10, 38, 44, 55–9, 69–77.

25 Beddoes to Davies Gilbert, 12 September 1792, Cornwall Record Office, DG 41/19 (I owe this reference to Larry Stewart); Fowler, 'Experiments', 83–91; Wilkinson, 'Elements', 1: 147–50 and 161–2. See also S. Jacyna, "Galvanic influences: themes in the early history of British animal electricity," in M. Bresadola and G. Pancaldi, eds, *Luigi Galvani international Workshop Proceedings*, (Bologna 1999), 167–85 and M. Bresadola, "Galvanismo senza Galvani: la ricezione dell'electtricità animale in Inghilterra, 1792–4," in A. Gatti and

Fowler's work had as an appendix a short letter from the Edinburgh natural philosopher John Robison, who had heard from his son about Fowler's experiment involving the eye-tongue circuit. Robison extended the experiment so that he kept a piece of zinc on his tongue and brought a piece of silver into contact with it, the silver being in contact with other parts of the body including the urethra and anus. In every case, the tongue was able to detect when the circuit was made. In another experiment Robison created an open wound in his toe and applied zinc to the affected part while placing silver on his tongue. In this case some irritation was felt at the site of the wound. Next he applied a zinc point to a hole in his tooth, placing a piece of silver on the inside of his cheek. When the metals were brought into contact, he experienced excruciating toothache. In general Robison found that increasing the number of pieces of zinc increased the effect, and told Fowler that he was now able to detect by taste alone the elements that were used in gold and silver 'trinkets'.[26]

The tongue test was soon repeated across Europe, and researchers developed a number of ingenious variations on the theme. Unquestionably, the most innovative galvanic self-experiments were performed by German physicians and natural philosophers, many of which made the original trials appear tame. In 1795, for example, Franz Karl Achard created a circuit designed to test the effects of galvanism on the roles of the intestines and the sphincter muscle. To the applause of commentators, Achard placed a zinc coating in his mouth and inserted a silver probe in his anus, and when the metals were connected, noted a sharp pain in his stomach followed by involuntary bowel contractions. Alexander von Humboldt duly repeated the experiment on a linnet near the point of death. At the moment when the circuit was established, to his amazement the creature opened its eyes, stood erect and fluttered its wings. It breathed for a further six or eight minutes before expiring "tranquilly". Further experiments convinced him that even if it were not wholly appropriate for resuscitating the genteel, the technique could be used to revive small animals that had drowned in urban apartments, thus repaying some of the debt owed to nature by the "massacres" exacted on it by humans.[27]

P. Zanardi, eds, *Filosofia, Scienza, Storia. Il Dialogo fra Italia e Gran Bretagna* (Padova, 2005), 215–32.

26 Fowler, 'Experiments', 169–74 (Robison's letter was dated 28 May 1793); Jacyna, 'Galvanic influences', 175–7.

27 Blondel, 'Animal electricity', 190; J.B. Trommsdorff, 'Geschichte', 11–12; Humboldt, *Versuche über die gereizte Muskel- und Nervenfaser nebst Vermuthungen über den chemischen Process des Lebens in der Thier- und Pflanzenwelte*, 2 vols (Berlin 1797–9) 1: 334 and Wilkinson, 'Elements', 1: 299–300.

Humboldt performed the most extensive, as well as the most extreme, experiments carried out in the eighteenth century. He confirmed the findings of Galvani and Aldini, showing that it was possible to create nerve-muscle circuits without any metals, and that organic (muscle) tissue possessed animal electricity. The most spectacular experiments were performed on his own body, often as a result of opening sores that lasted for several days. The accounts of previous writers were starting points for a number of exciting and innovative experiments in which he determined the conditions under which organic tissue acted most strongly as a conductor or insulator.[28]

Humboldt tested his teeth to see whether they were true insulators, and did a number of experiments on a gaping hole in his jaw left after a tooth extraction. Other experiments, which constituted pioneering exercises in the physiology and phenomenology of pain, gained him the respect of researchers across Europe. He opened up wounds in each of his deltoid muscles and made a connection between zinc and silver metals placed on each, an action that immediately gave rise to a 'violent smarting sensation'. Putting alkaline solutions on the wounds increased the visible reaction along with the pain. Following this, he made an opening in his wrist that exposed the radial artery, thereafter placing a zinc coating on the wound and touching the zinc with a silver coin. A peculiar but extreme sensation shot along his hand to the ends of the fingers and engulfed his entire palm, a feeling that was renewed every time the two metals were brought into contact. When he needed someone to critically check his researches Humboldt drew upon the services of Johann Wilhelm Ritter, who was more inclined than was Humboldt to understand galvanism in terms of the doctrines of *Naturphilosophie* that were then ascendant in Germany. Ritter argued that living beings were made up of connected galvanic circuits, and that all of nature was a cosmic animal. When Volta's pile became available, Ritter's philosophical commitments drove him to make the most extreme self-experiments that anyone had ever undertaken.[29]

The Pile

The publication of Volta's paper on the electric pile attracted the attention of medics, chemists, physicists and electricians. As with Galvani's initial discoveries, news of the pile travelled astonishingly quickly around Europe.

28 Wilkinson, 'Elements', 1: 203 and 261–306, esp. 284–5, 295.

29 Wilkinson, 'Elements', 1: 284–5, 295–7. For more details of Humboldt's experiments on his own body, see Steigerwald's essay in this volume.

This was thanks to personal contact but also to reports found in various letters, newspapers, books, translations, magazines, encyclopedias and journals. A number of histories of galvanism appeared at the start of the nineteenth century, their size both a testament to the work already done to that point as well as to the promise offered by the application of the pile in physics, chemistry, medicine and animal physiology. The greater power, continuous current, and the ease of use of the instrument made it a much more amenable research tool than anything that had gone before it. Apart from its potential as a multipurpose piece of equipment for the exact sciences, it provided experimenters with an unrivalled device for assessing the effects of a powerful electric current on the living organism. 'Galvanism' henceforth referred not to animal electricity but to both the natural causes that underlay the operation of the pile, and also to the effects that could be achieved by using it. Originally viewed as merely a more intense version of ordinary galvanism, it became apparent that the chemical and physiological effects that could be obtained with the pile were very different from, and much more striking than what could be achieved with ordinary circuits. The effects of the pile also appeared to be easier to measure, its power apparently proportional to the number of plates it contained. Finally, the battery allowed demonstrators to display to large audiences the extraordinary power of the galvanic current, using a wide variety of creatures such as domestic animals and humans.

Carl Johann Christian Grapengiesser, physician to the Crown prince of Prussia, was one of the first researchers to use the pile on patients, particularly in cases of blindness and deafness. Grapengiesser claimed some success in these trials, but reported difficulties in balancing the need for a powerful battery alongside the deleterious side effects that accompanied many of his experiments. Other parts of the body did not require such delicate control. For example, he tested the battery on a man in a military hospital with a scrotal hernia and an ileum that was so exposed that it hung to his knees. Although the patient's peristaltic movements had ceased to function, enhanced galvanic stimulation of the intestinal canal caused them to be restarted. Nevertheless, the patient felt great pain, and large secretions flowed from the intestines onto the metals. Grapengiesser's techniques and research problems were taken up by many researchers in France, Italy, England and Scotland, confident that here was a tool that was sufficiently powerful to effect positive change in the organism. The surgeon and galvanic physician Charles Henry Wilkinson, who was based in London, claimed that the application of the pile for curing paralysis was occasionally successful. However in cases of deafness and blindness – and especially when the pile was connected to the urethra, where it had been used with some success in the case of an enlarged prostate – any advantages arising from its use might be offset either by the terror felt by the patient or by debilitation due to overuse.[30]

The work of Giovanni Aldini was crucial to shaping the way that the battery was used across Europe to investigate the role played by electricity in the living organism. The nephew of Galvani, Aldini played the role of ambassador performed a decade earlier by Valli. Having spent much of the 1790s rebutting Volta's metallic account of the galvanic circuit, Aldini became the most enthusiastic adopter of the Voltaic battery in the field of experimentation on the "human animal machine". Continuing his attack on Voltaism, he argued that his experiments showed that all life was founded on an "animal pile", which was analogous to the artificial version discovered by Volta (thereby reversing Volta's own analogical argument). In the first year of its operation he tried various experiments on animals, subjecting hundreds of reptiles, mammals and birds to the power of the Leyden Jar and the pile. The 'Account' he published of his spectacular galvanic performances is most famous for the report he gave on the terrifying demonstration of galvanic vitality meted out on the corpse of a London criminal in January 1803. However, most of the book consisted of a series of anatomical and physiological experiments on animals. In Bologna, Turin and London he provided spectacular trials demonstrating what happened when the pile was applied to recently killed oxen, lambs and chickens, and he "declared war" against dogs, though hundreds, and probably thousands of felines were also dispatched. It was a logical progression to investigate "that noble being man, the sole object of my researches".[31]

In January 1802 Aldini procured for study "near the axe of justice" in Bologna, body parts from two recently decapitated criminals. Soon afterwards he participated in a more detailed galvanic investigation on the remains of a Bolognese assassin. This was led by the anatomist Carlo Mondini, whose preparation of the human subjects Aldini described in some detail. He praised Mondini's dexterity in the highest terms but disparaged himself as too inexperienced in anatomical dissections. However this experience was undoubtedly useful in his later galvanic performances, and was essential in establishing his credibility as a serious and moral individual when he showed his galvanised humans to large crowds. The Bolognese victims were all "robust" men and still

---

30 Grapengiesser, *Versuche den Galvanismus zur heilung einiger Krankheit anzuwenden*, (Berlin, 1801); Wilkinson, 'Elements', 1: 301–2; 2: 189, 449–50, 454–7.

31 Aldini, *An Account of the late Improvements in Galvanism*, (London, 1803), 8, 14, 47–51, 54–60, 63 (for the remark on dogs), 67 and 89–92; Sleigh, 'galvanism', 228–9.

possessed the sort of "vitality" that was necessary for his experiments. The pile produced a series of bodily contortions and 'horrid' grimaces in the faces of decapitated humans that could not be observed with a weaker current. Confirming his earlier views about the lack of response to galvanic stimulation exhibited by involuntary muscles, Aldini was unable to trigger any response in the heart but he emphasised that he was open to having the experiment repeated. In his renewed defence of animal electricity, he was perversely ingenious in the way he devised complex organic circuits, some of which involved creating arcs between the remains of two humans, or even between humans and animals, and others that connected various areas of the brain to different parts of the face. The "disagreeable sensations" excited by the "melancholy scenes" caused by working on humans were somewhat mitigated, he argued, by the belief that a new discovery might soon follow.[32]

News of Aldini's exploits circulated around Europe. In the summer of 1802 he worked in Turin with the anatomists Anton Maria Vassalli-Eandi, Carlo Giulio and Francesco Rossi, emboldened by the prize offered in June by Bonaparte for major new discoveries in electricity. Aldini and the others graduated from initial experiments on the head of oxen to work on the bodies of criminals, carried out before a large number of stunned spectators. Against the experience and beliefs of a number of previous researchers, including Bichat, Volta and Aldini himself, they obtained *post mortem* contractions in the heart and arteries (Aldini had shown galvanic influence on the involuntary organs only in the case of cold-blooded animals). This was initially achieved 'mechanically', that is, without the pile, but in time they had success in effecting galvanic stimulation of the heart. Contractions continued for up to 40 minutes after the pile had been disconnected. It was essential, Giulio commented in his report, to obtain cadavers as soon after death as possible, and for that reason the experimenters obtained a hall right next to the place of execution. The delay between death and experiment, and the insufficiently low temperature of the setting in which he had done his work were both invoked as reasons for Aldini's previous failures. The heart, Giulio concluded, was particularly susceptible to losing its vitality soon after death.[33]

---

32 Aldini, 'An Account', 67–86, esp. 67–8, 76–85. See also Sleigh, 'Life', 229–31 and 233, and for general issues surrounding human preparation see Cunningham, 'Anatomy', 231–46.

33 Aldini, 'An Account', 87–9 and 204–16. The experiments were originally reported in C. Giulio, "Rapport presenté à la Classe des Sciences Exactes de l'Académie de Turin le 27 Thermidor Sur les experiences galvaniques faites les 22 et 26 du même mois, sur la tête et le tronc de trois homes, peu de temps après leur decapitation," *Journal de Physique*, 55 (1802), 286–96 (translated in *Philosophical Magazine*, 15 (1803), 38–44), and 'Galvanism',

In the third section of his work on galvanic experiments, Aldini mentioned a number of cures of various illnesses and testified that many individuals had been revived having suffered 'artificial' death. He claimed that galvanism had the potential to improve the welfare of the entire human race. In 1801 he had carried out numerous experiments on patients, both alive and dead, at St. Ursula's Hospital in Bologna. Like many others Aldini seems to have had remarkably free rein to test the battery on patients suffering from deafness, blindness, paralysis, inflammation, and various forms of madness. Some of the most spectacular experiments were tried on people in the Bologna insane asylum, though not before he had tried a long series of "painful and disagreeable experiments" on himself. His work in the asylum apparently resulted in the cure of two people who had suffered from 'melancholic' madness, though Aldini admitted that the application had failed in many other cases, being positively dangerous in the case of 'raving' madness. In one case, he described in detail how a melancholic farmer acquiesced to the treatment and gained some degree of relief from his condition once he had been attached to the pile. Other comments imply that there was substantial resistance to the treatment from patients and even from the assistants, "who will even wish to proscribe it before it has been tried". For this reason, Aldini concluded, it was vital to simplify the technique as much as possible, and perhaps to hide the apparatus from the patient in case he or she was unduly petrified.[34]

The Guillotine and the Pile

Towards the end of 1802 Aldini travelled to Paris, where he performed experiments on guillotined criminals and vast numbers of 'large' animals. These shows drew huge crowds and his performances soon eclipsed those of Etienne-Gaspard Robertson, a showman who had been the first to use the battery in Paris (in the summer of 1800). In Paris, medics had immediately seen the possibilities

The Philosophical Magazine, 15 (1803), 188–9. Giulio in particular was motivated by the account given in Sue's *Histoire du Galvanisme*, which had ignored experiments done in the previous decade by Giulio and other Italians successfully proving that involuntary muscles were sensitive to galvanic action.

34 Aldini, 'An Account', 114–21, esp. 120; compare with Wilkinson, 'Elements', 1: 71. Aldini's work is now seen as a pioneering use of electricity in brain stimulation; see A. Parent, "Giovanni Aldini: From animal electricity to brain stimulation," *Can. J. Neurol. Sci.*, 31 (2004), 576–84, esp. 580–2 and T.G. Bolwig and M. Fink, "Electrotherapy for melancholia: the pioneering contributions of Benjamin Franklin and Giovanni Aldini," *Journal of Electroconvulsive Therapy*, 25 (2009), 15–18.

provided by the new machine, although there was antipathy to the idea of animal electricity, and a lingering suspicion of medical electricity. As head of the new Commission de l'Ecole de Médicine, Jean-Noel Hallé supervised a number of experiments on victims of paralysis. With the assistance of Philippe Pinel, Aldini attempted various galvanic cures of the insane at the Hôpital des Foux, and became a founding member of the Société Galvanique. As Christine Blondel has shown, physicians made up a substantial portion of its membership and it concentrated on the physiological and medical implications of the new research. Drawing from recently published reports of experiments by Grapengiesser and others in Germany, they performed a wealth of experiments on patients with various complaints but as a result of all this activity there remained little consensus regarding the medical efficacy of galvanism.[35]

The great promise represented by the pile, coupled with its ease of use, encouraged the physician Pierre-Hubert Nysten to continue the research programme started by his friend and colleague Bichat. Bichat, who had died in 1802, had performed an extensive and systematic examination of over 600 bodies by the winter of 1801–2, during which he had applied the new machine to test the vitality of various organs after death. Nysten proceeded to test the role of the heart in the general functioning of the body by assessing its susceptibility to *post mortem* galvanic stimulation. Noting that Bichat's experiments on the involuntary organs of higher animals and humans had been inconclusive, Nysten was guided by the experimental investigation of the heart given by the Turin anatomists. He set out to determine whether the results of the Italians, who had claimed to have successfully stimulated the heart through the application of a current, had really been obtained by galvanism, or by "mechanical irritability" only. Following Bichat, he wanted to know the exact order the various organs and body functions shut down, and like Bichat and Hallé, Nysten sought to test the responses of as many animals under as many conditions of death as experimental conditions permitted. Although he restricted himself to testing creatures that had died from strangulation, his extraordinary account of his single human anatomical dissection was on a victim of the guillotine.[36]

In his report of the procedure Nysten juxtaposed a relation of the dissection with a macabre, if jaunty story of his efforts to get permission to perform it. At 10 am on 14 Brumaire (4 November 1802) he had been going to work (carrying a vertical pile) when he heard the announcement of an imminent execution.

35 Aldini, 'Account', 91–2; Sue 'Histoire', 2: 384–94; Blondel, 'Animal electricity', 192–3, 195–6 and 203–4; M. Sirol, *Galvani et le Galvanisme. L'électricité animale*, (Paris, 1939), 202–8.

36 Nysten, *Nouvelles expériences galvaniques faites sur les organs musculaires del'homme et des animaux à sang rouge* (1803), 1–8 and 11–13.

He got permission first from the director of his school (reminding him that the intended research object was a criminal), and then from a series of officials, to perform experiments on the deceased. As the Italians had cautioned, every effort had to be made to get a body that was freshly killed and that still displayed vitality – thus Nysten did what he could to make sure that the carriage bearing the corpse did not tarry on its way to the graveyard. Despite the protestations of the curator, the dissection was started within five minutes of death in a ditch within the cemetery. Nysten recorded precisely the times when he submitted organs to galvanic stimulation, and also those times when the same organs no longer showed any "susceptibility". An hour after death, the vena cava and the pulmonary ventricle were allegedly still capable of contracting as a result of galvanic stimulation, although the intestines, stomach, oesophagus and bladder no longer did so. Indeed, he claimed that the heart retained its sensitivity to galvanism longer than the voluntary muscles, maintaining this power even after the 'vital heat' had disappeared. As a result, he placed the heart at the head of the first 'rank' of organs, in opposition to the conclusions drawn by the Italians.[37]

The Treaty of Amiens of March 1802 permitted a much greater exchange of scientific and medical personnel between Bonapartiste and other nations. Having taken on board the techniques of Mondini and the Turin anatomists, in the winter of 1802–03 Aldini travelled to England and performed experiments on animals and humans in front of fascinated spectators at venues such as Guy's and St. Thomas's hospitals. As in Italy and France, the most important contexts for selling galvanism to wider audiences was its potential for offering a precise criterion for distinguishing true from apparent death. The battery also offered the promise of actually resuscitating those who were partially drowned or asphyxiated, and it was in front of an audience at the Royal College of Surgeons concerned with these issues that Aldini performed his notorious series of experiments on a hanged criminal, George Forster, in January 1803. When the rods were applied to Forster's mouth and ear various facial muscles began to twitch, and the left eye opened. On inserting one of the rods into the rectum the entire body went into convulsive movements, so much so that Aldini reported that the effect "almost gave the appearance of re-animation". It was not, he reassured his readers, the aim of the experiments to "produce re-animation" but this was almost certainly the goal of other researchers. In any case, Aldini was again unable to produce convulsions in the heart.[38]

---

37 Nysten, 'Nouvelles expériences', 17–23 and 29–47.

38 Aldini, 'Account', 189–203, esp. 193–4 and 200–1; and Sleigh, 'Galvanism', 235–6.

Across the Channel, the guillotine was a much more precise piece of equipment for galvanic researches, and German medics and natural philosophers proved to be as enthusiastic as the French in using the new tool. In February 1803 an executed criminal at Breslau (Wroclaw) was subjected to battery tests performed by researchers who were informed by the account given by the Turin investigators. The experimenters attached the cut end of the spinal cord to a probe, but with the difference that the executed man's lips appeared to speak in response to prompting. Horrified-fascinated spectators were convinced that the criminal was still alive, but further tests on guillotined humans were inconclusive on the question of whether individuals could survive their own beheading. It was because such experiments might corrupt the moral sensibilities of spectators and also because they seemed unlikely to deliver definitive results, that the King of Prussia warned local authorities to be cautious licensing experiments on the victims of the guillotine.[39]

Nevertheless, the notion that – in the name of science – criminals could actually perform some useful service to the state *post mortem* justified the continuation of such trials. The most celebrated experiments took place at Mainz on 21 November 1803 on the bodies of members of the gang of one of Germany's most famous outlaws, Johann Bückler. Comprising 20 members, this group, known as the Räuberhauptman Schinderhannes gang, was guillotined in a matter of minutes in front of about 40,000 spectators. Students assessed the decapitated heads for consciousness immediately following the descent of the blade, after which the human remains were taken to a log cabin adjacent to the place of execution. Here the chemist Nicolaus Carl Molitor and members of the Medizinischen Privatgesellschaft of Mainz undertook controlled experiments on the decapitated individuals. In order to contrast the effects of galvanic influence and normal electricity, one room was equipped with Leyden Jars while another had a number of batteries. None of these experiments detected any consciousness in cadavers subjected either to high voltage or current. The finding was consistent with the government view that guillotining was a more ethically acceptable form of execution than hanging.[40]

39 Kevorkian, 'Brief history', 221–2. Wilkinson discussed suspended animation in 'Elements', 2: 463–9 and recommended techniques for revival at ibid. 2: 468–9. For the wider contexts see T. Blanning, *The French Revolution in Germany: Occupation and resistance in the Rhineland, 1792–1802* (Oxford, 1983) and R.J. Evans, *Rituals of retribution: Capital punishment in Germany, 1600–1987*, (Oxford, 1996).

40 See Borgards, 'Kopf ab', 130–2; Anon., *Galvanische und elektrische Versuche an menschen- und Thierkörpern. Angestellt von der medizinishen Privatgesellschaft zu Mainz* (Frankfurt am Main 1804). More generally, see G. Mann, "Schinderhannes, Galvanismus und die experimentelle Medezin in Mainz um 1800," *Medizin Historisches Journal* 12 (1977), 21–80

As Aldini had reported in his own case, the pile offered numerous opportunities for self-experiment, though the much greater current of the instrument meant that one had to be more circumspect than before. Robertson, Humphry Davy and others did a number of trials on their own bodies but the self-experiments carried out by Johann Ritter became notorious in the nineteenth century. Following his work with Humboldt, Ritter continued to perform galvanic experiments on frogs and other animals in the last years of the eighteenth century, as Joan Steigerwald explains in the next chapter. Contemporaries admired but also condemned Ritter for the way he experimented so relentlessly on his ravaged body. His experiments were brilliant, and even revolutionary, but inextricably personal and thus unrepeatable. Ultimately they exacted a hideous toll and he died, having sacrificed himself on the altar of scientific knowledge, at the age of 33 in 1809. As he took Humboldt's objectification and disregard for the self to its logical conclusion, so ironically his efforts to provide 'objective' accounts of *Nature* disintegrated into obscurantist subjectivism.[41]

Conclusion

The development of the galvanic circuit and the pile were quintessentially revolutionary events, occurring at a time when Europe was experiencing monumental convulsions in the social and political spheres. Specific political events directly intervened in galvanic enquiries and for a decade, the guillotine proved to be an excellent, if crude means of preparing human subjects for experiments. In the first instance, it was unclear whether these trials were being performed on humans who were in any sense 'alive'. The early years of battery-guillotine research witnessed an increasingly unseemly rush to move the sites of experimental enquiry as close as possible in time and space to the scene of despatch. Whatever religious beliefs or scruples attended the ritual of execution, the practical investigation and then the display of guillotined humans was in essence no different from the treatment accorded to brute

---

and J. Kevorkian, "A brief history of experimentation on condemned and executed humans," *J. National Medical Association*, 77 (1985), 215–22, esp., 221–22.

41 Ritter's *Beiträge zur nähern Kenntnis des Galvanismus und der Resultate seiner Untersuchung; herausgegeben von JW Ritter*, 2 vols, (Jena 1805); see S. Strickland, "The ideology of self-knowledge and the practice of self-experiment," *Eighteenth Century Studies*, 31 (1998), 453–71, esp. 456–61; Steigerwald, 'Subject as instrument' (this volume); Schaffer, 'Self-evidence' and L.K. Altman, *Who goes first? The Story of Self-Experimentation in Medicine* (Berkeley, 1988).

beasts, particularly the larger mammals. Secondly, it served to further erase distinctions between humans and animals that were already being eroded.

Galvanic practices were thus at the centre of major changes in religious and ethical sensibility, and galvanic experiments took place in areas where secularising processes were already operating. In two areas galvanism expanded liminal zones where once watertight boundaries were already being corroded. In the first place it complicated the already perplexing states that collectively existed under the term 'suspended animation', and it further muddied the difference between life and death. It is only a coincidence of sorts that galvanic work was exactly contemporaneous with discussions about the 'rights' of men, women, and even animals, and such concerns are not explicitly present in the many texts written on galvanic research during the period. Vivisection had long been common in European medical schools, but most voices raised against it were concerned about the descent into depravity that overcame the experimenter inured to wanton cruelty. It was in 1789 that Jeremy Bentham famously argued that the capacity to *suffer* should be a criterion of whether animals should be accorded the same rights as humans, a notion that was treated more fully – if ironically – three years later by Thomas Taylor. Although Taylor and others ridiculed Bentham's position, it was not long before legislation was introduced in European countries to reduce gratuitous cruelty to animals. The justification for allowing animal experimentation emphasised the positive utility that would arise from experimentation, and balanced this against the potentially baneful effects this could have on the researcher as well as the cruelty inflicted on animals.[42]

What of the ethical implications of early galvanic research? The physician (and later euthanasia advocate) Jack Kevorkian maintained that the research on guillotined victims was useful and indeed ethical by the standards of the day. In the middle of the eighteenth century, Denis Diderot, Pierre Maupertuis and others had argued that criminals ought to be operated upon *in vivo*. This was, not surprisingly, unacceptable to any European authority at the time, though one should note, as this volume shows, the routine use of patients for experiments both with and without their permission. As the cases of Nysten and others demonstrate, a large proportion of learned opinion believed that deceased criminals could serve a utilitarian purpose by involuntarily offering

42 Bentham, *Introduction to the Principles of Morals and Legislation*, (London 1789), ch. 17 and T. Taylor, *A Vindication of the Rights of Brutes*, (London, 1792). More generally see A.H. Maehle, "The ethical discourse on animal experimentation, 1650–1900," in A. Wear, J. Geyer-Kordesch and R. French eds, *Doctors and Ethics: The Earlier Historical Setting of Professional Ethics*, (Amsterdam, 1993), 203–51.

up their bodies for scientific experiment. In his account of the experiments on Forster, Aldini attacked "the unenlightened part of mankind" for their prejudice against those who, "however laudable their motives" performed experiments on the deceased, and he noted that the vulgar were so senseless as to condemn any sort of anatomical dissection. On the other hand, British laws, "founded on the basis of humanity and public benefit", had "wisely ordained" that the bodies of criminals could be devoted to a purpose "which might make some atonement for their crime, by rendering their remains beneficial to that society which they offended". As enlightened men and women discussed extending rights to women, slaves, Native Americans, the poor and even animals, one wonders how Aldini's audience simultaneously experienced the visceral horror of the "melancholy scenes" they observed, supported their utilitarian beliefs in the value of such research, and retained any residual religious sensitivity they might have had about the fate of the victim.[43]

---

43 J. Kevorkian, 'Brief history', 222–4; Diderot, 'Anatomie' in Encyclopédie, 1: 409–10; Arasse, 'Guillotine', 33; Aldini, 'Account', 189–90. More generally, see Bynum, 'Reflections' and Foucault, 'Discipline and Punish'. For London shows see R. Altick, *The Shows of London* (Cambridge, Mass. 1978).

CHAPTER 3

The Subject as Instrument: Galvanic Experiments, Organic Apparatus and Problems of Calibration

Joan Steigerwald[1]

What was being studied in galvanic experiments, among its simple chains of frog legs and metals? Galvanic experiments promised a new technique for investigating the phenomena of muscular contraction, and the respective roles of nerves and muscle fibers and the action of stimuli in effecting contractions. Organic parts, frog legs capable of reacting to stimuli, were also used as sensitive instruments for detecting weak forms of electricity. But galvanic experiments seemed also to indicate new forms of electricity, an electricity generated in organic material or through metallic contact, that appeared to be related to and yet distinct from the artificial electricity generated by electrical machines and atmospheric electricity. Galvanic experiments were also productive of chemical changes in organic parts and metals, and thus suggestive of relations between chemistry, electricity and organic processes. Finally, they promised medical applications, as a new useful instrument for the treatments of certain ailments and for distinguishing merely apparent from actual death. But because galvanic experiments intersected with such a variety of phenomena and interests, it was not always clear in these experiments what was being studied. In galvanic experiments, what constituted the phenomena being investigated, what was the apparatus generative of the phenomena, and what was the instrument reading the phenomena?

This entanglement of phenomena, apparatus and instruments was at the centre of investigations of galvanic experiments in German settings. Reports of Luigi Galvani's remarkable experiments in German periodicals from 1792 framed them in the terms of Galvani's dispute with Alessandro Volta, and the question of whether Galvani's experiments had disclosed a new form of electricity, animal electricity, or whether the frog legs used in the experiments were

1 Earlier versions of this paper have been presented at the Human Experiment Workshop at the University of Saskatoon in 2009, at the History of Science Society annual meeting in Montreal in 2010, and at the University of Toronto Reading Group in the History of Science in 2011. I thank participants in those meetings for stimulating feedback, as well as Erika Dyck, Larry Stewart and anonymous reviewers for very helpful comments, all of which have greatly improved this paper.

 | DOI 10.1163/9789004286719_005

simply a particularly sensitive instrument for detecting the small amounts of electricity generated by new apparatus. Discussions of the significance of Galvani's experiments occurred in German periodicals already engaged in an intense debate over organic vitality in the context of new applications of chemistry to vital processes, and questions of whether or not organic processes possessed a distinct vital principle. Investigations of Galvani's experiments also took place in the context of a developing attention to comparative physiology, to the differences in the organic processes in different species, and to the question of whether or not the results of experiments on the effects of different drugs on animals could be transferred to humans. The galvanic experiments of Alexander von Humboldt and Johann Wilhelm Ritter are particularly interesting, because they intervened directly into this complex of issues. They both positioned galvanic phenomena at the intersections of organic, electrical and chemical phenomena, and raised questions regarding the tools used to generate and to detect the phenomena appearing in their experiments.

Humboldt and Ritter inserted the experimental subject as an additional instrument into galvanic trials, at least in part, because it was not always clear from the entanglement of organic parts, metals and liquids how to distinguish the phenomena investigated from the tools of their investigation. But in using their own bodies, their organic parts and sense organs [*Werkzeuge*], as instruments [*Werkzeuge*] for reading nature alongside other modes of instrumentality, they reconfigured the relations between the apparatus and the subject and object of experiments. The human subject intervened into galvanic trials not only as the subject trying to make sense of the phenomena, both empirically and conceptually, but also as a corporeal part of the experiments. These self-experiments acted as further witnesses to their trials on frogs, confirming their reading of those results and their calibration of frog legs as instruments. But they also enabled a reading of phenomena that the frog apparatus could not, from within the phenomena. Humboldt and Ritter claimed the subject is doubly positioned inside the experiment, both as instrument and as subject matter, and thus uniquely able to apprehend and make sense of galvanic processes. The subject, as a corporeal instrument, acted as a tool of mediation between conceptual and sensory apprehension and the material phenomena, and thus provided a continuum between the subjective and objective aspects of experience. In galvanic self-experiments, instrumental mediations between subjective apprehension and its phenomenal object, between the agent and matter of experiment, were reciprocal, with not only the object of inquiry being decentred and infolded into the tools of inquiry, but also the subject being conditioned by the tools and becoming part of the products of experiments. If the matter of experiment often led to unanticipated excesses deflecting the subject's purposes and thus remaining recalcitrant to theoretical coherence, representations of matters of fact were often embodied in the apparatus of their production. This relation was also recursive, as the subject, and his conceptions, purposes and actions, became informed and instrumentalized through experiments, and these operations were then incorporated back into the phenomena.[2] In these galvanic trials there was thus no exteriority to instrumentality. Humboldt's and Ritter's galvanic self-experiments have significance in making explicit the instrumentality of not only the object but also the subject of experiment implicit in the scene of inquiry.

Self-experiments appear to be widespread in the history of science—from the use of taste to differentiate the qualities of substances in the seventeenth century, through experiments with electricity and opium in the eighteenth century, to experiments with infectious diseases, radiation and drugs in the nineteenth and twentieth centuries. Yet the practice of self-experiment has received surprisingly little attention from historians of science. Recent studies have highlighted the instability of self-experiments outside established contexts of public legitimization, and the narratives of self-discipline and heroic self-sacrifice invoked by scientists engaged in such practices.

Simon Schaffer provides an important framing of the problematic of self-experiment in his essay "Self-Evidence," taking the example of electrical experiments in the mid-eighteenth century. He argues that the experimenters' bodies were proffered into evidence by being integrated into the collective body of practitioners that provided the power and authority of public warrant. These self-experiments relied on bodily skills and bodily conditions that nevertheless required transmission and repetition. Credit was built up by suggestions of structural similarities between the instruments and materials investigated. But Schaffer argues calibration worked because the cultural settings supplied the stable relations between the phenomena in question and its surrogates. Bodily evidence was credited only through enculturation into the established practices of the authoritative community of natural philosophers.[3] Lucia Dacome

---

2 See: Donna Haraway, *When Species Meet* (Minneapolis: University of Minnesota Press, 2008); Karen Barad, *Meeting the Universe Halfway: Quantum Physics and the Entanglement of Matter and Meaning* (Durham: Duke University Press, 2007); Hans-Jorg Rheinberger, *Toward a History of Epistemic Things: Synthesizing Proteins in the Test Tube* (Stanford: Stanford University Press, 1997); and Andrew Pickering, *The Mangle of Practice: Time, Agency and Science* (Chicago: University of Chicago Press, 1995).

3 Simon Schaffer, "Self-Evidence," *Critical Inquiry* 18 (Winter 1992): 327–335. See also H.M. Collins, *Changing Order: Replication and Induction in Scientific Practice* (London: SAGE Publications, 1985); Collins discusses problems of replication and calibration, and their social warrant, but not self-experiments.

draws on Schaffer's framing of the public settings in which credibility and authority are established to show how weight watching was legitimated as medical practice in the seventeenth century. To gain medical authority and public consent for new practices of monitoring corporeal well-being, through comparing the measure of food ingestion against bodily discharges, weight watchers integrated tables of numerical evidence into personal narratives modeled on case histories. Such narratives allowed the adaptation of techniques of static medicine to shifting medical theories. Dacome contrasts the techniques of early weight watchers to William Stark's self-experiments in the late 1760s, in which Stark recorded the effects of different foods on his body through inner perception. His death shortly after the completion of his experiments enshrined them within a story of tremendous self-sacrifice, but also excessive zeal. By turning away from established practices of measuring bodily discharges to introspection, Dacome argues Stark was not able to establish a public context for his bodily evidence.[4]

Historians interested in self-fashioning have extended their analyses of the cultural settings of the production of scientific identity to consider the regimens of care of self deemed necessary for the production of reliable evidence. These regimens involved not only the disciplining of the passions but also of the body.[5] Michel Foucault has argued that self-knowledge is a product of practices of discipline—of the disciplines and technologies of knowledge, of the discipline of bodies and populations, and of the practices of self-discipline.[6] Jan Golinski draws on such analyses to consider the new discipline of chemistry, the new instruments for chemical investigations, and the new experimental self forged in the years around 1800. Taking as his example Humphrey Davy, he highlights Davy's self-experiments with nitrous oxide. Golinski argues that Davy employed an array of techniques in attempts to bring these highly subjective experiences into the collective body of evidence to yield credible scientific knowledge, but that the experiments evaded such attempts at management.

4 Lucia Dacome, "Living with the Chair: Private Excreta, Collective Health and Medical Authority in the Eighteenth Century," *History of Science* 39 (2001): 487–488.

5 Jan Golinski, "The Care of the Self and the Masculine Birth of Science," *History of Science* 40 (2002): 125–145; Christopher Lawrence and Steven Shapin, eds., *Science Incarnate: Historical Embodiments of Natural Knowledge* (Chicago: University of Chicago Press, 1988); Adrian Johns, "The Physiology of Reading," in *Books and the Sciences in History*, eds. Marina Frasca-Spada and Nicolas Jardine (Cambridge: Cambridge University Press, 2000), 291–314; and Loraine Daston and Peter Galison, *Objectivity* (New York: Zone Books, 2007).

6 Michel Foucault, *Discipline and Punish: The Birth of the Prison*, trans. Alan Sheridan (New York: Random House, 1977 [1975]); and *The History of Sexuality: Volume 1: An Introduction*, trans. Robert Hurley (New York: Random House, 1978 [1976]).

Davy, however, then explored and fashioned his identity as an "experimental self," positioning his own body and subjectivity as uniquely capable of investigating chemical powers. His self-experiments allowed Davy to display a heroic capacity for suffering in the cause of science.[7] Rebecca Herzig's study of narratives of self-sacrifice in American science at the end of the nineteenth century explores more broadly the willingness of scientists to endure extraordinary suffering for their science—submitting to endless hours of labor, deprivations and hardships, and even the pain and risk of self-experimentation. She finds that enthralled investigators were willing to endure any manner of pain in a dedication at once reasonable and compulsive. That it was largely only privileged subjects who could devote themselves to science adds to the paradox, in that it was the purportedly self-possessed who were compelled to suffer. Moreover, she documents scientists who represented themselves as prepared to sacrifice for science without clear reward or compensation in the public domain.[8]

Nicolas Langlitz traces the entanglements of objectivity and subjectivity in his study of neuropsychopharmacology and human experiments with psychedelics. As the name suggests, neuropsychopharmacology studies the effects of drugs not only on neurophysiology of the brain but also on the mental experiences of human subjects. Langlitz shows how these investigations are complicated by many researchers' personal experiences with drugs, through youthful experiments in the context of a cultural fascination with psychedelics and through their participation as test subjects in their own lab settings. That researchers become their own test subjects is an ethical requirement, as ethic boards insisted that experimental subjects be informed as to what to expect from the drugs. It is also an epistemic requirement, as researches argue they need to know effects first hand to understand the descriptions of experimental subjects and to design appropriate experimental settings. Langlitz's ethnography is also complicated by his own youthful experiences with psychedelics and his participation as a test subject in the labs he is studying. The looping or recursive interplay of objectivity and subjectivity in Langlitz's study raises fundamental questions about self-experiment in neuropsychopharmacology, as it at once produces and naturalizes its own preconditions.[9]

---

7 Jan Golinski, "Humphrey Davy: The Experimental Self," *Eighteenth-Century Studies* 45 (2011): 15–28.

8 Rebecca M. Herzig, *Suffering for Science: Reason and Sacrifice in Modern America* (New Jersey: Rutgers University Press, 2006).

9 Nicolas Langlitz, *Neuropsychedelia: The Revival of Hallucinogen Research since the Decade of the Brain* (Berkeley: University of California Press, 2013), especially 83–131. Langlitz draws on: Michel Foucault, *The Order of Things: An Archaeology of the Human Sciences*

Stuart Strickland's excellent essay on Ritter's galvanic self-experiments is the most directly relevant to this chapter.[10] His essay highlights a romantic trope of heroic self-sacrifice in the self-experiments of Ritter and Humboldt.[11] It also elaborates on how Ritter's self-experiments blurred the boundaries between the object and subject of study. Not only did Ritter use his own body as a laboratory instrument in the investigation of nature and other electrical instruments, the phenomena he was investigating was now within his own body. Analogies between the voltaic battery and his body seemed to give credibility to his self-experiments. But Strickland contends that by withdrawing from the discordant forces of the contested public realm, Ritter forfeited public warrant for his experimental results. Indeed, he characterizes Ritter in terms customary in portrayals of romantic figures—seeking refuge in personal experiences, his self-experiments became a personal odyssey that soon led to excess. Strickland thus subscribes to the common ideology of the ideology of romantic science—that it conflated knowledge of nature with knowledge of the self, and claimed for a disembodied intellect dominion over the natural world.

But, as indicated above and indeed by some of the essays in this volume, preoccupations with self-experiments and self-knowledge and excessive suffering for science are not unique to the Romantic period. Moreover, the extent of the experimental investigations by Humboldt and Ritter—their intense engagement with the material practice of experiment and their immersion of their own bodies in these experiments as instruments—demonstrate that embodied modes of knowing were central to their science. Self-experiments did become common in galvanic investigations. But given the entanglement of organic, chemical and electrical phenomena in galvanic trials, and given that

(London: Tavistock, 1970); Ian Hacking, "The Looping Effects of Human Kinds," in *Causal Cognition: A Multidisciplinary Approach,* ed. D. Serper, D. Premack and A. Premack (Oxford: Oxford University Press, 1995), 351–383; and Lorraine Daston and Peter Galison, *Objectivity* (New York: Zone Books, 2007).

10 Stuart Strickland, "The Ideology of Self-Knowledge and the Practice of Self-Experimentation," *Eighteenth-Century Studies* 31 (1998): 453–471.

11 Other historians using this characterization include: Michael Dettelbach, "Romanticism and Administration: Mining, Galvanism, and Oversight in Alexander von Humboldt's Global Physics" (Unpublished doctoral thesis, University of Cambridge, 1992), 87–89; and Michael Shortland, "'Was He or Wasn't He?' Eros and Kosmos in the World of Alexander von Humboldt", Lecture at York University, 7 October 1997. Contemporaries of Humboldt and Ritter also used this language: Johann Bartholoma Tromsdorff, *Geschichte des Galvanismus oder der galvanischen Elektricität, besonders in chemischer Hinsicht* (Erfurt: Henning, 1803), 106–107; and [Arnim] Review of *Fragmente aud dem Nachlasse eines jungen Physiker,* ed. Johann Wilhelm Ritter, in *Heidelbergische Jahrbücher der Literatur* III 1: 9 (1800): 122.

there was no single reliable instrument for reading galvanic effects, self-experiments were welcomed for offering an alternative reading of those effects. Humboldt and Ritter did take galvanic self-experiments further than their contemporaries, but through improved techniques and attention to detail. Indeed, sensations such as the bright flashes and bitter tastes produced in their self-experiments were striking and distinctive, while at times the frog instruments provided only unstable or ambiguous results, so that the empirical evidence produced by the former could often be more readily replicable and hence more credible than the latter. Self-formation was a predominant preoccupation, and thus became a widespread form of self-representation, in the romantic period. But it was a form of self-fashioning that occurred within the context of highly reflexive and critical awareness of the extent and limitations of the subjective contributions to acts of judgment, whether instrumental, cognitive or aesthetic. As shown in the following, Humboldt developed a view of nature in which aesthetic response, the instrumentality of the body and scientific instruments all contributed to a total view of phenomena. Ritter, rather than regarding these as complementary tools, regarded each as infolded in the other. It was not that he deemed nature as in some sense an unfolding of the mind, but rather that he held his representations of natural phenomena required tracing, with him, how they unfolded in his experiments and conceptualizations. Moreover, contra Strickland, Ritter's withdrawal into his experimental work was not characteristically romantic; it not only contrasts sharply with Humboldt's public success, but also with the commitments of the romantic circle into which he was welcomed in Jena to communal exchange. Ritter's decline had much to do with his failure in clearly articulating and promoting his experimental work in the public domain, even as that experimental work formed the most important part of his identity. Golinski's account of Davy as an experimental self reminds us of both the possibilities and difficulties of self-fashioning in the years around 1800. Unlike Humboldt, Ritter was not born to it, and, unlike Davy, he never acquired the capacity for it. An examination of the self-experiments of Humboldt and Ritter, then, not only offers valuable insight into experimental practices and epistemic concerns during the romantic period, but also contrasting portraits of romantic scientists.

The German Reception of Galvanic Experiments

Before examining in detail the experiments of Humboldt and Ritter, it is important to give some sense of the issues that formed the backdrop to their investigations in the German context. Because galvanic experiments touched on electrical phenomena and the instruments of its study, and on the phenomena of organic vitality and chemical processes in both living and nonliving matter, it intersected with several ongoing debates.

As has been well documented, Galvani, Professor of Anatomy in Bologna, attracted attention to his work by announcing in 1791 that he had discovered a new form of electricity, animal electricity. The initial experiments on which he based this claim produced contractions in a severed frog leg by the application of a bimetallic arc between the muscle and exposed nerve. Then in 1794 he claimed to able to produce a contraction in the frog leg by using a chain consisting solely of a nerve and muscle. Galvani's experiments were a development of those conducted earlier in the eighteenth century during the European wide debates over Albrecht von Haller's claim to demonstrate a vital property of irritability in muscle fibers, and over the possible uses of electricity in medical treatments.[12] When news of Galvani's experiments and claims for animal electricity began to be reported in German periodicals in 1792, they were already engaged in a renewed polemic over vital principles and irritability in

---

12 For details on the dispute between Galvani, Volta and the European debates regarding galvanism, see Robert Iliffe's paper this volume. Also see Marco Bresadola, "Animal Electricity at the End of the Eighteenth Century: The Many Facets of a Great Scientific Controversy," *Journal of the Neurosciences* 17 (2008): 8–32; Bresaldo, "At Play with Nature: Luigi Galvani's Experimental Approach to Muscular Physiology," in *Reworking the Bench: Research Notebooks in the History of Science*, eds. Frederic L. Holmes, Jürgen Renn, and Hans-Jörg Rheinberger (Dordrecht: Kluwer, 2003), 67–92; Marco Piccolino, "The Taming of the Electric Ray: From a Wonderful and Dreadful 'Art' to 'Animal Electricity' and 'Electric Battery'," in *Brain, Mind and Medicine: Essays in Eighteenth–Century Neuroscience*, eds. H.A. Whitaker, C.U.M. Smith and S. Finger (Boston: Springer, 2007), 125–143; Fabio Bevilacqua and Lucio Fregonese, eds., *Nuova Voltiana: Studies on Volta and his Times*, 5 vols. (Milan: Ulrico Hoepli Editore, 2000–2003); Paola Bertucci and Giuliano Pancaldi, eds. *Electric Bodies: Episodes in the History of Medical Electricity* (Bologna Studies in the History of Science, 2001); Marco Bresadola and Giuliano Pancaldi, eds., *Luigi Galvani International Workshop: Proceedings* (Bologna: Centro Internazionale per la Storia delle Università e della Scienza, 1998/1999); Francesco Moiso, "Magnetismus, Elektrizität, Galvanismus," in *Historisch-Kritische Ausgabe: Ergänzungsband zu Werke Band 5 bis 9. Wissenschaftshistorischer Bericht zu Schellings Naturphilosophischen Schriften. 1797–1800*, eds. Hans Michael Baumgartner, Wilhelm G. Jacobs and Hermann Krings (Stuttgart: Frommann-Holzboog, 1994), 165–372; Marcello Pera, *The Ambiguous Frog: The Galvani-Volta Controversy on Animal Electricity* (Princeton: Princeton University Press, 1992); Maria Trumpler, "Questioning Nature: Experimental Investigations of Animal Electricity in Germany, 1791–1810" (Unpublished doctoral thesis, Yale University, 1992); and W. Cameron Walker, "The Detection and Estimation of Electric Charges in the Eighteenth Century," *Annals of Science* 1 (1936): 66–100.

the face of new developments in chemistry, and suggestions of a chemical explanation of vitality by Christoph Girtanner.[13] Even Humboldt, in one of his first publications, a 1793 work on the physiology of plants, argued for a vital power countering chemical forces in living organisms.[14] Galvani seemed to provide evidence of a new form of vitality, and his frog apparatus promised to offer a new method for examining questions regarding vital principles experimentally.

Volta, Professor of Physics in Padua, contested Galvani's reading of his experiments, arguing in 1792 that contact between heterogeneous material provides an external electrical current that stimulates contractions in the frog leg. Volta's dispute with Galvani led him to physiological questions regarding the action of nerves, and the relations of nerve and electrical fluid, and he drew on analogies to electric fish in developing his voltaic pile. He also conducted self-experiments investigating the effects of electrical apparatus on his organs of taste, touch and sight. But he was primarily concerned with the physical phenomena and theory of electricity, and the development of electrical apparatus. Volta resisted claims that the frog leg was generative of electrical phenomena, instead was interested in it as an instrument for detecting weak electricity in his experiments on metallic chains. When further experiments on the frog leg suggested it engendered electrical effects, he developed alternative mechanical forms of electrometers. Volta developed his pile by bringing heterogeneous metals into contact.[15] But in 1796 the Italian Giovanni Fabbroni suggested that galvanic phenomena were actually chemical effects due to the oxidation of metals brought into contact in the presence of water. Indeed, after Volta's development of the battery in 1800, his theory of metallic contact had to contend with a growing interest in chemical interpretations of its action, a view supported by several German investigators.[16]

---

13 Christoph Girtanner, "Abhandlungen über die Irritabilität als Lebensprincip in der organisierten Natur," *Journal der Physik*, 3 (1791): 317–351, 507–537. See Joan Steigerwald, "Rethinking Organic Vitality in Germany at the Turn of the Nineteenth Century," in *Vitalism and the Scientific Image in Post-Enlightenment Life Science, 1800–2010*, eds. S. Normandin and C.T. Wolfe (New York: Springer, 2013), 51–75.

14 Alexander von Humboldt, *Aphorismen aus der chemischen Physiologie der Pflanzen*, trans. G.F.v. Waldheim (Leipzig: Voss, 1794 [1793]).

15 Giuliano Pancaldi, *Volta: Science and Culture in the Age of Enlightenment* (Princeton: Princeton University Press, 2003). See also note 12.

16 Fabbroni also contended that the acidic taste noted by Volta when touching different metals to the tongue indicated the role of oxygen in galvanic phenomena. Valeria Mosini, "When Chemistry Entered the Pile," in *Studies on Volta and his Times*, 5 (2003): 117–132; and

Galvani's experiments, and the debates surrounding them, were reviewed in detail by journals as diverse as the *Journal der Physik* and the *Medicinisch-chirurgische Zeitung* [*Medical-Surgical Periodical*] in the fall of 1792; both journals would continue to closely follow the debate over galvanism throughout the 1790s and into the 1800s, regularly publishing articles and reviewing works by Humboldt, Ritter, and other physicists and physiologists on galvanic phenomena.[17] By 1793, Galvani's work appeared in German translation.[18] The first substantive German work investigating galvanism was Christoph Heinrich Pfaff's *Über thierische Elektricität und Reizbarkeit* [*On Animal Electricity and Irritability*], a work that first appeared as a Latin dissertation in 1793, then in an abridged German translation in the *Journal der Physik* in 1794, and finally in a revised and an extended version in 1795.[19] In his galvanic experiments Pfaff used galvanic experiments as a tool to examine organic vitality, and investigated the effectiveness of different conductors and different exciters in different combinations on the circulation and direction of animal electricity. He also studied the effects of chemicals, of drugs and poisons, on irritability, and investigated how affinity to oxygen increased the effectiveness of exciters. But Pfaff was one of the strongest German advocate for Volta's metallic contact theory of the battery. In general German opinion remained divided on the composition of the materials in the chain generating galvanic phenomena, and on the question of whether the frog leg acted as a generator as well as detector of electricity. But there was widespread recognition of the significance of the experiments for investigating organic vitality.

Helge Kragh, "Confusion and Controversy: Nineteenth-Century Theories of the Voltaic Pile," in *Nuova Voltiana* 1: 133–157.

17 *Journal der Physik* 6.3 (1792): 371–415; and *Medicinisch-Chirurgische Zeitung* 4 (1792): 145–154. The *Magazine für der Neueste aus Physik und Naturgeschichte* and the *Journal der Erfindungen, Theorien und Widersprüche in der Natur- und Arzneiwissenschaft* also carried reviews and articles. On the first German publications on Galvani's experiments, see Trumpler, "Questioning Nature," 26–58.

18 Galvani's 1791 treatise "De viribus electricitatis in motu musculari Commentarius" appeared in German translation in 1793, as *Abhandlung über die Kräfte der thierischen Elektrizität auf die Bewegung der Muskeln, nebst einigen Schriften der H.H. Valli, Carminatus und Volta über eben diesen Gegenstand*, ed. D. Johann Mayer and trans. H. Ritschel (Prague: Calwe, 1793).

19 Christoph Heinrich Pfaff, "De electricitate sic dicta animali" (Inaugural dissertation at Karlsakademie, 1793); *Journal derPhysik* 8.1 (1794): 196–270, 280–284, 8.3 (1794): 377–386; and *Über thierische Elektricität und Reizbarkeit* (Leipzig: Sieggreid Lebrecht Crusing, 1795). See the reviews in *Medicinisch-Chirurgische Zeitung* 2 (1794): 170–175, and 2 (1796): 242; and the *Göttingische gelehrte Anzeigen* 167 (18 October 1794): 1665–1669, and 153 (4 September 1795): 1529–1532. On Pfaff, see Trumpler, "Questioning Nature," 99–126.

Humboldt's Galvanic Experiments: Frog Apparatus and Human Instruments

Humboldt witnessed his first galvanic experiments in Vienna in the fall of 1792. From a wealthy family, but not yet able to claim his inheritance, the 23 year-old Humboldt had just begun working as a mine inspector. After restlessly moving between studies in cameralism, engineering, literature, botany and other sciences at universities in Frankfurt, Berlin, Göttingen and Hamburg, and travels to Holland, England and France, Humboldt finally completed ten months of study at the Freiberg Mining Academy, if leaving without obtaining a degree. He would work as inspector for the Prussian Mining Services in Ansbach-Bayreuth until 1797, when the death of his mother brought him into his inheritance and he began preparing for travels to Spanish America. Humboldt began galvanic experiments in earnest in 1794, finding the simple experiments easy to conduct despite the long hours and extensive travel demanded by his work; the galvanic apparatus of a few metal bars, forceps, glass plate and anatomical knife could be easily carried in a satchel on horseback. The appearance of Pfaff's work in the spring of 1795 with similar experimental work took him by surprise, prompting him to delay publishing but also to recast his own work. Then, in August of 1795, delayed in Como by foul weather, he had the opportunity to do galvanic experiments with Volta, who demonstrated how the application of a potash solution instead of water to the nerves increased the responsiveness of a frog leg.[20] But exploring the implications of this technique took Humboldt in different directions than Volta, to a detailed examination of the influence of chemicals on the effectiveness of galvanic chains and to the calibration the frog leg as a reliable instrument for detecting galvanic effects. As Humboldt's investigations developed, he became less interested in galvanic phenomena and increasingly interested in using galvanic experiments to explore organic vitality. Fabbroni's chemical interpretation of galvanic phenomena was also taken

20 See Humboldt's letters, 12 December 1792, 29 July 1795, and 26 August 1795, in *Die Jugendbriefe Alexander von Humboldts 1787–1799*, eds. Ilse Jahn and Fritz G. Lang (Berlin: Akademie, 1973), 219–223, 438–440, 454–456; Humboldt, "Ueber die gereizte Muskelfaser, aus einem Briefe an Herrn Hofrath Blumenbach," *Neues Journal der Physik* 2.2 (1795): 115–116; "Neues Versuche über den Metallreitz in Hinsicht auf die verschiedenartige Empfänglichkeit der thierischen Organe," *Neues Journal der Physik* 3.2 (1796): 172–184; and *Versuche über die gereizte Muskel- und Nervenfaser, nebst Vermuthungen über den chemischen Prozess des Lebens in der Thier- und Pflanzenwelt,* 2 vols. (Berlin: Rottman, 1797), 1: 3, 8, 31–32. On Humboldt's galvanic experiments, see Moiso, "Magnetismus, Elektrizität, Galvanismus," 330–338; Trumpler "Questioning Nature,"129–157; and Dettelbach, "Romanticism and Administration," 63–89.

up by Humboldt. Indeed, Humboldt modified his 1793 stance on the need for a special vital power, and instead began to explore a vital chemistry. In 1797 he published the results of his galvanic investigations in a massive thousand page work, *Versuche über die gereizte Muskel- und Nervenfaser* [*Experiments on Stimulated Muscle and Nerve Fibers*]. For Humboldt galvanic phenomena became the combined alterations of organic parts, metals and chemicals. The frog leg was not only a means for testing effective chains, but also the object in which to study receptivity to stimulus and vital chemistry in organic bodies. Self-experiments acted as further witnesses to his experiments with frog legs, introduced to confirm the calibrations of frog legs as instruments, and to help disentangle the complex of alterations taking place in organic bodies and confirm his reading of these alterations through a vital chemistry.

If Humboldt's galvanic experiments began by addressing the main questions asked in earlier investigations, he soon introduced a new set of concerns. Like Pfaff, he carried out exhaustive experiments examining the contested conditions for effective galvanic chains—testing both heterogeneous and homogenous combinations of metals; probing into the purity of metals; doing trials on the role of moisture; and determining which metals were effective and in which combinations (Figure 3.1). He was particularly fascinated with Galvani's experiments with solely organic parts, which Pfaff had not explored, and concluded that the nerve and muscle alone can provide the requisite heterogeneity for an effective chain. But by 1796 he realized the import of the condition of organic parts, and their receptivity to stimuli, for the success of experiments. It had become commonplace for experimenters to attend to the quality of frog specimens, to ensure that they were always fresh and lively. But Humboldt concluded that negative results in experiments demonstrated nothing, as long as the experimenter could not prove his experiments were made with specimens which possess maximum animal excitability. The significance of Volta's technique for increasing the excitability of a frog leg with potash solution now became clearer. Nonsuccess only gives a result for a specific test under specific conditions; one positive result under conditions of increased excitability can overthrow a series of negative results. Humboldt expressed astonishment that this observation escaped Volta and others when they declared particular experiments ineffective. For Humboldt, galvanic effects were as dependent on the excitability or receptivity to stimuli of the organic parts as the strength of the stimuli.[21]

---

21 Humboldt, "Neues Versuche," 171–175; "Aus einem Briefe des Herrn Oberbergraths von Humboldt an Herrn Hofrath Blumenbach," *Neues Journal der Physik* 2.4 (1795): 471–473; and *Versuche* I: 22–27, 68, 98–99.

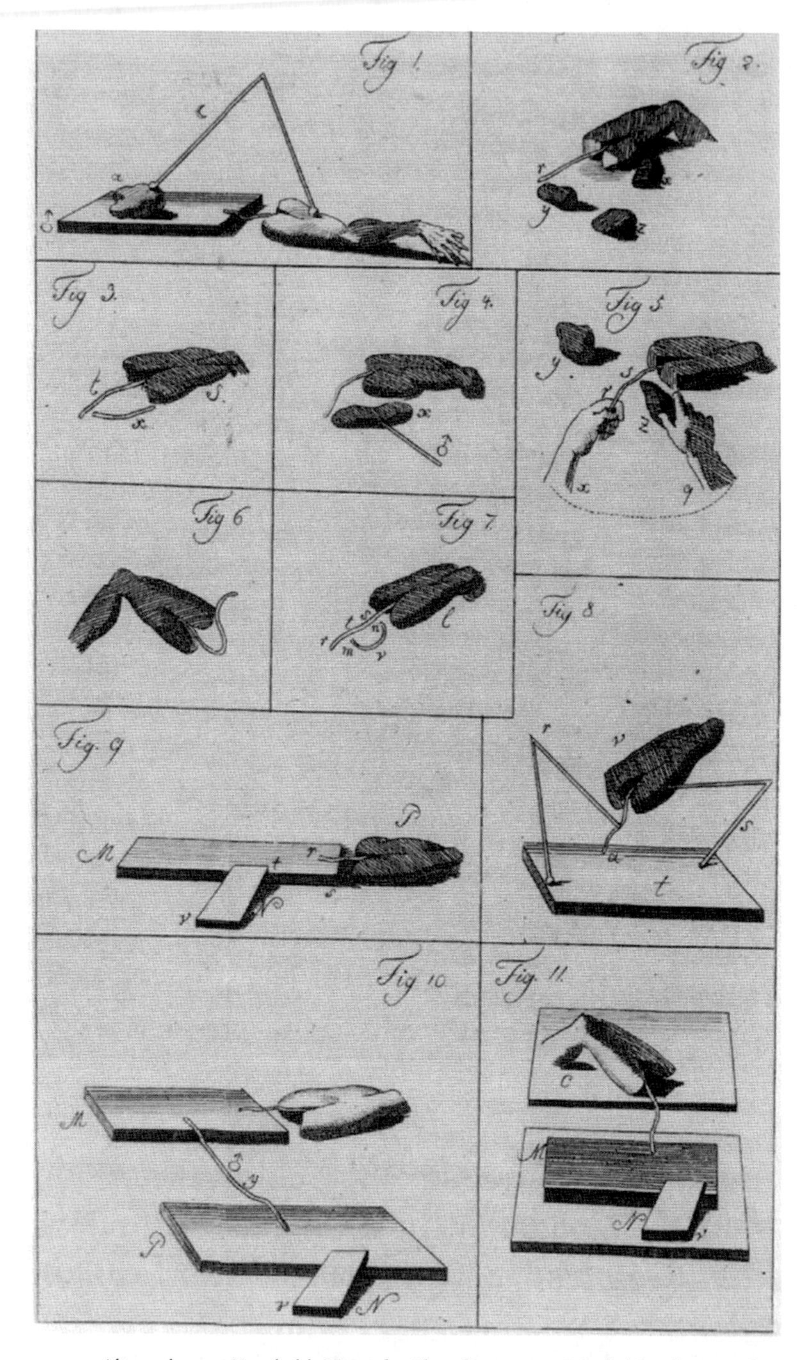

FIGURE 3.1 *Alexander von Humboldt,* Versuche über die gereizte Muskel- und Nervenfaser, nebst Vermuthungen über den chemischen Prozess des Lebens in der Thier- und Pflanzenwelt, *2 vols., volume 1.*
BERLIN: ROTTMAN, 1797

Accordingly, Humboldt turned his attention to the condition of the frogs used in his experiments. Volta had been unhappy with the frog as an electrometer because he found it unstable and difficult to standardize, preferring to work with inorganic materials. Humboldt set out to calibrate the frog to make it a more effective instrument. Building on Volta's technique with potash solution, Humboldt found he could increase the excitability of the frog leg chemically by the application of alkaline solutions to the nerves and acidic solutions to the muscles. He further explored the depressing or exciting effects of opiates, alcohols and other drugs as well as different gases. He extended concern with the quality of specimens to a wide variety of factors that he determined affected their responsiveness—the species, age, sex, strength, nutrition and health of the frog; the mating season, time of year and even day; and the climate, humidity, temperature, light and air quality of the experimental setting. He continually reminded his reader, at each place where he introduced a new series of experiments in his text, how to prepare frog legs for maximum effectiveness.[22] Using natural historical, physiological, environmental and chemical considerations, Humboldt thus calibrated the frog leg on a scale from low to high excitability. He came to regard it as a living instrument that could be used to examine not only different metals, but also the composition of material. It could be used as an *Anthrakoskope* to detect the presence of coals effective in galvanic chains, or a *Hygrometer* to measure moisture, much as our own bodies could be used as thermometers or electrometers.[23]

Humboldt routinely compared experiments with frog legs to experiments on himself, confirming his calibration of frogs as instruments, but also confirming his reading of changes in excitability through chemical alterations to the organic body. Indeed, his first publication in 1795 remarked that although self-experiments were the most interesting they were the least studied, a circumstance he set out to rectify.[24] In his self-experiments, as in his first frog experiments, Humboldt used a simple apparatus of a few metal bars to generate effects. He began by galvanizing his tongue, following Volta, noting that tongue experiments had already been made in Germany thirty years previously by Johann George Sulzer in his study on pleasure; Humboldt recorded the now

22 *Versuche*, I: 22–27, 34–36, 68–87, 242–248; and II: 173–175.

23 Ibid., I: 189–191; and II: 175. On contemporary uses of the human body as an aid to thermometers and meteorological measurements, see Hasok Chang, *Inventing Temperature: Measurement and Scientific Progress* (Oxford: Oxford University Press, 2004), 8–56; and Jan Golinski, *British Weather and the Climate of Enlightenment* (Chicago: University of Chicago Press, 2007), 108–109, 150–153; and "Care of the Self," 138–139.

24 Humboldt, "Ueber die gereizte Muskelfaser," 119.

familiar unpleasant acidic or bitter taste and burning sensation. He galvanized his eyes, teeth and nose, repeating experiments already undertaken by Pfaff and others, recording sensations of a bright flash, pain, pressure in the head and an impulse to sneeze. He concluded each sense organ has a characteristic reaction, according to its characteristic energy.[25] Humboldt continued with a series of original self-experiments of remarkable sophistication. He galvanized the exposed nerves in wounds, at first incurred accidentally and then produced intentionally. In his most striking experiment, he prepared a wound on his back, by forming and opening a blister, and applied different galvanic chains. He recorded which of his back muscles contracted, the sensations of shock, pain and burning, and the dark-red and caustic serous-fluid that formed and inflamed the wound. Humboldt then directly compared the effects produced in himself to those of experiments made on frogs, by laying a frog leg in his back wound, and, on galvanizing the wound, having assistants observe the leg "hop" at the same time as he noted his sensations. Humboldt even included his whole body in the galvanic chain, by placing a zinc disc in his mouth and a silver rod in his anus. He then compared the hefty convulsions produced in these experiments to ones on animals, by similarly galvanizing a bird which had just stopped breathing and thus reviving the signs of life, if only briefly. He also tried a few experiments on the sense organs of animals, reading sensations similar to his own into their reactions.[26] In his self-experiments, Humboldt carefully recorded his sensations of increased pulse, shock or throbbing; of heat, acidic taste or flash of light; and of the alterations to fluids in his wounds as well as to his excrement; recording how these effects varied with different experimental arrangements. Reading within his own body different degrees of receptivity to stimulus, he used these results to evaluate his observations in his frog experiments. As in his frog experiments, the focus of his self-experiments shifted from the calibration of the galvanic apparatus to its effects on organic vitality. These self-experiments offered a host of new observations of physiological and chemical alterations in the organic body in response to stimuli to complement those made in experiments with the frog leg and its singular response of muscular contraction.

From the beginning of his engagement with self-experiments, Humboldt raised concerns about the subjectivity of his judgments based on them, and variations in the bodies and impressions of experimenters. Indeed, Humboldt emphasized the importance of having witnesses to his experiments, and when

25 *Versuche*, I: 195, 206, 308–322.

26 "Ueber die gereizte Muskelfaser," 119–120; "Neues Versuche," 166–170; and *Versuche*, I: 33–35, 307–336.

possible enlisted established authorities, such as Christian Wilhelm Hufeland, professor of medicine at the University of Jena, to this end. He cautioned that "experiments on sensation must be made with great care, for the fantasy and the expectant opinion with which one proceeds with the experiment easily lead to deception."[27] He accordingly introduced techniques designed to reduce the subjectivity of these experiments. The back experiments were devised to avoid a subject's expectant response when observing apparatus being brought into contact with his body. Witnesses to the experiments on Humboldt's back could testify to the contraction of the frog leg, the alterations in the fluids in the wound and the muscle contractions of the back muscles correspondent to his sensations. To compare the sensations generated by a particular experiment, he had a wire run from the wound on his back over the tongue of several persons. As he felt "burning and throbbing" on his back, a second person claimed to taste acid.[28] Michael Dettelbach has suggested an erotic connotation to these experiments, as Humboldt galvanized his anus and conducted experiments in the company of a young Lieutenant with whom he had a romantic friendship, Reinhard von Haeften.[29] But these self-experiments were described as more painful than pleasurable, and conducted on different occasions, before different witnesses and over several years. Humboldt had found his tongue difficult to domesticate, both physically and morally, and turned to back experiments as more reliable.[30] But his frog legs were also unruly instruments, despite his careful attention to calibration and experimental conditions; he found each experiment altered the receptivity to stimulus of a specific specimen, and each specimen was unique. Humboldt thus turned to self-experiments as further witnesses to his experiments on frogs, experiments that offered a reading of those phenomena independently of the experiments with frog legs and against which he could evaluate those experiments. Indeed, the results of his self-experiments were often very striking and produced distinct sensations that could be read with less ambiguity than the movement of the frog leg.

---

27 *Versuche*, I: 307; and "Ueber die gereizte Muskelfaser," 119.

28 "Neues Versuche," 168; and *Versuche*, I: 332–335.

29 Dettelbach, "Romanticism and Administration," 87–89. On Humboldt's sexuality, see Shortland, "'Was He or Wasn't He?';" and Joan Steigerwald, "Figuring Nature, Figuring the (Fe)male: The Frontispiece to Humboldt's *Ideas Towards a Geography of Plants*," in *Figuring it Out: Science, Gender and Visual Culture*, eds. A. Shteir and B. Lightman (Hanover, NH: University Press of New England, 2006), 54–82.

30 Humboldt, *Versuche*, I: 197.

Both his experiments with frog legs and his self-experiments helped Humboldt come to the conclusion that nerves were an essential component of galvanic phenomena; galvanic phenomena, as phenomena of stimulus, required nerves, whether the resultant alterations were to muscles or to sensory organs. Humboldt thus conceived his galvanic studies as a contribution to the ongoing German debate over organic vitality and the role of nerves in irritability. He regarded one of his most important results to be the demonstration of an effective circle around nerves [*Wirkungskreise*], giving them a form of action at a distance and enabling them to be effective where anatomical dissection found few traces of nerves, such as in the heart.[31] Humboldt's interpretation of his experiments was that a galvanic fluid flows from the nerve into the muscle to produce a contraction, with the influx of galvanic fluid back into the nerve stimulating it to this action. The chain used in an experiment affects the flow of the galvanic fluid—in chains with heterogeneous metals the flow of galvanic fluid is hindered, accumulating to a point of break through, to produce strong contractions; in chains with solely animal parts the galvanic fluid moves easily and hence only produces weak contractions.[32]

Although referring to a galvanic fluid, Humboldt expressed hesitation in speculating over the nature of such a fluid, and became increasingly convinced that organic vitality is too complex to be reduced to a single substance or force. His attention to techniques of chemically altering the excitability of frog legs and to the chemical alterations generated in his self-experiments convinced him of the significance of the chemistry of vital processes, a proposal he developed in the second volume of his 1797 work. In making this proposal, Humboldt sought to distance himself not only from his 1793 position, which argued for a *Lebenskraft* separating organic from inorganic processes, but also from Girtanner, who emphasized too singularly the similarities between combustion and life; both positions Humboldt now argued hindered insight into organic phenomena. Instead, he introduced a conception of "an important and *new* branch of natural science," vital chemistry [*vitale Chemie*], "*the investigation of the chemical alterations of composition* [chemische Mischungsveränderung] *that occur in the excitable matter* [erregbarer Materie] *during the vital functions*." Excitability is the common condition of organic matter, with each organic fibre having a specific excitability dependent on its specific form and composition, and each also able to preserve its specific excitability despite being uninterruptedly

31 Humboldt presented this result as a significant contribution in his first publication on galvanism ("Ueber die gereizte Muskelfaser," 122), and as one of the main results of his book (*Versuche*, I: 205–231, 483–486).

32 *Versuche*, I: 390–421.

stimulated and in a continual exchange of material with the external world. Vital functions are alterations of form and composition that are the common result of a complex entanglement of phenomena.[33]

Humboldt's *Versuche über die gereizte Muskel- und Nervenfaser* appears to suffer from a circularity, which is however mitigated by recognizing the entanglement of phenomena in his work. Having determined the constitution of effective galvanic chains and calibrated his frogs as instruments, Humboldt applied these instruments to the study of comparative physiology and vital chemistry. He used chains of metals to examine the variable excitability of different organic parts in a range of organisms, from plants through worms, insects and amphibians to birds and mammals. He also used the frog apparatus to examine the effects on the excitability of organic parts of a range of chemicals and gases, and to compare galvanic effects to those of electricity, magnetism, light and heat. But he then concluded that this frog apparatus was actually the cause of the galvanic phenomena and the objects he wished to study—stimuli, organic excitability and vital chemistry. As Humboldt pursued the study of the complex alterations of form and composition of excitable organic matter under varied stimuli, he implemented an increasingly diverse set of instruments to read these alterations, using physiological effects (such as increased or decreased pulse) and chemical changes to the organic fluids and fibres alongside the frog leg and his own sensations. In this entanglement of phenomena, there is no primary cause, and thus no single instrument for its examination—Humboldt used the tools of galvanism, vital functions and chemistry to study the relations of galvanism, organic vitality and chemistry. His self-experiments did not resolve this confusion of instruments, causes and effects, as the matter under investigation was now within his own body, the subject reading the phenomena also the object of study. But despite his caveats over their potential subjectivity, Humboldt presented his self-experiments alongside his experiments with frog legs and other tools, as respectable independent witnesses to those results and as a confirmation of his reading of organic vitality through a vital chemistry.

Self-experiments became an established part of Humboldt's experimental repertoire. His galvanic self-experiments were a development of the self-experiments he conducted as a mine inspector to study the airs in mines. He had compared both chemical effects and the effects on plants and on miners' health of different gases to effects on his own body. The techniques he developed in his galvanic self-experiments also prepared him for his travels through Spanish America from 1799 to 1804. Humboldt prepared for his trip by traveling

33 Ibid., II: 41–42. See I: 483; and II: 52–75, 100–147, 430–436.

to European centers of science collecting physical instruments, including different versions of the same instrument made by different manufacturers according to different principles, and collecting the necessary expertise to use each properly. He hauled a vast array of these instruments through America, from thermometers and electrometers to hygrometers and eudiometers, using them to record all aspects of the environments he encountered as they varied in time and place.[34] Fascinated by the exuberant vegetation and wildlife he encountered in the tropics, he began to study how they varied with the specific physical conditions of specific regions that he measured with his instruments. He was particularly eager to examine the giant electric eels of South America,

FIGURE 3.2 *Alexander von Humboldt and Aimé Bonpland in their jungle hut.*
IMAGE FROM PAINTING BY E. ENDER IN POSSESSION OF THE INSTITUTE FÜR GEOGRAPHIE AND GEOÖKOLOGIE LEIPZIG DER AKADEMIE DER WISSENSCHAFT DER DDR.

34 Michael Dettelbach, "The Face of Nature: Precise Measurement, Mapping, and Sensibility in the Work of Alexander von Humboldt," *Studies in History and Philosophy of Biological and Biomedical Sciences* 30.4 (1999), 475–481; and "Global Physics and Aesthetic Empire: Humboldt's Physical Portrait of the Tropics", in *Visions of Empire: Voyages, Botany, and Representations of Nature*, ed. D.P. Miller and P.H. Reill (Cambridge: Cambridge University Press, 1996), 261–272.

renowned for their powerful shocks, which he attributed to their tropical environment. But despite his extraordinary investment in physical instruments, Humboldt continued to record the effects of the different environments he studied on his own body. He even used his body as an instrument to investigate the specific characteristics of the shock of the giant eels of the new world under varied conditions, and on his return would use it to compare those effects to the electric rays of the old world.[35] Humboldt also privileged aesthetic affect, under the influence of Johann Wolfgang von Goethe and the landscape artist Joseph Anton Koch. He regarded an aesthetic view of a landscape,

FIGURE 3.3 *Alexander von Humboldt in his library (1856).*
LITHOGRAPH AFTER A WATERCOLOR BY EDUARD HILDEBRANDT, IN POSSESSION OF THE STAATLICHE MUSEEN ZU BERLIN, PREUSSISCHER KULTURBESITZ, KUNSTBIBLIOTHEK.

35 Humboldt, "Versuche über die elektrischen Fische," *Annalen der Physik* 22.1 (1806): 2–13; and "Beobachten über den Elektrische Aal des Neuen Welttheils," in *Beobachtungen aus der Zoologie und vergleichenden Anatomie, gesammelt auf einer Reise nach den Tropen-Ländern des neuen Kontinents* (Tübingen: Cotta, 1806) II.2: 49–125.

of its characteristic forms and vegetation, as a cultivated sensibility only found in developed cultures. Indeed, he regarded aesthetic perception as part of the array of instruments he brought to America to characterize its geography, as an instrument of judgment that he could use together with his scientific instruments and the instrument of his own body, these instruments working together to provide a total view of a region.[36]

During his travels through America Humboldt immersed himself in the object of his study, the physical geography of a region and its characteristic vegetation, using his whole person, corporeal and sensible, aesthetic and intelligent, alongside his physical instruments, to apprehend phenomena. But on his return to Europe, Humboldt moved away from his immersion in the natural world. He instead embarked on an ambitious writing career, producing volume after volume recounting his studies in America and his views of the nature, in a beautiful prose, writing increasingly at a remove from his subject matter. Already well established in the scientific communities of Europe before his departure, through the positive reception of his works on galvanism, plant physiology and the airs of mines, and through his family connections and the opportunities they presented, the reports on his travels through America were eagerly anticipated. Humboldt, however, increasingly turned from active scientific investigation to writing for a wide reading public, writing to educate them on his vision of the cosmos and the methods for its study.

Ritter's Galvanic Experiments: The Subject of Experiment

While Humboldt was at the University of Jena in 1797 to study anatomy in preparation for his trip to Spanish America, he met the nineteen year-old Ritter. Ritter had initially trained in pharmacy, then came to Jena in 1796 to pursue chemistry further. From a poor family, lacking a genteel upbringing and education, he never felt comfortable in the elite intellectual and scientific circles surrounding him. He nevertheless soon acquired a reputation for his observational skill and untiring occupation to galvanic experiments. Impressed, Humboldt invited Ritter's comments on the first volume of his work on galvanism, comments he then included as an appendix to the second volume. In 1798 Ritter offered his own contribution in *Beweis, dass ein beständiger Galvanismus den Lebensprocess in dem Thierreich begleite* [*Proof that a Continuous Galvanism Accompanies the Process of Life in the Animal Kingdom*],

36 See Steigerwald, "Figuring Nature."

the first of many publications on galvanism.[37] He began by examining the experiments of Humboldt, Volta and Galvani, arriving at a definition of galvanism as dependent on the experimental arrangement of heterogeneous substances rather than a special fluid or substance. Ritter's skill lay in experimental investigations of the relations between galvanic, organic, chemical, electrical and other processes. What he lacked was the ability to articulate and conceptualize the results of his investigations. While Humboldt's talents as a writer became as important to his success as his experimental investigations, Ritter's comparative failures and excesses as a writer hindered his success despite the recognition of his experimental work. Ritter made his self-experiments central to his achievement as a scientist, understanding them not only as one tool amongst others for the investigation of galvanic phenomena, like Humboldt, but also as providing unique insight into the inner workings of this activity by enabling him to trace them within the instrument of the human body.

Ritter began his galvanic experiments by repeating, revising and extending earlier experiments, investigating exhaustively the varied conditions under which a galvanic chain could be demonstrated to be effective (Figure 3.4). Beginning with the basic galvanic chain of a nerve-muscle preparation and two heterogeneous metals, silver and zinc (experiment 1), he produced a series of transformations of this basic chain. Ritter repeated Humboldt's experiments increasing and decreasing the excitability of nerves through the application of exciting or depressing solutions, as well as applying cuts or ligatures to nerves (6–8), but rejected Humboldt's conclusion that nerves were essential to exciting contraction of the frog leg. Varying the combinations of exciters and conductors in the chain, he confirmed Volta's claim of the need for heterogeneous materials or asymmetrical arrangements (9–25). But he also confirmed Galvani's experiments, producing effective galvanic chains with solely organic parts (26–28). Combining these results, he concluded each organic part provided the requisite heterogeneity to produce galvanic phenomena. Ritter then explored the effects of different exciters and conductors with different chemical qualities on the effectiveness and direction of galvanic chains, claiming to demonstrate what others had only suggested, that substances with a high affinity to oxygen effected

---

37 For biographical details on Ritter, see Klaus Richter, *Das Leben des Physiker Johann Wilhelm Ritter* (Weimar: Hermann Böhlaus, 2003); and Walter D. Wetzels, *Johann Wilhelm Ritter. Physik im Wirkungsfeld der deutschen Romantik* (Berlin: Walter de Gruyter, 1973). On Ritter's galvanic experiments, see Joan Steigerwald, "Figuring Nature: Ritter's Galvanic Inscriptions," *Bulletin de la Société d'Histoire et d'Epistémologie des Sciences de la Vie*, 2 (2008); Moiso, "Magnetismus, Elektrizität, Galvanismus," 365–372; Strickland, "Ideology;" and Trumpler, "Questioning Nature," 159–188.

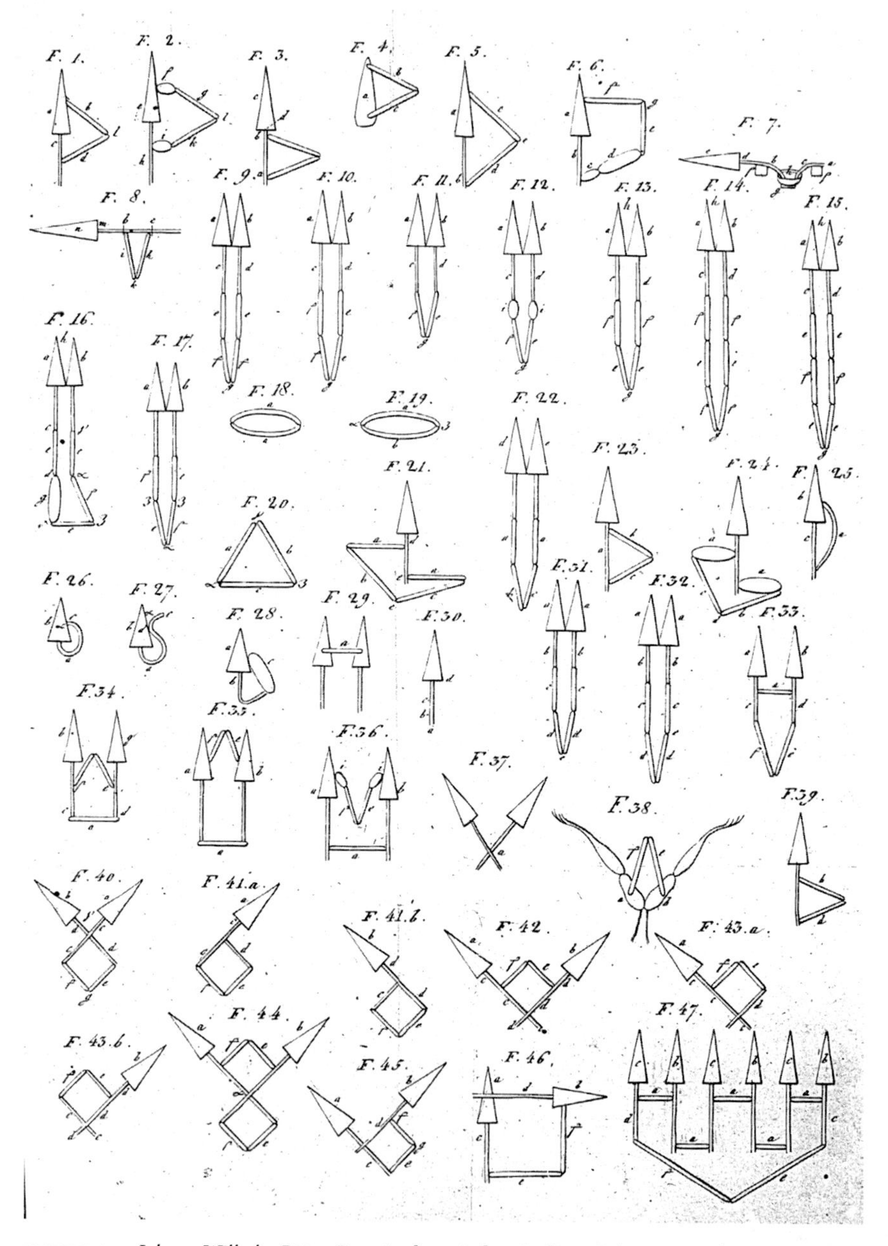

FIGURE 3.4 *Johann Wilhelm Ritter,* Beweis, dass ein beständiger Galvanismus den Lebensprocess in dem Thierreich begleite.
WEIMAR: INDUSTRIE COMPTOIRS, 1798

stronger contractions than oxidized substances.[38] His pages of figures and experiments presented these results with far greater systematicity and detail than previous investigators.

Bringing together an array of approaches to galvanic experiments, in subsequent works Ritter explored the relations between organic, chemical, electrical and galvanic processes. Humboldt had focused attention on the chemical alterations effected in galvanic experiments, developing a conception of a vital chemistry based on these alterations in the chemical composition of excitable organic matter. Ritter used Humboldt's conception of vital chemistry to account for the effectiveness of each organic part in galvanic chains; the more complex their organic composition, "the more they are vitalized [*belebt*]," the greater their heterogeneity and thus effectiveness.[39] But he also used the chemical changes produced by different galvanic chains, such as water becoming milky white and zinc showing signs of oxidation, as instruments to replace the frog leg to indicate the presence of galvanic activity and to explore purely inorganic galvanic chains.[40] He used physiological phenomena to examine chemical, electrical and galvanic effects; galvanic phenomena to explore physiological and chemical effects; and chemical phenomena to examine galvanic and electrical effects. He also further extended his interests to include magnetism, light and heat.[41] In the end, he offered experimental definitions of these different phenomena—chemical interactions are when two substances are transformed into a new substance; electrical interactions when a tension is created between two substances; and galvanic interactions when a continuous action is produced between three heterogeneous substances, with each working on the others both directly and indirectly.[42] For Ritter, "all galvanism is

38 For example, in Figure 3.4, experiment 11, if (e) is zinc and (f) silver, muscle (b) will contract more strongly than (a), where (c) and (d) are nerves. Ritter attributed this result to the greater affinity of zinc to oxygen. Ritter, *Beweis, dass ein beständiger Galvanismus den Lebensprocess in dem Thierreich begleite* (Weimar: Industrie Comptoirs, 1798), 52–56.

39 Ritter, "Versuche und Bemerkungen über den Galvanismus der Voltaischen Batterie," *Annalen der Physik* 7 (1801): 443–445.

40 Ritter, "Beweis, dass die Galvanische Action oder der Galvanismusauch in der Anorganischen Natur möglich und wirklich sey," in *Beyträge zur nähern Kennntniss des Galvanismus und der Resultate seiner Untersuchung*, 2 vols. (Jena: Frommann, 1800–1805) I.2: 111–284.

41 See the articles collected in Ritter, *Beyträge*; and *Physisch-Chemische Abhandlungen in Chronologisher Folge*, 3 vols. (Leipzig: Reclam, 1806).

42 Ritter, *Beweis*, 172–173. Ritter continued to use this experimental definition of galvanism even while working with Volta's battery and studying the electricity of different bodies. See *Das Electricische System der Körper* (Leipzig: C.H. Reclam, 1805), 49 and 326.

one," whether the constituents of the chain are organic or inorganic; galvanic phenomena are those produced by galvanic experimental arrangements.[43]

Ritter also inserted himself into his experiments as another instrument alongside others, using his own body to read galvanic action, like Humboldt, extending instrumentality into the subject as well as the objects studied. Ritter used self-experiments to complement experiments with frog legs, like Humboldt, directly comparing the results of experiments on his sense organs to those with the frog leg. He even devised a variation of Humboldt's back experiment, incorporating a frog leg into a galvanic chain applied to his tongue, and recording the twitching of the frog leg correspondent to his sensations of taste.[44] But Ritter also claimed self-experiments enabled him to present new results. The human being, as "the pinnacle [*Spitze*] of organization on earth," as the most complex and vitalized being, stands at "the pinnacle of all possible galvanicity," and thus can demonstrate phenomena not apparent in other chains.[45] Ritter used self-experiments to demonstrate what he regarded as his most important result, the continuation of galvanic action throughout the closure of a chain, phenomena that eluded the frog apparatus, which only contracted on the opening or closing of a chain. Although Humboldt, amongst others, had mooted this possibility, Ritter claimed to establish it as a matter of fact through the instruments of his sense organs [*Werkzeuge*]. He then used this result to account for the loss of excitability of frog legs through continued enclosure in a galvanic chain.[46] He claimed his body provided a more sensitive instrument than the "frog apparatus," offering more nuanced readings of variations in the strength and character of galvanic effects. After Volta's development of the battery in 1800, Ritter used it to study the effects on his sense organs in greater depth, fine tuning his body as an instrument, and then further used his senses to calibrate different columns, measuring the differences in the quantity and quality of sensations produced by columns of different materials and numbers of layers.[47]

Ritter was cognizant of worries over the subjectivity of self-experiments, over the difficulty of repeating them and how impressions might vary between

---

43 "Versuche," 441.

44 *Beweis*, 82.

45 Ibid., 444–446.

46 Ritter, 78–119. Compare Humboldt, *Versuche*, 1: 200.

47 "Wirkung der Galvanischen Batterie auf die verschiedenen Sinne des Menschen, beym Eintritt, Seyn, und Austritt in und aus der Ketter jener," *Beyträge* 2 (1802): 3–13. See Strickland, "Ideology," 453–471; and Roberto Martins, "Ørsted, Ritter and Magnetochemistry," in *Hans Christian Ørsted and the Romantic Legacy in Science*, eds. Robert M. Brain, Robert S. Cohen and Ole Knudsen (New York: Springer, 2007), 339–385.

experimenters. Unlike Humboldt, as an impoverished student, Ritter conducted his experiments alone in his rented room, having little access to credible witnesses. But he positioned his own body as authoritative, claiming that the extent of his investment of his own body in his studies gave him greater expertise and that his own senses were thus uniquely perceptive. The experiments required considerable practice, not only to steady the hand in applying armatures to sensitive organs but also to develop the ability to distinguish phenomena that at first appeared confused. In his experiments on the organ of sight, for example, Ritter claimed to detect not only a positive or enhanced illumination with an effective chain, but also a negative or decreased illumination from normal when he reversed the metals in the chain; he also detected the appearance of colours that altered with the alteration of the chain.[48] He contended that results that were ambiguous with simple galvanic chains became clearer with the enhanced effects produced by the galvanic battery. Strickland suggests Ritter had an erotic relationship with his voltaic columns,[49] although, as in the case of Humboldt, it would need to be a masochistic one. Indeed, Ritter expressed shock at the strength and painfulness of his first experiments with the voltaic battery, concluding his account of his self-experiments with the battery with a list of the ill effects incurred and a warning to others.[50] Ritter figured his experiments with the battery through narratives of self-authority and self-sacrifice rather than self-pleasure. The singularity of his self-experiments, Ritter contended, came from his mastery of this experimental art, rather than the subjectivity of his interpretations.

In contrast to his widely cited talent in experimentation, when Ritter attempted to provide theoretical generalizations of the results of his investigations articulation often failed him. His inability to express his meaning was in part a rhetorical failure, but also a failure of conceptualization. In presenting his galvanic experiments, he moved systematically between different variations, first trying one order of metals and then its inverse, first opening and then closing the chain, and repeating each variation with different sense organs. In this systematicity, however, it can be seen how a growing preoccupation with polarity and analogy began to direct the order of experiments as well as the form of individual experiments. If Humboldt worried over the contemporary fixation on polarity, Ritter revelled in signs of polarity, and found through them a rhythm for his experiments and textual narratives. In experiments on light, this preoccupation proved productive in directing his investigations

48 Ritter, "Versuche," 451–459.

49 Strickland, "Ideology," 454–455.

50 Ritter, "Versuche," 476–482.

on ultraviolet light; in his self-experiments, it seems constraining as he tabulated phenomena into endless oppositions of positive and negative, red and blue, heat and cold, acid and alkaline.[51] He was often unable to give clear conceptualization to the analogies and relations he pursued experimentally between organic and inorganic processes, and between galvanism, chemistry, electricity, light, magnetism and heat. Indeed, he lamented that although he had the materials for a battery in advance of Volta's announcement of his discovery, he did not apprehend it.[52] Ritter's theoretical speculations were largely confined to the margins of his texts, to concluding *Winke* (signs or gestures to future significances), and to notes in diaries or letters to friends, often the products of reveries late at night after a long day of experiments. It is in these attempts to theorize that language fails him; a poor writer at best, lacking a genteel education, these passages spiral into incoherence, a circling series of clauses, their excessiveness shattering articulation. In these passages, Ritter tried his hand at philosophizing on the relations of man, God and nature, following the philosopher and theologian Johann Gottfried Herder, or his hand at figurative forms of expression, following the romantic writer Novalis, but he lacked the skills of these masters of these forms. His inaugural lecture at the Bavarian Academy of Science in 1806, on *Die Physik als Kunst* [*Physics as Art*], reveals most strikingly this failure of expression. The lecture to the Academy was a genre in which a scientist would be expected to relate his work to the history of his field and suggest its larger significance for humanity. But Ritter's attempt to portray the relations between the sciences, the creative aspects of scientific work and the transformative potential of physics for humanity left many in his audience bewildered.[53] The contrast to Humboldt's successes with such forms is striking.

Ritter lacked the skills of self-promotion that came naturally to privileged figures like Humboldt or at which less privileged figures like Davy or Volta learned to excel. Despite his early successes, his frequent publications in leading journals, and the praise of his talents as an experimentalist until the end, he remained desperately poor, lacking even sufficient funds to pay for his doctorate. He struggled in Jena, borrowing money from friends and collecting some funds from teaching. He also entered into a publishing venture in an

51 Jan Frercks, Heiko Weber and Gerhard Wisenfeldt, "Reception and Discovery: The Nature of Johann Wilhelm Ritter's Invisible Rays," *Studies in History and Philosophy of Science* 40 (2009): 143–156; and Martens, "Ørsted," 355–360. Compare Humboldt, *Versuche*, I: 385.

52 Richter, *Leben des Physiker*, 36.

53 See the letter from Karl Erenbert Moll, at the time Vice-President of the Bavarian Academy of Science, in K.E. Moll, *Mitteilungen aus seinem Briefwechsel*, 4 vols. (Augsburg, 1829–1835), II: 596.

effort to earn money, producing thousands of pages on his experimental work. But these unrefined works never captured a significant audience, in a stark contrast to Humboldt's remarkable success as a writer. It would be a false image, however, to cast Ritter in his poverty and isolation as a romantic figure. The romantic circle in Jena was self-consciously communal, both conceiving and practicing "romanticizing" as critical exchange and collective philosophizing or "*Symphilosophie*." Ritter was welcomed into this circle—first befriended by Novalis and then invited as a regular guest to the home of Friedrich and Dorothea Schlegel—while these friends also lamented the young scientist's periods of withdrawal and compulsive study. He was also generously supported by Herder, and by his publisher, Carl Friedrich Ernst Frommann.[54] Humboldt was only one of several scientists expressing interest in his work, and introduced him to the scientific community.[55] Most famously, his former student, Hans Ørsted promoted his work in the international community by translating it into French, and as a result, Ritter was invited to apply for the Galvani Prize in Paris in 1805. But some of Ritter's recent experimental work could not be confirmed—that magnetic phenomena arise through the contact of metals, of zinc and silver; and that an electric compass can be made from a gold needle to demonstrate the electrical polarity of the earth. The prize was not awarded to Ritter.[56] In Munich from 1805, Ritter became ever more isolated and depressed, spending hours and even days continually galvanizing parts of his body, experimenting with a variety of drugs, pursuing the relations between varied natural phenomena until his death in 1810.[57] Ailing, damaged and

---

54 See Friedrich Schlegel's letter to Novalis, which also attempts to include Ritter in their "Symphilosphie," in Novalis, *Novalis Schriften*, eds. Paul Kluckhorn and Richard Samuel, 3rd ed. (Stuttgart: W. Kohlhammer, 1977–1988), IV: 490–491, 500–501; and Novalis on "Romantisieren", in Novalis, *Novalis Schriften*, II: 545. See Ritter, *Fragmente aus dem Nachlasse eines jungen Physikers*, in *Key Texts of Johann Wilhelm Ritter (1776–1810) on the Science and Art of Nature*, ed. and trans. Jocelyn Holland (Boston: Routledge, 2010 [1810]), 37–55; Wetzel, *Ritter*, 15–17, 29–30; and Richter, *Leben des Physiker*, 47–52.

55 See the letter from Adolph Ferdinand Gehlen to Novalis, 25 Feb. 1810, expressing failed attempts to help Ritter establish himself, in Else Rehm, "Über den Tod und die Letzten Verfügungen des Physikers Johann Wilhelm Ritter," *Jahrbuch des freien deutschen Hochstifts* (1974): 304–311. On the reception of Ritter's work, see Gerhard Wiesenfeldt, "Eigenrezeption und Fremdrezeption: Die galvanischen Selbstexperimente Johan Wilhelm Ritters (1776–1810)," *Jarhrbuch für Europäische Wissenschaftslehre* 1 (2005): 207–232; and Heiko Weber, "J.W. Ritter und J. Webers Zeitschrift 'Der Galvanismus'," in *Naturwissenschaft um 1800. Wissenschaftskultur in Jena-Weimar*, eds. Olaf Briedbach and Paul Ziche (Weimar: Hermann Böhlaus, 2001), 217–247.

56 Wetzel, *Ritter*, 36–40; and Martens, "Ørsted," 372–374.

57 Strickland, "Ideology;" and "Circumscribing Science: Johann Wilhelm Ritter and the Physics of Sideral Man" (Unpublished doctoral thesis, Harvard University, 1992).

depressed, his body no longer a reliable instrument, Ritter became increasingly lost in the material of his experiments.

In a final attempt at validation, Ritter published a curious autobiography, narrated in the third person and presented as an obituary. Appearing as *Fragmente aus dem Nachlasse eines jungen Physikers* [*Fragments from the Estate of a Young Physicist*], an editor introduced the life and work of a gifted physicist who had passed away prematurely. But the text was authored by Ritter, who published it shortly before his death in 1810. This prologue was followed by 700 pages of fragments from his scientific papers and diaries. This doubling of his self, as subject and editor of the fragments, enabled self-study, as Ritter used the (auto)biography to attempt to make sense of his life and work. But he also used it to attempt to construct a new image of himself. Through the figure of the editor, Ritter tried to imagine himself as if an other were telling his story, and thus imagine himself as other. The prologue was a device to displace himself from the text in an attempt to correct the record of his achievements as a scientist. In fact, he falsified his biographical data, exchanging the events of his life with those of others, creating a new genealogy and a new significance for his life and work.[58] Ritter acknowledged his mentors and friends, but also laid claim to an instinctive understanding formed through self-education; he claimed he was misunderstood because he possessed a higher perspective. The *Fragmente* were to offer an education in the art of physics for aspiring scientists, but also to document early insights he failed to publish before others and insights not yet come to fruition. Although he claimed not to care about his lack of professional success, focused on higher purposes, his bitterness is evident throughout.[59] The purported authenticity provided by another authoring his biography, the purported honesty of the candid autobiography, was read as too biased, too personal, too arrogant and too unformed. The attempt at romantic "jest [*Scherz*]" backfired and left former friends and colleagues insulted.[60] In the

58 As Jocelyn Holland argues in her analysis of the prologue, the third person autobiography gives expression to a tension between difference and identity. Ritter's belated reflection on his life sought to give his life purpose, but also incorporated that reflection into the life, while at the same time reflecting on the activity of this belated reflection. Through his (auto)biography he gazed into a mirror, but at a distorted image. Jocelyn Holland, *German Romanticism and Science: The Procreative Poetics of Goethe, Novalis, and Ritter* (New York: Routledge, 2009), 116–125.

59 Ritter, *Fragmente*, 20–122.

60 See the reviews of the *Fragmente* in *Allgemeine Literatur Zeitung* 193 (19 Jul. 1810): 393–397; and *Heidelbergische Jährbuch der Literatur* 3.1.9 (1810): 116–125. The latter review by Ludwig Achim Arnim, Ritter's former friend, responded with a brutal assessment of Ritter's flawed character and betrayals.

end, the *Fragmente* only reinforced the image of Ritter as obscure and self-absorbed.

What Ritter did acquire from his association with the romantic circle was a belief in a method of spontaneous fragmentary writing, the recording of thoughts and experiments as they unfold in the physical and mental "workshop of the physicist [*Werkstätte des Physiker*]," so that even insights not fully formed could be revealed in works. He thus enlisted others, in reading this work, to work with him and to assist in making sense of both the parts and the whole. As Ritter argued in *Die Physik als Kunst*, insight into the inner workings of a phenomenon arises from being pulled into its activity, internalizing it rather than viewing it externally, being immersed within it and recreating it within oneself. Only thus can one apprehend it wholly. In his self-experiments, he understood this inner apprehension as not only an intellectual process, but also as a sensory and corporeal one.[61] Self-experience becomes the experience of nature, as the self becomes at once subject and object of study, and the instrument of their mediation. In leaving fragments of his experiments he hoped others would be able to trace that experience with him. If in the end Ritter had failed as a scientist, unable to ascend to clear articulation, and isolated and lost within his experiment materials, he hoped the *Fragmente* might write his life differently by providing experiments and reflections that might yet be given significance. The *Fragmente* demonstrate the porous boundary between Ritter's life and work, and demonstrate that it was in his experimental work, especially in his self-experiments, that he forged his identity.

"Experimental Selves"

Self-experiments helped Humboldt and Ritter make sense of the entanglement of phenomena and instruments in galvanic experiments. Galvanic experiments intersected with organic, chemical and electrical phenomena. Moreover, no single reliable instrument existed for reading galvanic effects—the frog leg provided unstable or ambiguous results; the chemical effects were not yet fixed; and mechanical apparatus were insufficiently sensitive. By inserting themselves into the experiments Ritter and Humboldt attempted to calibrate more precisely the other instruments utilized in galvanic experiments and to disentangle the phenomena studied from the tools generating and reading that phenomena. But the striking and distinctive sensations produced in self-experiments also provided an important perspective on these effects. With

61 Ritter, *Fragmente*, 96–97; and *Die Physik als Kunst*, in *Key Texts*, 575–583.

the phenomena investigated occurring within their own bodies, the subject was doubly positioned as an instrument for reading galvanic activity and as an object of investigation. By extending the instrumentality of their experiments into the subject as well as the object studied, self-experiments offered a means for mediating between conceptual and sensory apprehension and the material phenomena. But this mediation was reciprocal and recursive. By incorporating their own sensory selves into the experiments, not only was both the subject and object affected by the tools of inquiry, but these instrumentalized objects and subjects in turn affected the tools and methods of investigation. Humboldt was interested in galvanic experiments as a means to study organic vitality and vital chemistry. He used self-experiments to confirm his trials with frog legs, and developed them as one of the many tools enlisted in his investigations of nature. He used the instrument of his body to complement physical instruments and the instrument of aesthetic judgment to provide a total view of natural phenomena. Ritter was interested in galvanic experiments as a means to explore the relations between organic, chemical, electrical and other phenomena. He used self-experiments not only to complement his trials with frog legs, metals and chemistry, but also to read phenomena these instruments could not. He contended self-experiments provide a more nuanced apprehension of galvanic effects from inside the experimental material, arguing that insight into the inner workings of phenomena can arise only by internally recreating it. Ritter as well as Humboldt developed practiced skills and sophisticated techniques of self-experimentation that gave them credibility alongside trials with frog legs and other instruments. The self-experiments of both Humboldt and Ritter thus formed an important site for forging their identity as scientists and as "experimental selves."[62] They also provided a means of self-examination; in instrumentalizing themselves they externalized themselves as objects. But the two proffer strikingly distinct figures of the romantic scientist, challenging a common ideology of the ideology of romantic science. Humboldt in the end distanced himself from this experimental space, successfully writing himself into one of the most widely read theorists of a general view of nature in the nineteenth century. Ritter, in contrast, unable to ascend to clear articulation, remained immersed within this experimental space, his self-experiments becoming, in effect, his true autobiography.

62 Golinski, "Humphrey Davy: The Experimental Self."

CHAPTER 4

Shocking Subjects: Human Experiments and the Material Culture of Medical Electricity in Eighteenth-Century England

Paola Bertucci[1]

In contemporary Western societies medical patients are accustomed to being tested or treated by means of electrical instruments. Their presence is so familiar that it would be unsettling to enter a hospital or a medical laboratory unfurnished with the high tech apparatus through which research, diagnoses and therapies are routinely carried out. The technologization of medicine has produced systems of trust that rely on black boxed instruments, which profoundly influence contemporary perceptions of the human body and of the self.[2] However, the applications of scientific instruments for medical purposes have a history of debates and controversies.[3] In the eighteenth century, when the medical profession was regulated by the guild system, the intersections between experimental philosophy and medical practices created uncharted territories that blurred disciplinary divides and gave rise to conflicting epistemologies of medical efficacy. The early applications of electricity as a medical remedy offer a striking case of the tensions that such intersections engendered.[4]

---

1 For comments on earlier versions of this paper I am grateful to: participants in the workshop "The makers' universe" at McGill University, participants in the symposium "Alternative therapies of the 18th century" at the Wellcome Institute in London, Sally Romano, Lucia Dacome, Jan Golinski, and the editors of this volume.

2 On the black boxing of scientific instruments see Bruno Latour, *Science in Action : How to Follow Scientists and Engineers through Society* (Cambridge, Mass.: Harvard University Press, 1985); on the social construction of technological systems: Wiebe E. Bijker, Thomas P. Hughes and Trevor Pinch eds. *The Social Construction of Technological Systems: New Directions on the Sociology and History of Technology* (Cambridge, Mass.: MIT Press, 1987); on medical technology see Joel Howell, *Technology in the Hospital : Transforming Patient Care in the Early Twentieth Century* (Baltimore: Johns Hopkins University Press, 1995); on medical imagining and self perception: Renée van de Vall and Robert Zwijnenberg eds. *The Body Within: Art, Medicine, and Visualization* (Leiden: Koninklijke Brill NV, 2010).

3 Jeffrey P. Baker, *The Machine in the Nursery* (Baltimore: Johns Hopkins University Press, 1996); Joel Howell, "Early Perceptions of the Electrocardiogram," *Bulletin of the History of Medicine* 58 (1984), 83–98; Bettyann Kevles, *Naked to the Bone : Medical Imaging in the Twentieth Century* (New Brunswick, N.J. : Rutgers University Press, 1997).

4 Paola Bertucci and Giuliano Pancaldi eds. *Electric Bodies. Episodes in the history of medical electricity* (Bologna: CIS, 2001).

 | DOI 10.1163/9789004286719_006

Beginning in the 1740s, electrical treatments were often offered by practitioners who did not make a mystery of their ignorance of medical theory. Their claims to authority were based on the success of their new methods as assessed by patients themselves, rather than on theoretical grounds. Often working at the fringes of the medical system, electrical healers attracted at best skepticism, but more often indifference, on the part of the medical establishment.[5] They were not the only "irregulars" who advertised unusual therapies riding the wave of the self-help ethos of the time. Yet electrical treatments were quite exceptional in their employment of instruments that came from the domain of the experimental sciences.[6] Electrical machines and Leyden jars were not simply aiding tools for surgical operations or for the preservation of health; they were understood as the means to collect an "electric fluid" that, when applied to the human body, elicited physiological responses. It was the action of such a fluid on the body that caused the healing process, even though no theory explained exactly how. In 1780, Fellow of the Royal Society Tiberius Cavallo explained that "hitherto it has not been discovered that the electric fluid acts within the human body by any chymical property, as other medicines do"; the prevailing view in England was that electricity exerted a mechanical action upon muscles and nerves, even though recent cases indicated that it could have "some other action upon the human body besides that of mere stimulus."[7] If there was no agreement on the theory of medical electricity, electrical healers were all aware that their treatments could only be applied by means of scientific instruments. The strategies they devised to build trust in electrical apparatus differed greatly. In the 1750s John Wesley, one of the first advocates of the healing virtues of electricity, strove to naturalize the electric matter: although it was produced by means of man operated machines, electricity was for him a natural power, whose healing properties bore testimony to divine benevolence. Wesley attempted to make the electrical machine transparent to his patients: what mattered most to him was the fact that electricity was

5 On the early applications of electricity to medicine see Bertucci and Pancaldi, *Electric Bodies* (ref. 4); Stanley Finger, *Doctor Franklin's Medicine* (Philadelphia: University of Pennsylvania Press, 2006); Harry Whitaker et al., eds. *Brain, Mind, Medicine: Essays in the History of Eighteenth-Century Neuroscience* (Boston: Springer, 2007).

6 On the medical marketplace in 18th-century England see Roy Porter, *Health for Sale: Quackery in England* (Manchester: Manchester University Press, 1989); Roy Porter, ed., *Patients and Practitioners: Lay Perceptions of Medicine in Pre-Industrial Society* (Cambridge: Cambridge University Press, 1985); Dorothy and Roy Porter, *Patient's Progress: Doctors and Doctoring in Eighteenth-Century England* (Oxford: Basil Blackwell, 1989).

7 Tiberius Cavallo, *An Essay on the Theory and Practice of Medical Electricity*, (London, 2nd ed., 1781) London, 1781, p. 7 and note.

a "primitive remedy" created by God before the Fall, which the machine only revealed.[8] This paper will show that instead, in the later part of the century, electrical practitioners attracted attention to the electrical apparatus that afforded the therapy, with an emphasis on their own ability to master its performance. Trust in the electrical machine could not be taken for granted: electrical treatments were lengthy and often painful, and even though the eighteenth-century upper classes played with shocks and sparks in their salons, it was widely known that electrical imbalances in the atmosphere were responsible for life-threatening phenomena, such as lightning, thunderstorms, and even earthquakes.[9] What strategies did electrical healers employ to build trust in their therapies? The technologies of consensus building devised and employed by seventeenth-century experimental philosophers in the validation of experimental results have been widely explored.[10] I will argue that late eighteenth-century electrical healers attempted to adapt such technologies to their own practice: they presented accounts of medical cases as reports of experimental results, providing trustworthy testimonies and calling for virtual witnessing.[11] The unpredictability of the human body's responses to electricity, however, destabilized their attempts.

The essay follows the trajectory of John Fell, a Quaker surgeon in the English province who, in his mid-career, decided to specialize in electrical treatments. The de-centered view offered by this case provides rare insights into the pragmatic demands of provincial audiences that fed on the public culture of science.[12] Fell was neither physician nor quack. His career as an electrical

---

8 Paola Bertucci, "Revealing sparks: John Wesley and the Religious Utility of Electrical Healing," *British Journal for the History of Science* 39 (2006), 341–62.

9 Simon Schaffer, "Natural Philosophy and Public Spectacle in the Eighteenth Century," *History of Science*, 21 (1983), 1–43; on spectacular demonstrations: Paola Bertucci, "Sparks in the Dark: The Attraction of Electricity in the Eighteenth Century," *Endeavour* 31 (2007), 88–93. For a general overview on the history of electricity: John Heilbron, *Electricity in the 17th and 18th centuries : a Study of Early Modern Physics* (Berkeley: University of California Press, 1979).

10 Steven Shapin and Simon Schaffer, *Leviathan and the Air Pump; Hobbes, Boyle, and the Experimental Life* (Princeton: Princeton University Press, 1985); Steven Shapin, *Social history of Truth : Civility and Science in Seventeenth-Century England* (Chicago: The University of Chicago Press, 1994).

11 On virtual witnessing see Steven Shapin, "The House of Experiment in Seventeenth-Century England," *Isis* 97 (1988), 373–404.

12 Larry Stewart, *The Rise of Public Science: Rhetoric, Technology, and Natural Philosophy in Newtonian Britain, 1660–1750* (New York: Cambridge University Press, 1992); Jan Golinski, *Science as Public Culture : Chemistry and Enlightenment in Britain, 1760–1820* (New York: Cambridge University Press, 1992).

healer did not result from the institutionalized paths of medical education or apprenticeship, or from heroic competition in the medical marketplace. His self-directed apprenticeship derived from the public lectures he attended and from the experiments he carried out on patients and instruments. As we shall see, he held high expectations about the new direction he took, keeping records of his expenses, treatments, experiments and correspondence.[13] These documents offer insights on the material culture of electricity and its role in fostering the cultural ambitions of provincial practitioners. By examining the range of skills Fell strove to acquire in order to establish himself as a trustworthy medical electrician, I will argue that the management of the electrical apparatus was the most relevant aspect of the training and practice of electrical healers. The combination of practitioner, patient and instruments constituted an unstable experimental system that turned medical treatments into trials aimed at testing the performance of electrical instruments.

Manufacturing Safety

The earliest applications of electricity for medical purposes dated back to the early 1740s. Fashionable salon demonstrations that employed the human body indicated that sparks and shocks caused physiological reactions, which several electricians throughout Europe hoped to convert into a new way of treating diseases. Although authoritative experimenters such as Jean Jallabert in Geneva claimed that electricity could restore voluntary movement to paralyzed limbs, controversial results undermined the credibility of medical electricity. The infamous controversy over the "medicated tubes," a miraculous electrical treatment hastily supported by the Bologna Institute of Science and subsequently discredited by the abbé Nollet, became a well known cautionary tale of the dangers that learned societies could encounter when endorsing the new remedies.[14] However, in the 1770s, studies on fishes such as the torpedo and the *gymnotus electricus* (or electric eel), whose "electric organs" were shown to produce sparks and shocks in the same way as the electrical instruments

---

13 Wellcome Library, London: Ms 1175, *Miscellanea Electrica* (hereafter Ms 1175). This notebook contains John Fell's expenses, income, experimental records and correspondence with a number of London electricians.

14 Paola Bertucci, "Sparking Controversy. Jean Antoine Nollet and Medical Electricity South of the Alps," *Nuncius. Journal of the History of Science* 20 (2005), 153–187; id. *Viaggio nel paese delle meraviglie. Scienza e curiosità nell'Italia del Settecento* (Turin: Bollati Boringhieri, 2007).

employed in fashionable entertainments, boosted new interest in the role of electricity in the animal economy. The new subject of "animal electricity" captivated Georgian Londoners, who crowded the displays of live eels brought from South America. The eel literally shocked audiences and inspired a number of erotic poems that played with the fish's shape and electric vitality.[15]

Such demonstrations pointed to a close relationship between electricity and life and revived interest in the medical applications of electricity. In 1780 Tiberius Cavallo, the author of the successful *Treatise of Electricity in Theory and Practice* (London 1777), published *An Essay on the Theory and Practice of Medical Electricity*, reprinted in a second edition the following year, in which he acknowledged that in recent years the medical applications of electricity had progressed enormously. In the same period in Bologna, the physician Luigi Galvani began experimenting on dissected frogs with the aim to understand the role of electricity in animal motion, even though his work became widely known only a decade later.[16] In 1782 the instrument maker Edward Nairne envisaged enough potential in the medical applications of electricity as to petition for a patent – the first in the class of electricity – for his improved electrical machine.[17] According to the surgeon John Birch, who had established the Electric Department at St. Thomas Hospital, since 1777 "the many improvements of the Electrical Machine...have furnished the practice of medical electricity with a variety of accuracy and application 'till then unknown": the technical improvements of electrical apparatus enhanced accuracy in electrical treatments to the point that the year marked a clear divide between the "old" and the "new" medical electricity.[18]

Birch's emphasis on instrumentation was widely shared, as the eighteenth-century term for the medical applications of electricity indicates. Calling themselves "medical electricians," those who offered electrical treatments wanted to

15 Stanley Finger and Marco Piccolino, *Shocking History of Electric Fishes: From Ancient Epochs to the Birth of Modern Neurophysiology* (New York: Oxford University Press, 2011); Christopher Plumb, "The 'electric stroke' and the 'electric spark': Anatomists and Eroticism at George Baker's Electric Eel Exhibition in 1776 and 1777," *Endeavour* 34 (2010), 87–94.

16 Marco Piccolino and Marco Bresadola, *Rane, torpedini e scintille: Galvani, Volta e l'elettricità animale* (Turin: Bollati Boringhieri, 2003); Marco Bresadola and Giuliano Pancaldi eds. *Luigi Galvani Proceedings* (Bologna: CIS, 1999).

17 On Nairne's patent electrical machine see Paola Bertucci, "A Philosophical Business: Edward Nairne and the Patent Electrical Machine," *History of Technology* 23 (2001), 41–58.

18 John Birch, *Considerations on the Efficacy of Electricity in Removing Female Obstructions* (London: Cadell, 1780), p. iv (note).

be perceived as practitioners with skills and expertise in the theory and, above all, in the practice of electricity. A number of them were instrument makers who specialized in electricity and subsequently devoted themselves to its medical applications. This was the case of William Swift, a turner who made electrical machines at Greenwich, or John Read, a cabinet maker who invented a portable electrical machine for medical uses, and of a number of fellows of the Royal Society who made or designed instruments, such as William Henly, the inventor of the homonymous electrometer, the public demonstrator James Ferguson, and the apothecary Timothy Lane. The association between instrument makers and medical electricians was so strong that Nairne had "respectfully to inform the public" that his "other avocations make it impossible for him to attend" to the many "patients desirous of receiving the benefit of medical electricity."[19] In his *Course on electricity*, the instrument maker George Adams made a visual statement on the relevance of instruments to the practice of medical electricity: the frontispiece represented a young girl undergoing electrical treatments to her paralyzed forearm at the presence of her mother; while the image portrayed the apparatus in full detail, the human figures were only sketched (Fig. 4.1).

Late eighteenth-century audiences had come to learn that electricity's relationship with life was a double-edged one: the electric fluid seemed to carry a vivifying principle that disappeared in dead bodies, yet it was also a natural power with potentially lethal effects. The case of the St. Petersburg electrician Georg Richman, who died struck by lightning while incautiously experimenting on the electricity of the atmosphere, showed that lack of expertise in the management of the apparatus could prove lethal. Salon demonstrations showed that strong shocks could kill small animals, so advocates of medical electricity attempted to persuade patients that there was no danger involved in electrical therapies.[20] The concern with "safe" electricity boosted invention and provided new business opportunities. If, in the 1750s, the earliest advocates of medical electricity described three methods for treatment and did not devote too much attention to the description of the apparatus, three decades later medical electricians emphasized that there were many more ways in which

19 Edward Nairne, *Description and Use of Nairne's Patent Electrical Machine* (London, 1786), 79. The association between medical electrical practice and expertise in the management of instruments also characterized the work of the itinerant demonstrator James Dinwiddie. See Larry Stewart and Jan Golinski's papers in Bernard Lightman, Gordon McOuat, and Larry Stewart eds., *Circulating Knowledge: East and West* (Leiden: Brill Publishers, forthcoming).

20 On lightning and electricity, see, P. Heering, O. Hochadel and D. Rhees eds., *Playing with Fire: Histories of the Lightning Rod* (Philadelphia: The American Philosophical Society, 2009); on eighteenth century electricity in general, see Heilbron, *Electricity* (ref. 8).

FIGURE. 4.1 *Frontispiece from George Adams's* An Essay on Electricity *(London, 1785). Notice the detailed representation of the electrical apparatus, in contrast with the human figures that are only sketched.*
COURTESY OF THE BAKKEN MUSEUM AND LIBRARY OF ELECTRICITY IN LIFE.

electricity could be made to act as a healing agent and several new instruments with which to offer "safe" treatments.[21] They launched a vocal campaign against the administration of strong shocks and described at length the soothing virtues of "gentle" electrification. In 1767 the London apothecary and FRS Timothy Lane presented to the Royal Society his invention of an electrometer especially designed to measure the intensity of electric shocks.[22] The "medical electrometer," as the instrument came to be called, could be used to set an upper limit to the intensity of electric shocks and met with success among medical electricians. Cavallo, for his part, designed special instruments to apply electricity in the form of "gentle streams" of sparks in order to reduce the pain associated

21 The three methods of applying electricity for medical purposes were: the electric bath, in which the insulated patient was connected to the electrical machine by direct contact or through a metallic chain; the drawing of sparks from the affected part through pointed metallic "directors;" the electric shock, administered by discharging a Leyden jar through the affected part. Richard Lovett, *Subtil Medium Prov'd* (London, 1756).

22 Timothy Lane, "Description of an electrometer invented by Mr Lane; with an account of some experiments made by him with it," *Philosophical Transactions of the Royal Society*, 57 (1767), 451–460.

with electrical treatments. Instrument makers differentiated their offer of electrical machines and marketed a variety of "directors" especially designed for treating toothaches and ailments of the eye or the ear.[23] The sheer quantity of new instruments demonstrates more forcefully than any textual source that in spite of controversy and skepticism by the 1780s medical electricity had become a thriving business and a tempting investment – one that required training in electrical experimentation. As Cavallo pointed out, the ability to manage electrical instruments was key to the success of medical electricity:

> That Electricity has been of great benefit in many cases, where the application of other medicines has failed, is beyond doubt, ...its inefficacy in several cases is in great measure to be attributed to the injudicious application of it, indeed more than to any other cause.[24]

It is no surprise, then, that when Fell decided to become a medical electrician, his first preoccupation was to equip himself with good electrical apparatus. We do not know much about Fell's education or training, except that he worked as a surgeon at Ulverston in Lancashire (now Cumbria) when, at the beginning of 1783, he asked his brother-in-law, who lived in London, to seek the advice of some expert electrician. At the time, London was Europe's electrical capital. No other city hosted as many experts in the science of electricity active at the same time. London's reputation in matters electrical extended beyond the Channel: even well-reputed physicists such as Alessandro Volta embarked on long journeys across Europe in order to become acquainted with electricians whose expertise ranged from instrument making to meteorology, physiology, and the art of healing.[25] London's electric microcosm was diversified and widespread. The Fleet Street area was home to the workshops of electrical instrument makers such as Jesse Ramsden, Edward Nairne and George Adams, who attracted commissions from Europe and beyond. Fellows of the Royal Society who had greatly contributed to the science of electricity lived not far from Somerset House, in the Strand, the venue of the Society's meetings. The notorious quack James Graham had erected his Temple of Health and Hymen, an electrical extravaganza that combined electricity and sexual therapies, in Pall Mall. In Southwark, the surgeon John Birch was responsible for the newly founded Electric Department at St. Thomas Hospital. In Moorsfield, the

---

23 Tiberius Cavallo, *Essay* (*ref.* 7).

24 Tiberius Cavallo, *A Treatise of Electricity in Theory and Practice* (London, 1777). 88.

25 Giuliano Pancaldi, *Volta: Science and Culture in the Age of Enlightenment* (Princeton: Princeton University Press, 2003).

FIGURE. 4.2 *Electricity in London A. Miles Partington, Cavendish Square. B. Tiberius Cavallo, 8 Little St. Martin's Lane. C. Timothy Lane, Aldersgate Street. D. John Wesley's Electrical Dispensary, Moorsfield. E. Edward Nairne, 20 Cornhill. F. Mr. Long, Soho. G. James Graham, Pall Mall. H. George Adams, Fleet Street. I. William Henly, Borough, Southwark. L. John Birch, St. Thomas Hospital. M. Henry Cavendish, Great Marlborough Street. N. John Canton, Spital Square Academy. O. John Read, Knightsbridge. P. Michael Underwood, Cavendish Square. Q. William Swift, Greenwich* (*out of map*).

Dispensary founded by John Wesley offered free electrical treatments to the poor. In addition to such institutional or semi-institutional premises, a number of self-styled electrical healers practiced in their own homes all over the city.[26] Eighteenth-century sources mention a number of them who achieved some notoriety: Miles Partington at Cavendish Square, John Read at Knightsbridge, a Mr. Long in Soho, Michael Underwood at Cavendish Square, William Swift at Greenwich.[27] The electric topography of the city reveals that electrotherapeutics constituted the most common practice through which Londoners became familiar with the new science of electricity (Fig. 4.2).

On the other hand, in the market town of Ulverston no highly-skilled instrument maker or medical practitioner who lived on the healing virtues of sparks

26 On Graham see Roy Porter, "The Sexual Politics of James Graham," *British Journal for Eighteenth-Century Studies*, 5, (1982), 201–6; on Wesley: Bertucci, "Revealing Sparks" (ref. 8).

27 I found the medical electricians' addresses in: Tiberius Cavallo, *Complete Treatise on Electricity, in Theory and Practice* (London: Dilly, 1786), 186; [anon,], *New Thoughts on Medical Electricity* (Sevenoaks: Clout, 1782), 9; James Ferguson, *An Introduction to Electricity* (London, 1770), 127; relevant entries in *Oxford Dictionary of National Biography*, online ed., ed. Lawrence Goldman, Oxford: OUP, http://www.oxforddnb.com.

and shocks could help Fell set up his new business. However, Cumbria was not too far from the West Midlands of the English Industrial Revolution and it was a convenient stop for itinerant demonstrators heading to Scotland.[28] Lecturers travelling between London and the provinces offered courses of experimental philosophy that spread the cultural novelties that inflamed audiences in the capital.[29] Provincial audiences did not participate in these kinds of cultural activities simply as spectators: they expected to acquire practical knowledge to start new businesses and to become cultural protagonists on their turn.[30] Fell's self-guided apprenticeship in the domain of electricity was firmly grounded in such courses. In the 1780s he attended the lectures on electricity offered by a Mr Long, who arrived at Ulverston after working as a mechanical assistant for Adam Walker, a successful demonstrator in London.[31] A medical electrician by the name of Long was active in London roughly at the same time, and it is possible that they were one and the same.[32]

Fell's brother-in-law, Morris Birckbeck, was friends with Lane, whose reputation in matters electrical had escalated after the invention of his electrometer. In the 1770s Lane worked together with other fellows of the Royal Society on crucial electrical affairs, such as the experiments on the artificial torpedo conceived by Henry Cavendish and the controversy on lightning rods.[33] Birckbeck passed on to him Fell's commission. Being acquainted with Nairne, Lane set out to examine the new patent machine "with critical attention" and, after conversing "with several first-rate electricians on the new construction, who all acknowledged its superiority," he warmly recommended it to Fell, "in preference to all others." The examination of the machine took nine months and, since London makers worked on commission, Fell had to wait upon another two to have his machine at Ulverston. Once Fell decided to order the machine, Lane personally attended to its construction, visiting Nairne's workshop several times: he "tried the power of the cylinder, & examined every part of the apparatus, pronouncing the whole excellent."[34] The careful selection

28 Peter Jones, *Industrial Enlightenment: Science, Technology and Culture in Birmingham and the West Midlands, 1760–1820* (New York: Manchester University Press, 2008).

29 Stewart, *Rise of Public Science* (ref.).

30 Larry Stewart and Paul Weindling, "Philosophical Threads: Natural Philosophy and Public Experiment among the Weavers of Spitalfields," *The British Journal for the History of Science* 28 (1995), 37–62.

31 Long was "manager to the mechanical parts" of the Eidouranion, a spectacular equipment that Walker employed in his scientific lectures in London theatres: Ms 1195, f. 84.

32 Lancashire Record Office, DDX 317/83 (John Fell to unknown, 15 February 1798).

33 Henry Cavendish, "An Account of Some Attempts to Imitate the Effects of the *Torpedo* by Electricity," *Philosophical Transactions of the Royal Society* 66 (1776), 196–225.

34 MS 1175, ff. 2–3.

and testing of the machine is yet another instance of the crucial relevance of the apparatus in the practice of medical electricity.

The arrival of Nairne's patent electrical machine at Ulverston in October 1783 marked the beginning of Fell's career as a medical electrician. His initial investment is revealing of his expectation to profit from electricity, as the amount he spent on Nairne's machine complete with medical apparatus was on the higher end of the machine's price range. Being one of the first purchasers of the newly invented instrument, Fell had to learn how to use the machine without the *Directions and use of the patent electrical machine* that Nairne published in 1784. Since Fell engaged in medical treatments soon after the arrival of Nairne's electrical machine, it is reasonable to believe that he was not new to electrical instruments and experiments. He purchased eight of the most popular texts on electricity only the following year, and made a present of two of them, so his knowledge of electricity likely did not derive from texts, but mostly from the courses of experimental philosophy offered by travelling lecturers.[35]

The Proper Subjects for Electric Trials

Building trust in medical electricity at Ulverston required a trustworthy electrical practitioner. As Roy Porter pointed out, patients of fringe remedies often trusted therapies as a result of their trust in the practitioner, yet electrical treatments presented the additional challenge of being painful and not widely available.[36] In a small sized provincial village, trust was first gained through personal connections. The son of a surgeon, Fell enjoyed a reputation among his fellow Quakers, and it is likely that his religious affiliations brought him at least a few patients.[37] As scholars have shown, medical electricity spread quickly through religious communities and word of mouth.[38] Even if we cannot be sure about his patients' confessions, his list of "electrified patients" – as

35 MS 1175, f. 16.

36 See ref. 4 above.

37 Evidence about Fell being a Quaker in Backhouse Papers, Durham University Library, Archives and Special Collections, BAC/58 3 May 1764: Certificate of the marriage of John Fell of Ulverston, Lancashire, son of Stephen Fell, Practitioner in Physic, of Ulverston and Margaret his wife, and Sarah Birkbeck, daughter of William Birkbeck, merchant, of Settle, Yorkshire and the late Sarah Birkbeck, at the Quaker Meeting House in Settle.

38 Jonathan Barry, "Piety and the Patient: Medicine and Religion in Eighteenth Century Bristol," in Porter, *Patients and Practitioners*; Bertucci, "Revealing Sparks" (ref. 8); Simon Schaffer, "The Consuming Flame: Electrical Showmen and Tory mystics in the World of Goods," in J. Brewer and R. Porter eds., *Consumption and the World of Goods in the Eighteenth Century* (London: Routledge, 1993).

he called them – suggests that family bonds and connections were crucial to the making of his authority as a medical electrician. Scrolling down the list, we find groups of people with the same family name, or patients indicated simply as somebody's sons, daughters, wives, brothers or maid-servants.[39] Yet, when he started his career in medical electricity, Fell was already known in learned Quaker circles even beyond Ulverston. The natural sciences figured prominently in the education of Quaker children, and they constituted one means of cultural exchange among communities living in different provinces.[40] Fell and his wife were interested in botany and exchanged specimens with fellow Quakers in other villages.[41] They were friends with the Manchester chemist John Dalton, himself a Quaker, with whom Fell also shared membership in the Manchester Literary and Philosophical Society.[42] Fell, as a non-resident, was a honorary member, a position held also by Alessandro Volta, Joseph Priestley, Jean Hyacinthe Magellan, and John Lettsom.[43]

This set of connections certainly played a role in establishing Fell's reputation as a cultivated surgeon. Yet it was not obvious that patients would accept undergoing electrical treatments as experimental subjects. Fell was not the first to confront this problem. Previous medical electricians had defined the "proper subject" for medical electricity as a patient who had nothing to lose, who had not benefited from the various traditional therapies, and who had been referred to electricity by somebody authoritative, possibly a physician. Medical electricians indulged in the descriptions of the miserable state of the patients who were sent to them, with the double result of presenting electricity as a harmless last-resort remedy, and acclaiming it as the one that succeeded where others had failed.[44]

Fell carefully selected his first case. The patient was a 19 year old woman who "contracted 'typhus nervosus' " several months earlier and had gradually lost her appetite, becoming "pale and emaciated." She was "seized with frequent faintings, and at last with the most dangerous debility," associated with the interruption of her menses. The woman had been treated by the London

39 Ms 1175, ff.21–25.

40 Geoffrey Cantor, *Quakers, Jews, and Science: Religious Responses to Modernity and the Sciences in Britain, 1650–1900* (Oxford: Oxford University Press, 2005).

41 John Fell to unknown, 15 Feb. 1798. Lancashire Record Office, DDX 317/83.

42 Arnold Thackray, *John Dalton: Critical Assessments of His Life and Science* (Cambridge, Mass.,: Harvard University Press, 1972): Letter by Fell to Dalton (dated 5 April 1801) published at pp. 149–50. On the Manchester Lit and Phil: p. 53.

43 *The Medical Register for the Year 1783*. London: printed for Joseph Johnson, No 72, St. Paul's Church-Yard, [1783].

44 Birch, *Considerations* (ref. 18); Cavallo, *Essay* (ref. 7); Miles Partington, "A Cure of Muscular Contraction by Electricity", *Philosophical Transactions of the Royal Society* 68 (1778), 97–101.

physician William Heberden who, after trying "preparations of the bark, bitters and iron," declared her incurable. After five months of failures, her father, himself a physician, lost hope of recovery and turned to Fell. Upon an "attentive examination of the case," Fell declared the girl "a proper subject for electric tryals." She spent four days at Fell's, during which she underwent five rounds of electric operations, consisting of "six very slight shocks from Os pubis to the upper part of the Os sacrum, and the same number from each Os ilium to the knee or foot of the contrary side." Her menses returned, she recovered completely and went back home "in perfect health," with her father calling the event "a resurrection from the dead."[45]

Fell's choice of a case of "suppression of the menses" points to his hope to start his new career with a success. The disease was often treated successfully by electricity. In 1779 John Birch published *Considerations on the efficacy of electricity in removing female obstructions*, a booklet that went through a second edition the following year and that was in Fell's library. Birch was a surgeon at St. Thomas Hospital, where he established an Electric Room in which he routinely received patients who had failed to respond to other, more traditional therapies. In his promotion of medical electricity, Birch repeatedly underscored the method's "certainty" against the suppression of the menses. Along with other advocates of medical electricity, he explained that shocks should never be applied to pregnant women.[46] This frequent warning – together with the emphasis on the method's infallibility – has been interpreted as an implicit advertisement for procuring abortions.[47] There are several elements that might support this thesis. Birch's accounts of successful cases show that medical electricians who offered remedies for the obstruction of the menses did not spend too much time on excluding the possibility of a pregnancy. As a disease unrelated to pregnancy, the "obstruction of the menses" was associated with debility, frequent fainting, and difficult digestion – symptoms that were very similar to those of early pregnancies.[48] The ambiguity of the diagnosis could certainly leave room for a don't-ask-don't-tell relationship between electrician and patient. In eighteenth-century England abortion was rarely prosecuted, even though it was legally a crime if performed after the quickening of the child: the proceedings of the Old Bailey of London do not record any indictment for abortion until 1823, when the laws started to become more

45 Fell's notes. Ms 1175, ff. 31–33.

46 Birch, *Considerations* (ref. 18), p. 3.

47 Porter, *Health* for sale, p. 148.

48 Birch, *Considerations* (ref. 18), p. 43 (Case v).

severe.[49] However, views on the subject varied greatly, and several English texts welcomed the severity with which Catholic countries condemned men or women who procured abortions. English audiences were familiar with the story of the Scottish "Doctor" Philip who, in 1770, was sentenced to be banished to the plantations for life after being found guilty of giving drugs and performing "desperate operations" to cause abortion in a pregnant woman.[50] The *Harrop's Manchester Mercury* listed several cases of people indicted for murder after forcing pregnant women to take abortifacients that caused death.[51]

Although it is possible that electricity was employed in this way, it would be too simplistic to conclude that the "removal of female obstructions" was synonymous with abortion. In line with recent studies on early modern conceptions of fertility and menstruation that have proposed a more nuanced interpretation of the "suppression of the menses," I believe that the employment of electricity in the treatment of this disease illuminates its function as a normalizing agent – a balancing power that, if managed properly, restored the body to its normal state.[52] Early-modern bodies were still understood within a humoral theory that prescribed a balance of bodily fluids to maintain physical as well as moral health. As Birch pointed out, women – much more frequently than men – were affected by diseases caused by their own body: "As the intention of nature in the formation of the female sex, was, among other things, for the nutrition of the child while in the womb," women were endowed with a "superfluous quantity of blood from the time they are capable of conception"; since blood was a stimulant, the periodical discharge through the menses defended women from the dangers of such accumulation.[53] Women's health was entirely dependent upon the regularity of the menstrual flux, whose obstruction immediately caused illnesses, infertility, and bad habits. Women affected by the suppression of the menses lacked the natural balance of bodily fluids that stood at the basis of physical as well as social health. Birch was not

49 *The Proceedings of the Old Bailey. London Central Criminal Court, 1674–1913*: www.oldbaileyonline.com; Richard Burn, *The Justice of the Peace, and Parish Officer*, London, 1764, vol. 1, 179–80.

50 *The Scots Magazine* 33 (1771), 499.

51 *Harrop's Manchester Mercury (and General Advertiser)*, issue 475 (1761–1763), 850, 899 (1767–1769), 1201 (1773–1775).

52 Jennifer Evans, "'Gentle purges corrected with hot spices, whether they work or not, do vehemently provoke venery': Menstrual Provocation and Procreation in Early Modern England," *Social History of Medicine* 25 (2012), 2–19; Wendy Churchill, "The Medical Practice of the Sexed Bodies: Women, Men, and Disease in Britain, 1600–175," *Social History of Medicine* 18(2005), 3–22.

53 Birch, *Considerations* (ref. 18), 2.

the first to suggest that electric shocks could be applied to the site of the disorder to remove such obstructions: authoritative physicians, such as Van Swieten, Cullen and Musgrave, had pointed to electricity, understood as a mechanical "deobstruent," as a possible remedy against the disease. Birch's main goal was to argue that, even though the obstruction of the menses usually fell under the care of physicians, the application of electricity, being "an operation of the hand," required the skills of surgeons.[54]

The role of electricity as a normalizing agent extended to other diseases caused by imbalance of bodily fluids. Melancholy, which resulted from a "depression of the spirits," affected most commonly young and studious men. Usual therapies for this disease included moving to places with better climate and breaking the ordinary study routine, but such remedies were tried to no avail on a 19 year old student at Cambridge who was then sent to Fell. The surgeon found that the boy's "intense application to his studies" changed his attitude from cheerful to gloomy and timid; he was "so inattentive that it was sometimes necessary to ask him a question 3 or 4 times before an answer could be exorted." Observing "an almost constant uneasiness in his head," Fell decided to pass 50 small shocks in every direction through the boy's head, every morning and evening, for eleven days, "which restored him to his usual state of health and spirit."[55]

The idea that the electric fluid could act on the nerves was relatively recent: because of the involuntary motion that electric shocks imparted to muscles, electricity was initially regarded mainly as a remedy against paralysis. Fell's "trial" indicated that it could prove effective also in the so-called "English malady."[56] With his application of electricity to nervous disorders, in particular to the brain, Fell placed himself at the forefront of medical electrical research. Five years later, the Windsor physician James Lind would speculate with the President of the Royal Society Joseph Banks that electricity might prove useful in treating the King's madness, even though he explained that the subject needed further experimental inquiries:

> From a scrupulous investigation into what Authors have written upon the subject of Insanity I do not find that among the variety of means which have been employed in the cure of that malady, that Electricity has

54 Birch, *Considerations* (ref. 18), 8.

55 Ms 1175, ff. 40–41.

56 Fell was by no means the first to suggest that electricity could be used against nervous diseases. See Heather Beatty, *Nervous Disease in Eighteenth-Century England* (London: Pickering and Chatto, 2011).

> ever been tryed. If we may credit the accounts of the state of the Brain of insane persons found upon dissection I think there is great reason to believe that it might be of service in that disorder, and appears to me to merit a fair Tryal. I should be glad if we could find out any body that would make the experiment in London, where there are such frequent opportunities of doing it.[57]

As infrequently as electricity was applied to the brain, the treatment soon elicited critical responses. The Reverend Edward Harwood, who underwent electrical treatment for his recurrent melancholy, warned against its possible dangers: "I would advise persons labouring under any violent headach [sic], that they would be extremely careful about permitting a fluid of such omnipotent energy to be darted and dashed through their brains."[58] Harwood's warnings were one more instance of the fear that electrical treatments caused on patients. Electricity's "omnipotent energy" needed to be governed by qualified practitioners who would be able to extend their trustworthiness to their apparatus. Fell was trying to do just that. Soon after the arrival of the patent electrical machine at Ulverston, he ordered another set of instruments to train himself in the material practice of electricity. Meanwhile, he prepared a letter with the account of "a few cases where electricity was applied with success after other means had failed" that he addressed to Lane in London.[59] The letter contained not only the descriptions of his "electric trials" on patients but also specific details about the performance of the patent electrical machine. Fell noted that Nairne's machine made it unnecessary, for example, to uncover the patient and made the administration of mild shocks more comfortable for both patient and practitioner.[60] The bodies of Fell's electrified patients put Nairne's new instrument to the test.

The Human Body and the Electrical Machine

The human body was an essential component of eighteenth-century spectacular electrical demonstrations. Starting in the 1730s, when Stephen Gray invented

---

57 Fitzwilliam Museum Library, Cambridge. Mss H 140 (Lind to Banks, 27 November 1788). There is no evidence that electricity was ever tried on George III.

58 Edward Harwood, *The Obstinate Case of the Rev. Dr. Harwood: An Obstinate Palsy of Above Two Years Duration, Greatly Relieved by Electricity*, London, 1784.

59 Ms 1175, f. 31.

60 Ms 1175, ff. 33–34.

the experiment of the "flying boy," the property of the human body to conduct electricity engendered a vast range of experiments designed to display the properties of the newly discovered power of nature. Electrical practitioners staged interactive performances that allowed spectators to feel the effects of electricity on their own bodies: when electrified their hair raised towards the ceiling, their fingers issued sparks, the precious embroidery on their clothes became luminous. Contemporary texts explained how to set spirits on fire with an electrified finger, how to make chains of people jolt together, how to make ladies proffer electric kisses. Illustrated texts gave evocative names to these experiments – the "Venus electrificata," "setting spirits on fire," "electric commotion" – thus presenting them as standardized, even though it was well known that the responses of the human body to electrification were unpredictable.[61] Electricians knew that the outcome of their performances was highly dependent on the weather, the status of the apparatus, and even on clothing. Human bodies responded differently to electrization.[62] Although a number of medical electricians claimed that electricity increased bodily perspiration, temperature, and the pulse by a fixed, measurable amount, experienced experimenters such as Cavallo pointed out that those who underwent electrization, even just for entertainment, were generally afraid of electricity, and all the physiological alterations resulted from fear or apprehension.[63] When such anxieties about electricity were dispelled, no physiological changes could be observed: Cavallo underwent electrization himself, without ever noticing any change in his pulse. He did not intend to use his own body as conclusive evidence, though. Rather, he underscored that the practice of electrical experimentation demonstrated that there was no standard electric body.

---

61 Bertucci, "Sparks in the dark"(ref. 9); Heilbron, *Electricity* (ref. 9); Simon Schaffer, "Self Evidence," *Critical Inquiry* 18 (1992): 327–362; Paul Elliot, "'More Subtle than the Electric Aura': Georgian Medical Electricity, the Spirit of Animation and the Development of Erasmus Darwin's Psychophysiology," *Medical History* 52 (2008), 195–220.

62 I am using here the term "electrization" in its eighteenth century meaning, to indicate all possible interactions between the human body and the electrical machine – not just the administration of electric shocks. A number of medical electricians believed that by connecting the human body (insulated from the ground) to the electrical machine's conductor, the electric fluid would run into the patient's body, accelerating his or her pulse, and increasing insensible perspiration. This method of applying electricity would still be indicated as electrization, even if it was by no means as painful as the drawing of sparks or the electric shock.

63 Cavallo, *Essay* (ref. 7), 2. On the increase of insensible perspiration as a medical therapy see Lucia Dacome, "Living with the Chair: Private Excreta, Collective Health and Medical Authority in the Eighteenth Century," *History of Science* 39 (2001), 467–500.

The variety of individual responses made it "impossible to prescribe the exact degree of electrization that must be used";[64] electrical treatments met with contrasting results because of the "variety of temperaments" and the "coincidence of circumstances": "not every disorder, nor every temperament…requires an equal, or perhaps any application of electricity."[65] Unlike other remedies, electricity was not a specific for particular disorders, with the possible exception of the suppression of the menses.

Medical electricity relied on an epistemology of efficacy grounded in integrated experimental settings where patient, practitioner, and instruments constituted an unstable unit. The operator could not rely on any electric pharmacopoeia that associated therapies to diseases: he had to "be instructed by experience," learning how to calibrate the intensity of electricity to the patient's constitution and temperament, and be able to apply the smallest degree of electricity that "the patient can conveniently suffer."[66] Experience could only be gained through the skillful use of electrical apparatus. As Birch emphasized, learning how to manage electrical instruments was as crucial to medical electricity as anatomy was to surgery, or chemistry to medicine.[67]

Fell was fully aware of the necessity of engaging in electrical experiments. A few months after the arrival of Nairne's electrical machine, he purchased several instruments from London. Although he remained a customer of Nairne and George Adams for eleven years, his "electrical expenses" show that his most consistent investments were made in the first two: in 1784, he spent on electrical instruments four times the amount he spent on the patent electrical machine, equipping himself with the most popular devices for electrical demonstrations.[68] He carefully studied the standard demonstrations of electrical lecturers: how to make a model house explode with an electric spark, how to make paper puppets dance on a metallic wire, how to compose words of (electric) fire in the dark (Figs. 4.3–4.6). For him, the instruments that arrived at Ulverston from London were objects of experimental enquiry in their own rights: he continuously modified and adapted them to his needs. As he progressed with his experiments, he realized that several practical details that were crucial to the success of electrical experiments did not find space in published accounts. So, he worked painstakingly on the making of lacquers,

64 Cavallo, *Essay* (ref. 7), 26.

65 ibid., 87.

66 ibid., 27.

67 Birch, *Considerations* (ref. 18), ix.

68 Ms 1175, ff. 12–20.

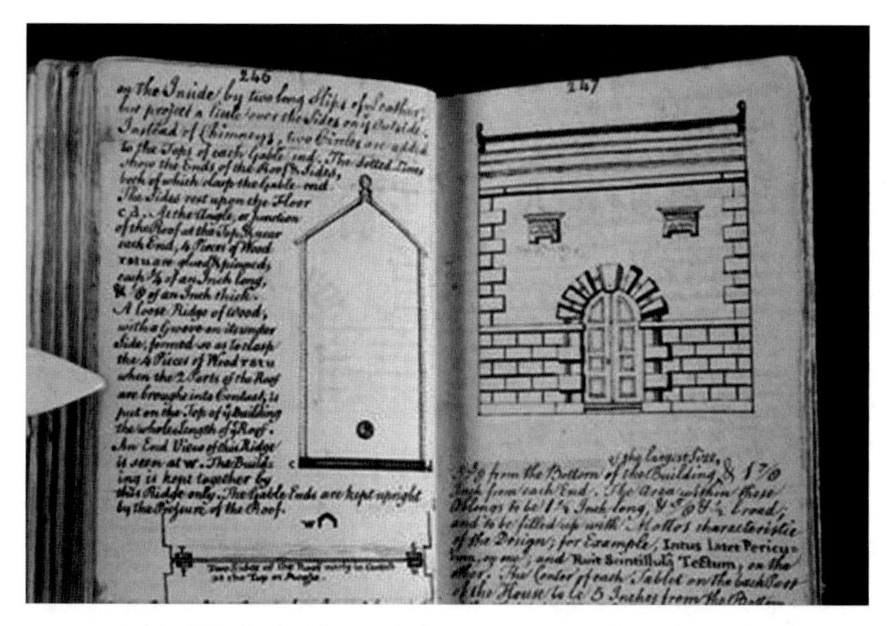

FIGURE. 4.3 *John Fell's sketch of the "powder house". Ms 1175,* Miscellanea Electrica. *Wellcome Library, London.*

varnishes, amalgams and other materials employed for the preservation of the apparatus, and studied the design of various electrical machines in order to optimize the laborious process of charging. The process of modification and adaptation led him to inventions, whose details he shared with Nairne and Adams.

Instruments created new channels of communication that made the experiments of a provincial surgeon relevant to metropolitan electricians: Nairne was impressed with Fell's "fire shooter," an amusing experiment that he found "very prettily devised" and started to review Fell's order personally, sending comments along with the apparatus (Fig. 4.7). He encouraged Fell to continue with experiments and explained that his patent electrical machine resulted from gradual improvements to the apparatus that he made while performing experiments similar to those described by the surgeon.[69] When Fell ordered special instruments for treating disorders of the ear, Nairne praised the choice, describing the successful test he had performed on himself, and did not charge the surgeon.[70] Adams was slower in his replies, but no less interested in Fell's

69 Nairne to Fell, Ms 1175.

70 Nairne to Fell, MS 1175 ff. 58–59.

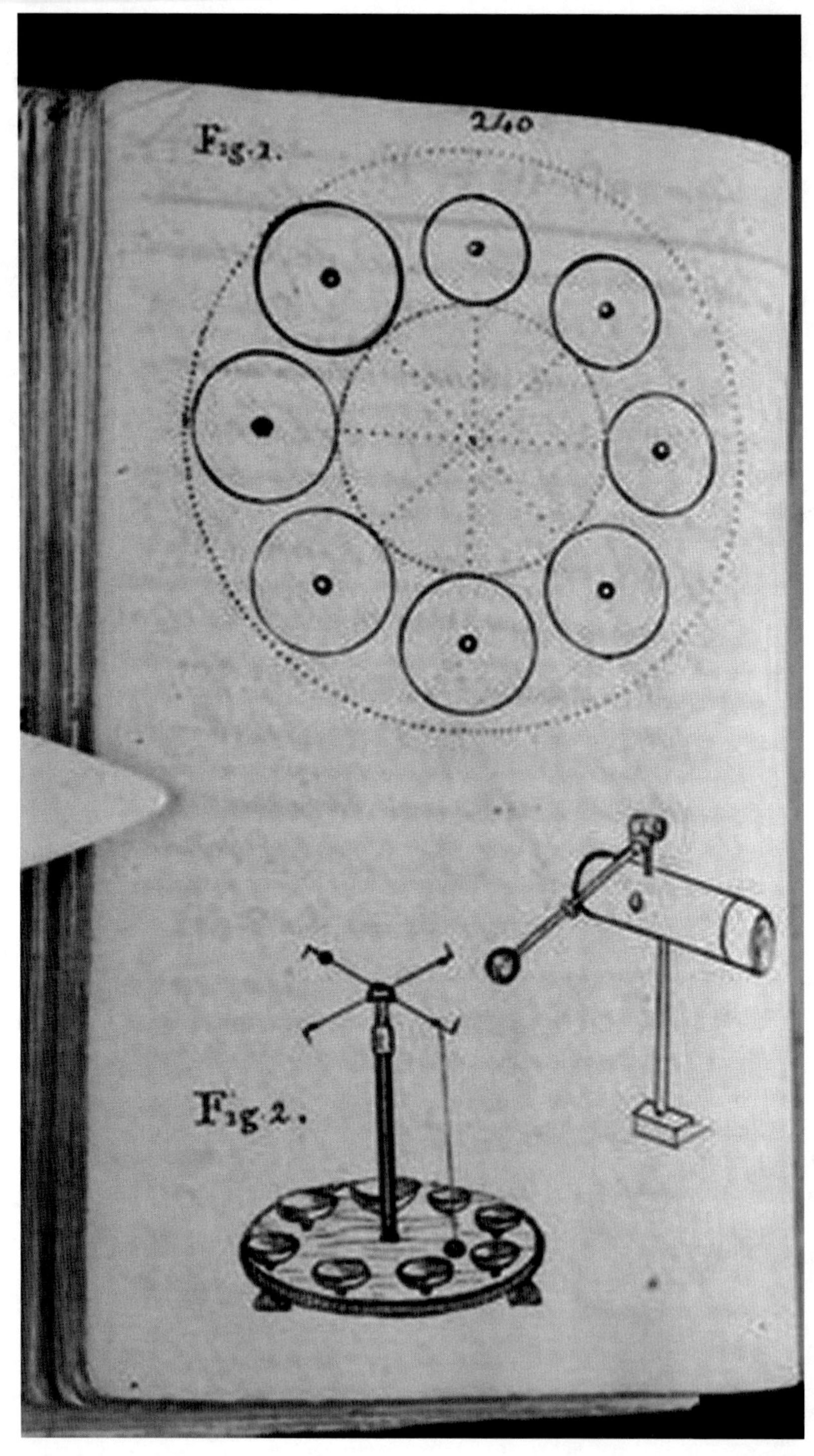

FIGURE. 4.4 *John Fell's sketch of "electric bells". Ms 1175,* Miscellanea Electrica. *Wellcome Library, London.*

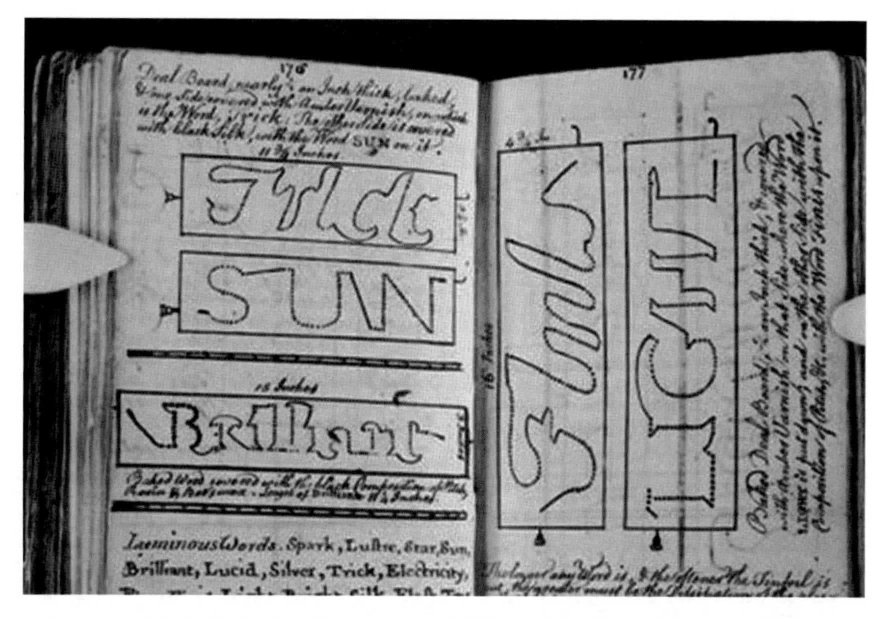

FIGURE. 4.5 *John Fell's sketch of "luminous words". Ms 1175,* Miscellanea Electrica. *Wellcome Library, London.*

original contributions: he was so enthusiastic about Fell's invention of a "double jar" as to publish it in the second edition of his *Essay on electricity*.[71]

Fell's engagement in what he called "rational recreations" was part of his self training to become a lecturer of electricity. He realized that the most effective way to spread interest in medical electricity at Ulverston was to familiarize his community with electricity: with time he noted that electric patients were hardly dependable and often discontinued treatment, with the exception of those who formed "a resolution to persevere, from an opinion of the practice."[72] In a few years, he became a sought after lecturer, who offered several courses to groups and individuals, such as John Dalton, and who played an active role in the scientific education of young Quakers in the area.[73] His self training raised his expectations on public lecturers coming from London, whose courses he attended with a critical eye, carefully observing the experimental demonstrations and pointing out their fallacies. Long, for example, added exotic drama to electrical shows by staging a "perpetual war" between an electric eel and a fly along the rivers of Surinam. He employed an artificial eel made of cork and

71 Adams to Fell, Ms 1175, f. 61; Fell's description of double jar: Ms 1175, ff. 75–76. George Adams, *An Essay on Electricity* (London, 1785), 128–29.

72 Lancashire record Office. Mss DDX 317/82 (Fell to Barton, 25 October 1787).

73 Lancashire record Office. Mss DDX 317/83 (Fell to unspecified, 15 February 1798).

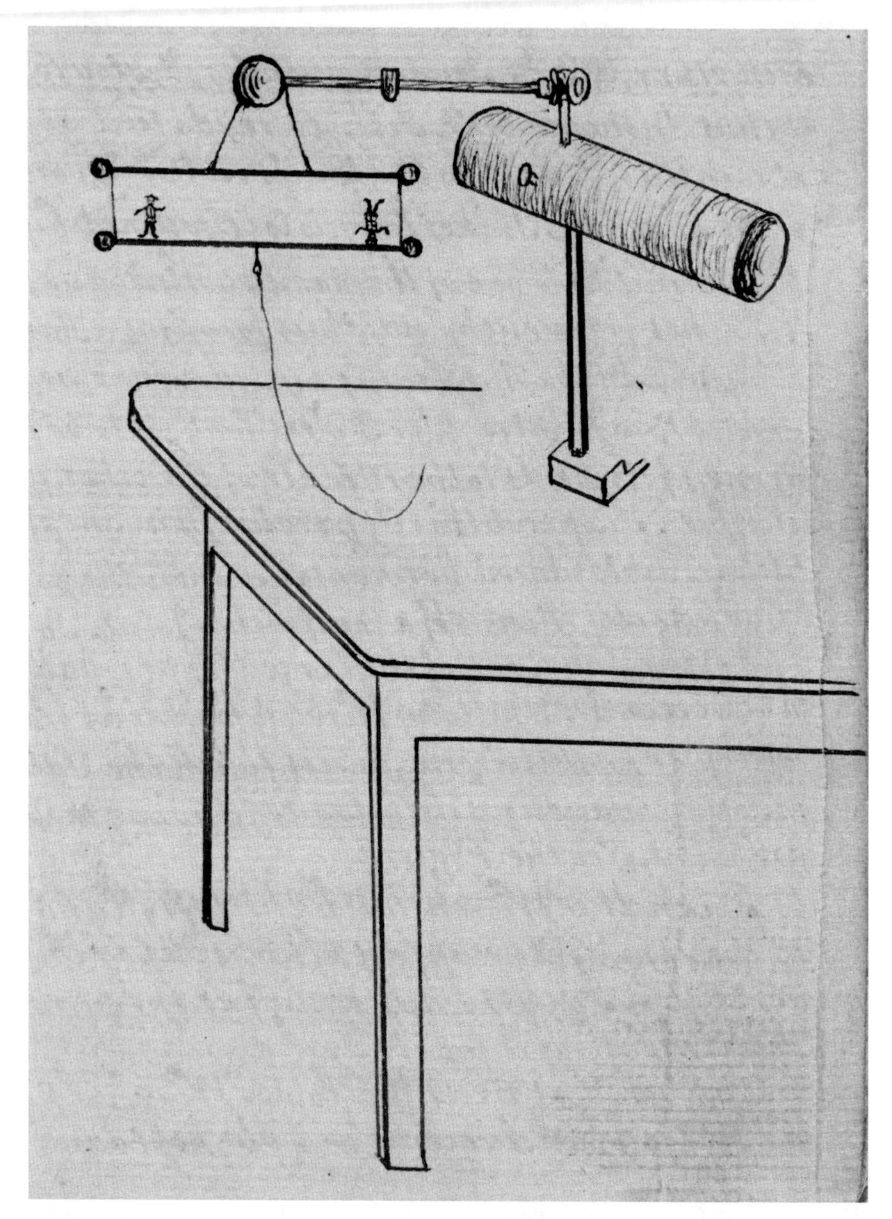

FIGURE. 4.6 *John Fell's sketch of the "electric dancers" apparatus. Ms 1175,* Miscellanea Electrica. *Wellcome Library, London.*

tinfoil together with a light feather to act as the fly; when charged, they attracted and repelled each other, simulating the chase. Fell sketched Long's apparatus and noted with great disappointment that the demonstrator failed to obtain shocks from the eel (Fig. 4.8). It was the ability to draw sparks from

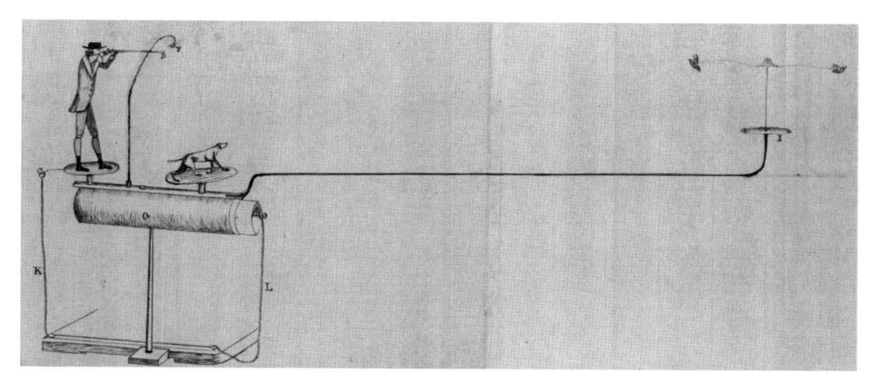

FIGURE. 4.7 *John Fell's sketch of the "fire shooter" apparatus. Ms 1175,* Miscellanea Electrica. *Wellcome Library, London.*

FIGURE. 4.8 *John Fell's sketch of Long's artificial eel. Ms 1175,* Miscellanea Electrica. *Wellcome Library, London.*

the fish that proved to London audiences that the eel was able to generate electricity. So, Fell contrived his own model that, he emphasized, never failed to give shocks.[74]

Fell's experimental activity elicited the attention of electricians in London. Lane read the letter with the description of Fell's "electric trials" to the Royal Society and, following his endorsement, other electricians became interested in Fell's medical experiments.[75] Inverting the direction of exchange between province and metropolis, Miles Partington, a London surgeon turned a successful medical electrician, introduced himself to his colleague at Ulverston, expressing hopes "to form a channel of communication for our mutual information."[76] Partington was the most visible metropolitan medical electrician. Praised by Birch as the practitioner who most effectively demonstrated

74 Fell's experiment with the electric eel are described in Ms 1175, ff. 82–86. On the history of electric fishes, see Stanley Finger and Marco Piccolino, *The Shocking History of Electric Fishes: From Ancient Epochs to the Birth of Modern Neurophysiology* (New York: Oxford University Press, 2011).

75 Ibid., 39.

76 Partington to Fell, Ms 1175, f. 49.

the efficacy of electricity in "the regulation of the animal economy," he was well known at the Royal Society.[77] In 1778 his report of a successful treatment of "a case of muscular contraction" by electricity was published in the *Philosophical Transactions* and his electrical cures were also described in Cavallo's *Essay on the theory and practice of medical electricity*.[78] Partington's business in London was so successful as not to leave him time to complete a treatise on medical electricity that he started to plan around 1780.[79] He was interested in Fell's activity as, he explained, "every new particular in this branch of practice may be productive of more important experiments."[80] The wording of this sentence reveals the epistemic functions that medical electricians such as Partington and Fell hoped their practice would serve. The "more important experiments" Partington referred to were new cases where electricity proved successful after other therapies had failed. Results of this kind could be achieved by studying how different experimental settings affected the treatment's outcome. In his own report of the successful treatment of a fistula lachrymalis – a disease that responded only occasionally to electricity – Partington detailed the "disposition of the apparatus": by arranging the instruments, the patient, and his own body in the way he described, "the effects of electrization are considerably increased, the pungency of the sparks is felt much deeper into the electrified part of the body: the heat occasioned by it is also greater, and therefore it seems more efficacious for internal complaints."[81]

In these accounts, medical electricity was presented as yet another kind of electrical experimentation, which resulted not only in new therapies, but also in the perfecting of instruments. Because of the uniqueness of each human body, the combination of patient, practitioner, and electrical apparatus could not be standardized. This instability gave practitioners the opportunity to experiment with the apparatus, trying new materials and new forms of applying electricity. Partington, for example, designed new "directors" for treating the fistula; the process of treatment thus became an experiment to test his instruments: "The short experience I have had with these directors does not enable me to determine how far they can be useful...they do not answer my most sanguine expectations; but yet in several instances they seem to have

77 Birch, *Considerations* (ref. 18), v–vi.

78 Miles Partington, "A Cure of a Muscular Contraction by Electricity," *Philosophical Transactions*, 68 (1778), 97–101.

79 Birch, *Considerations* (ref. 18), vii.

80 Partington to Fell, Ms 1175, f. 49.

81 Cavallo, *Essay* (ref. 7), 101.

afforded considerable relief."[82] Fell too regarded electrical treatments as unstable settings where instruments could be modified to improve the comfort and safety of patients and practitioners. As he explained to Nairne and Partington, the insulating stool that he designed with the collaboration of local artisans at Ulverston resulted from his work on electric patients (Fig. 4.9).[83]

Aware that electricity "was never seriously admitted as an article, by Practitioners of Physic, for the cure of diseases," Partington believed that the best way to advocate its medical efficacy was to compile a collection of experiments,

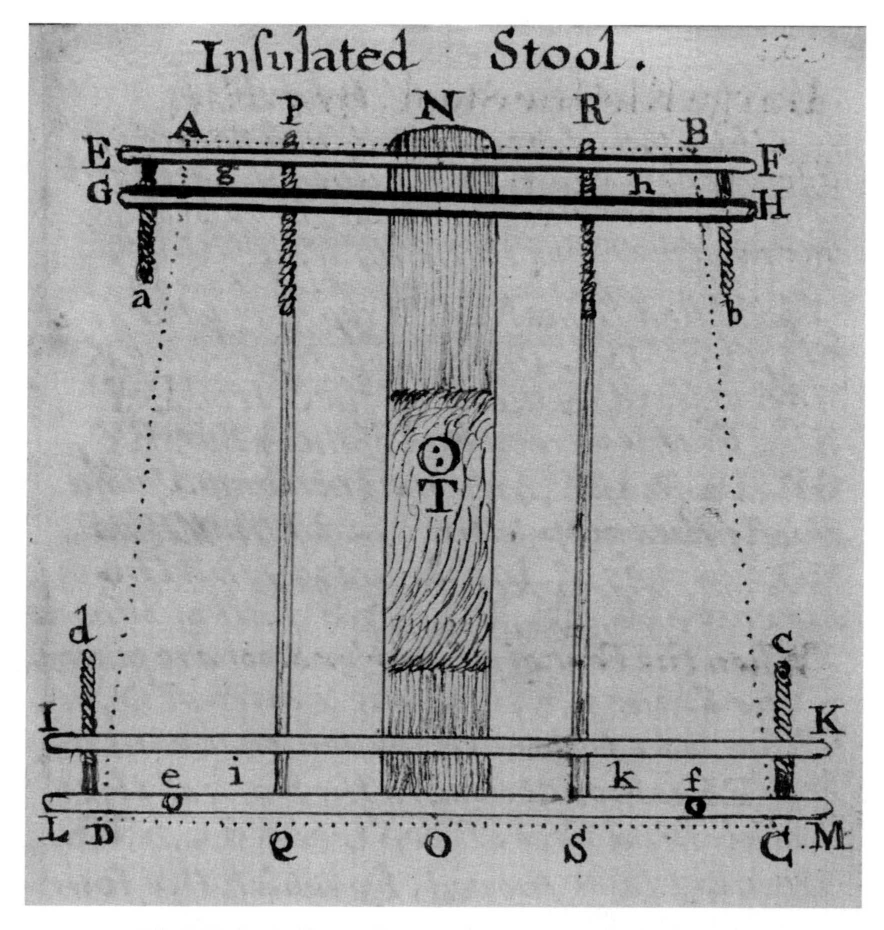

FIGURE. 4.9 *John Fell's sketch of his insulating stool. Ms 1175,* Miscellanea Electrica. *Wellcome Library, London.*

82 Cavallo, *Essay* (ref. 7), 102.

83 Ms 1175, f. 125–26.

in the style of the books of experimental philosophy. They would describe the disease, the apparatus, and how the practitioner used it. When put together, the sheer number of successful cases "would form a body of facts that would stamp conviction on all candid persons of the great use of this agent in the cure of diseases."[84] The numerous texts on medical electricity had already begun to conquer several practitioners who "have now machines in their possession" and contributed daily with accounts "of fresh advantages received from this agent, either as an auxiliary to medicine, or independent of other medical assistance."[85] Partington noted, however, that the physicians' resistance was hard to break:

> I am debarred too much from the necessary part of trying the effects of medicines combined with electricity, for want of zeal in our physicians with whom I generally have to act, and find not easy to persuade to experiments.[86]

As Cavallo explained, medical electricity, "different from other physical applications, requires quite a nicety of operation than a thorough knowledge of the disease."[87] The "nicety of operations" consisted in the skillful employment of "philosophical instruments," an approach that called for a new epistemology of medical expertise grounded in the material culture of electricity. Nairne emphasized that it was advantageous "that we are not under the necessity of waiting till a theory is established, before we can receive benefit from the powerful, though safe, application of electricity."[88] Partington, however, was aware that medical electricity's dependence on philosophical instruments constituted the source of its problematic status not only within the medical profession but also within the experimental community. The Royal Society, with its several electricians, was initially interested in publishing reports of electrical treatments. With time, however, and with the rise in number of controversial cases, the Fellows realized that discussions on medical electricity would inevitably turn to debates over the efficacy of treatments, which were best left to the medical profession. Partington felt obliged to warn Fell that the Royal Society would not publish any of the cases read by Lane: even though several fellows

84 Ms 1175, ff. 49–56.

85 Ms 1175, f. 138.

86 Ms 1175, f. 168.

87 Cavallo, *Essay* (ref. 7), 5.

88 Edward Nairne, *The Description and Use of Nairne's Patent Electrical Machine* (London, 1786), 66.

remained interested in the subject, reports of electrical cures were no longer published in the *Philosophical Transactions*.[89] Nonetheless, he was interested in any result the surgeon from Ulverston wanted to share. Although the numerous patients did not leave him time to engage in "rational recreations," he believed that "experiments of this kind indirectly furnish convenient & useful applications to the Human body." The usefulness of such applications did not so much concern the efficacy of the therapy as the perfecting of the apparatus: Partington explained to Fell that it was while he was performing "experiments for amusement" that he discovered that a wooden point on top of a metallic conductor worked miracles in the application of electricity to the eye.[90]

Conclusion

At the intersection of experimental philosophy and medical practice, medical electricity did not constitute a "trading zone" where objects could be exchanged independently of the various meanings that different groups attributed to them.[91] It was rather a liminal space where individual practitioners merged skills, traditions, and epistemologies, alienating established institutions. If Partington was convinced that the publication of numerous cases would eventually win the skepticism of the medical establishment, fellows of the Royal Society such as Cavallo realized that medical efficacy needed to be assessed in more general terms. The lack of a standard electric body made it impossible to stabilize therapies, matching diseases to treatments. Partington hoped to establish trust in medical electricity with a process of virtual witnessing, a practice that was commonly employed by experimental philosophers.[92] However, this strategy for building consensus could not be easily translated into the domain of medical practice. Cavallo explained that the account of a few successful cases "does by no means establish the reputation of the treatment, when a vast number of unsuccessful trials are concealed from the eyes of the public."[93] He claimed that in order to obtain "a proper estimate" of the

89 Ms 1175, ff. 50–51, 138.

90 Ms 1175, f. 136.

91 On the notion of the "trading zone" see Peter Galison, *Image and Logic: a Material Culture of Microphysics* (Chicago: The University of Chicago Press, 1997).

92 Shapin, "The House of Experiment" (ref. 11).

93 Cavallo, *Essay* (ref. 7), 54.

efficacy of any remedy, it was necessary to show the "proportion between the successful, and the unsuccessful trials."[94] So, he created a list of common diseases, indicating for each whether electricity was appropriate and which electrical treatment seemed most effective. In spite of all these efforts, the vast majority of physicians continued to ignore electricity. In 1787, Fell was dismayed that a man from Manchester, whose pain in the spermatic chord was so intense as not to allow hunting on horseback, had to be brought to Ulverston to receive electrical treatments: "there are 3 gentlemen of the faculty within 200 yards of him, yet, strange to tell, every one of them is totally ignorant of electricity."[95]

It was to dispel such ignorance that Fell engaged in public courses on electricity, familiarizing his audiences with electrical instruments. For him, as for Partington and other medical electricians, building trust in electrotherapeutics required building trust in the electrical apparatus. Not every patient was persuaded. In the span of a few years Fell became annoyed by the lack of perseverance of some of his electric patients who resolved to give up on electricity.[96] He noted, among his 65 electric patients, those who did not complete the entire planned treatment.[97] He did not note, with the exceptions of the cases he described to Lane, whether the others were cured. Instead, he filled his notebook with descriptions of experiments and new instruments that he invented in large measure as a result of his medical practice. His experimental activities brought him to the attention of various naturalists in the West Midlands, who asked his opinion to evaluate collections of instruments and requested his courses on electricity. At the turn of the century, Fell was more popular as an electrical expert than as a medical electrician.[98] By then, he had realized that his electric trials on humans did not bring the same kind of popularity or income as his other experiments. Medical electricity was an experiment that proved successful when it came to testing and perfecting instruments, or training oneself in the management of the apparatus. Yet, its therapeutic efficacy remained elusive, just as the electric body did.

---

94 Cavallo, *Essay* (ref. 7), 55.

95 Lancashire record Office, DDX 317/82 (Fell to Barton, 25 October 1787).

96 Lancashire record Office. Mss DDX 317/82 (Fell to Barton, 25 October 1787).

97 Ms 1175, ff. 21–24.

98 Lancashire record Office. Mss DDX 317/82 (Fell to Barton, 25 October 1787).

CHAPTER 5

Pneumatic Chemistry, Self-Experimentation and the Burden of Revolution, 1780–1805

Larry Stewart

The Politics of Airs

By the late 18th century, the Unitarian minister Joseph Priestley proposed democracy might be manufactured in a crucible, from an air pump, or by an electrical machine. His meaning implied, at the very least, that experimental philosophy should be so diffused that knowledge of the natural world could no longer be confined to the world of scholars or gentlemen but would achieve much with widespread practice. In the 1760s, Priestley had asserted "progress might be quickened, if studious and modest persons, instead of confining themselves to the discoveries of others, could be brought to entertain the idea, that it was possible to make discoveries themselves."[1] On the verge of the political upheavals of the previous half century, Priestley's manifesto carried an uncertain burden. Yet, by the time of the American and French Revolutions and England's own industrial transformation, it seemed possible to draw broad inferences from the spread of natural knowledge. This was ultimately the foundation in 1790 of Priestley's famous metaphor, in the clearest declaration of his natural philosophy, of the impact of an air pump or electrical machine that would shake the foundations of the British establishment.[2]

Priestley's mantra was not at all unique. The doctrine was common currency amid reformers and in republican circles. Engagement in experimental discoveries was particularly taken to assert democratic principles. Thus, the

---

1 Joseph Priestley, *The History and Present State of Electricity, with Original Experiments*, Second Edition, corrected and enlarged (London: J. Dodsley, J. Johnson, J. Payne, T. Cadell, 1769), 546.

2 Priestley, *Experiments and Observations on Different Kinds of Air, and Other Branches of Natural Philosophy, Connected with the Subject.* In Three Volumes; Being the Former Six Volumes Abridged and Methodized, with Many Additions (Birmingham: Thomas Pearson; and London: J. Johnson, 1790), Vol. 1, Preface, p. xxiii; See also Maurice Crosland, "The Image of Science as a Threat: Burke versus Priestley and the 'Philosophic Revolution,'" *British Journal for the History of Science* 20 (July, 1987): 277–307, esp. 282. On the politics of science in the British Enlightenment, see T.H. Levere, L. Stewart, H. Torrens, *The Enlightenment of Thomas Beddoes: Science, Medicine and Reform* (forthcoming).

 | DOI 10.1163/9789004286719_007

midland's chemical manufacturer James Keir, in his 1789 revision to the widely-circulated *Dictionary of Chemistry* of Pierre Joseph Macquer, famously declared the need for a "diffusion of a general knowledge, and of a taste for science, over all classes of men..."[3] Shortly thereafter, in his dismay at the worst excesses of the French Revolution, the democratic doctor Thomas Beddoes sought an antidote to Terror. He hoped the success of his own public lectures on chemistry in Bristol might usefully provide "a chance of preventing some acts of barbarity in the times that I fear are coming."[4] Many were the burdens of physics and chemistry.

I argue here that the promotion of experimental method was adopted in the late 18th century precisely on the notion that revisions to social and political structure demanded a new, and broadly-dispersed, epistemology. In the Enlightenment, this also explicitly came to transform the body itself as an experimental device. Indeed, it was the body as an ideal site and instrument of investigation that helped erode what we might assume as commonplace disciplinary boundaries in the laboratory – between the new gas chemistry, galvanic electricity, and even the many public adventures with hot air balloons. In all these, research depended on the function of the human – both as practitioners

FIGURE 5.1
Thomas Beddoes (1760–1808), English physician. From Memoirs of the Life of Thomas Beddoes, M.D. *by Stock, John Edmonds, M.D., 1811 (Wellcome Library, London).* HTTP://WELLCOMEIMAGES.ORG/INDEXPLUS/IMAGE/M0000717.HTML

3 [James Keir], *The First Part of a Dictionary of Chemistry, &c.* (Birmingham: Pearson & Rollaston, 1789), iii.

4 Birmingham Central Library [BCL], James Watt Papers [JWP], W/9/7. Thomas Beddoes to James Watt, April 21 [1796?].

and as subjects. Yet, early-modern experimenters did not accept disciplinary boxes. This may well have followed from the wide public enthusiasm for experimental practice. As lightning and balloons amply showed, much of this very experience proved as dramatic as it was exceedingly dangerous.

In the late 18th century, new gases demanded attention. Balloons were all the rage. Jean-François Pilâtre de Rozier was seemingly the first in Paris to ascend in a hot-air balloon. But it was Jacques-Alexandre-César Charles and Nicolas-Louis Robert who, in December, 1783, did so by tediously-produced hydrogen. After seventy hours that it took to fill the balloon, they rose before a crowd of up to 400,000 witnesses.[5] Such an enormous gathering on the Champ de Mars was compelling evidence of a public scientific arena. With rather less drama, the ultimately ill-fated chemist Antoine Lavoisier quietly tested the chemical differences between the air in the box at the top of the nearby theatre of the Tuileries compared to that lurking among the audience of the pit. Common health was one issue. He pointed to the need to freshen airs, otherwise "spectators would be exposed to the most fatal accidents before the conclusion of the performance."[6] Unlike Lavoisier, or the ultimately doomed Pilâtre de Rozier, Charles survived his balloon adventures and the Revolution, even later being employed by the Royalist regime in assessing the *cabinets de physique* of émigrés who had fled persecution in France. While Lavoisier would fall victim to politics, for some at least, a public science seemed to converge with "the revolutionary spirit" of the age, only to be constrained in the early 19th century by the rise of a professionalization and the denigration of anything that smacked of the popular appeal that had emerged from the Enlightenment.[7]

How far would experimenters go? Take, for example, the provocative Pilâtre de Rozier, who had once been criticized as "scatterbrained, dissipated, keen on

---

5 Dalhousie University Archives [DUA], James Dinwiddie Correspondence, MS 2–726, C 45. Aerostation, No. 1 (1793?), p. 5; Jean Pierre Poirier, *Lavoisier. Chemist, Biologist, Economist*, trans. Rebecca Balinski (Philadelphia: University of Pennsylvania Press, 1996), 145–147.

6 Lavoisier, 'Observations on the Alteration produced in the Air of Places Where Great Number of Persons are Assembled,' in Thomas Beddoes, *Letters from Dr. Withering, of Birmingham, Dr. Ewart, of Bath, Dr. Thornton, of London, and Dr. Biggs, Late of the Isle of Santa-Cruz; Together with Some Other Papers, Supplementary to Two Publications on Asthma, Consumption, Fever, and Other Diseases* (Bristol: Bulgin and Rosser; London: J. Johnson and H. Murray, 1794), 7–9.

7 Cf. Thomas Broman, "The Habermasian Public Sphere and 'Science IN the Enlightenment'," *History of Science* 36 (June, 1998): 139–142; Michael R. Lynn, *Popular Science and Public Opinion in Eighteenth-Century France* (Manchester and New York: Manchester University Press, 2006), 148; Marie Thébaud-Sorger, *L'Aérostation au temps des Lumières* (Rennes: Presses Universitaires de Rennes, 2009).

pleasure and unmentionable studies." He was, however, handsome – and it was at his school, the 'Musée de Monsieur,' that Madame Lavoisier once attended lectures. According to the radical Thomas Beddoes, who like many would have an urgent interest in the newly-identified gases, Pilâtre de Rozier had repeated experiments on inspiration "and when it was objected to him that he did not empty his lungs of atmospheric air, he proceeded in the following manner. He inhaled a large quantity of hydrogene [sic] air, and then, applying his mouth to a tube, he blew out the air and fired it at the end of the tube, so that he appeared to breathe flame. In this case, he observed, if there had been any admixture of atmospheric air in his lungs, a detonation must have taken place in his mouth and his chest." Such attempts became chemists' obsessions, notably at the Arsenal in Paris with Lavoisier and his assistant Armand Séguin.[8] Beddoes similarly attested to the effects of hydrogen on the pulse of the 'confectioner' and aeronaut James Sadler, once briefly his laboratory assistant. Sadler "inspired, for above a minute and a half at a time, pure hydrogene [sic] from steam and heated iron. From the rate of eighty-four in a minute [his pulse] increased to an hundred and ten in fifteen seconds, and became soft and weak."[9] Sadler survived the adventure and turned balloonist. Pilâtre de Rozier, risk taker that he was, ultimately became ballooning's first casualty when his own balloon caught fire over the Channel.

Experimenting with the Public

Public experience and sometimes a solitary laboratory life increasingly intersected. And human trials magnified the complexities of a spreading practice – as true of the rage for electrical research as much as it was of the chemistry of gases and their dramatic effects.[10] Joseph Priestley was particularly sensitive to

---

8 Cf. Lavoisier, Oeuvres, (November 17, 1790) quoted in Poirier, *Lavoisier*, 300; on assistants see also pp. 187, 189, 215. See Séguin and Lavoisier, "Premier Memoire sur la respration des animaux," *Ouvres de Lavoisier publiées par les soins de Son Excellence le Ministre de l'Instruction Publique et des Cultes*. Vol II (Paris: Imprimerie Imperiale, 1862; New York and London: Johnson Reprint Corporation, 1965), 688–703, esp. 703.

9 Thomas Beddoes, *A Letter to Erasmus Darwin, M.D.* ON *A New Method of Treating Pulmonary Consumption, and Some other Diseases Hitherto Found Incurable* (Bristol: Bulgin and Rosser; London: J. Johnson [1793]), 44–45. On Pilatre de Rozier, see Poirier, *Lavoisier*, pp. 126, 146, 324; and Michael Lynn, *Popular Science and Public Opinion in Eighteenth-Century France* (Manchester and New York: Manchester University Press, 2006), 82–90.

10 Cf. Paul Elliott, "'More Subtle than the Electrical Aura': Georgian Medical Electricity, the Spirit of Animation and the Development of Erasmus Darwin's Psychophysiology," *Medical History* 52 (2008): 195–220.

the management of the new realm of experimental devices that had emerged in the 18th century. The spread of electrical apparatus was already deemed especially promising – both in experiments and in electro-therapy, as explained in the previous chapters. Of gases, or pneumatic chemistry, Priestley was equally passionate. Yet, in advancing pneumatic chemistry, there were also cautions:

> It is known to all persons who are conversant in experimental philosophy, that there are many little attentions and precautions necessary to be observed in the conducting of experiments, which cannot well be described in words, but which it is needless to describe, since practise will necessarily suggest them; though, like all other arts in which the hands and fingers are made use of, it is only *much practise* that can enable a person to go through complex experiments, of this or any other kind, with ease and readiness.[11]

Experiment was a means of democratic diffusion. Hence, Thomas Beddoes' later dictum of gas chemistry suffocating the pretense of privilege in Europe's *ancien regime* was surely a remarkable notion. Natural knowledge dispersed would corrode the corrupt foundations of inequality. Beddoes echoed Jean-Jacques Rousseau at the heart of Enlightenment reform. Beddoes' optimism, as in Lavoisier, linked the discoveries of the laboratory with the promotion of the general good. To be conversant with experiments of many kinds was one of the most important markers of enlightened achievement – even more so when the laboratory explored the depth of human frailty. When human trials were widely reported (and even sometimes carefully sanitised) they were not easily reproduced. Impressions of effects were subject to all manner of distortion and influence. We encounter these throughout the century, in Carl Wilhelm Scheele's inspirations of inflammable air, Priestley's experiments on the non-existent phlogiston, Beddoes' accounts of the impact of the so-called 'factitious' airs, or of his assistant Humphry Davy's famous stupefaction with nitrous oxide. Even when widely reported, we see only obliquely into the private face of enlightenment.[12] The validity of enlightened experiment ultimately required broad evidence of utility.

11 Joseph Priestley, *Experiments and Observations on Different Kinds of Air*, vol. II (London: J. Johnson, 1775), 6–7; quoted in Trevor H. Levere, "Measuring Gases and Measuring Goodness," in Frederick L. Holmes and Trevor H. Levere, eds., *Instruments and Experimentation in the History of Chemistry* (Cambridge, Mass., and London: MIT Press, 2002), 105–135, esp. 111.

12 Thomas Beddoes, *Considerations on the Medicinal Use, and on the Production of Factitious Airs*, Part I (London, 1796), 29; Roy Porter, *Doctor of Society. Thomas Beddoes and the Sick Trade in Late-Enlightenment England* (London and New York, 1992), 44–45; Trevor H. Levere, "Dr. Thomas Beddoes: The Interaction of Pneumatic and Preventive Medicine

The willingness to venture into innovative trials, then without much in the way of constraint, may now seem remarkable. Beddoes' account, before he was driven by his democratic politics from royalist Oxford, of one particularly injudicious experiment is especially striking:

> At Oxford, in 1790, I proposed to a distressed negro, to try to whiten part of his skin with oxygenated marine acid air. He was to exhibit the appearance, if it should be curious, for the relief of his family. His arm was introduced into a large jar full of this air, and the back of his fingers lay in some water impregnated with it at the bottom of the vessel. It was perceived that he had ulcerations from the itch between his fingers: and this made me very cautious about the experiments. In 12 minutes he complained of severe pain from the ulcers, and the arm was withdrawn. The back of his fingers had acquired an appearance as if white lead paint had been laid upon them, but this did not prove permanent. A lock of his hair was whitened by this acid. Next day the ulcers became extremely painful, and the hand swelled from the inflammation; this deterred him from a continuance of the experiment after he was cured of his complaint. You cannot safely impute the effect of this powerfully stimulating acid to its oxygene [sic] alone.[13]

As appalling as this episode now seems, Beddoes' promise of a barbarity vanquished was obviously less enlightened than we may have expected of a radical democrat. The limits of experimenting were undetermined. For him, promoting chemical knowledge was sufficient reason for human trials. Indeed,

---

with Chemistry," in *Chemists and Chemistry in Nature and Society, 1770–1878* (Aldershot: Variorum, 1994), 1–17, esp. 2–3, 10; Thomas Beddoes, M.D. and James Watt, Engineer, *Considerations on the Medicinal Use, and on the Production of Factitious Airs*, Part III, Second Edition (London, 1796), 67. Robert Thornton to Beddoes, January 1, 1795; Michael Neve, "The Young Humphry Davy: or John Tonkin's Lament," in Sophie Forgan, ed., *Science and the Sons of Genius. Studies on Humphry Davy* (London: Science Reviews, 1980), 1–32, esp. 16–17; Susan C. Lawrence, *Charitable Knowledge: Hospital Pupils and Practitioners in Eighteenth-Century London* (London and New York: Cambridge University Press, 1996), 316; June Z. Fullmer, *Young Humphry Davy: The Making of an Experimental Chemist* (Philadelphia: American Philosophical Society, 2000), 211ff.

13 [Beddoes], *Considerations on the Medicinal Use, and on the Production of Factitious Airs*: Part I. By Thomas Beddoes, M.D. Part II. By James Watt, Engineer. Edition the Third. Corrected and Enlarged (Bristol: Printed by Bulgin and Rosser; London: For J. Johnson, in St. Paul's Church-Yard, 1796), 45; John Edmonds Stock, *Memoirs of the Life of Thomas Beddoes, M.D. with an Analytical Account of His Writings* (London: John Murray, 1811; Reprint edition, Bristol: Thoemmes, 2003), 66–67.

experiments on animals were secondary. As James Watt had already told his son, he was disturbed by hearing of tests on dogs, and was "not certain how far such experiments are necessary."[14] But, more broadly, it was from what the celebrated engineer once simply described in Enlightenment Birmingham as the "pleasure of experimenting" that we now may learn of the relation of early-modern chemistry with disease – even when, in Watt's case, experiment was forced to give way before the daily business of manufacturing.

Beddoes championed experiment as essential to breaking the icy grip of a bogus medical profession upon those they so easily duped. Amidst common accusations of quackery, which reformers hurled even at the medical establishment, the public promotion of chemistry required deft handling.[15] In this circumstance the physician John Ewart, of the Bath City Infirmary and Dispensary, told James Watt in 1794 that, as an engineer, Watt was "likely to see more clearly & to report more faithfully the result of your experiments, than most physicians are... But a few disinterested philosophers like you shall make the faculty ashamed of their supineness, & conscious of their defects, then may we hope for a new practice of medicine with healing on its wings."[16] Similarly, Beddoes argued that if Watt were able to assert as fact that Watt's own experimental subjects had "greatly benefitted" from the airs, "it would have much weight with the public."[17]

In the case of Beddoes, an apparent democratic ethos tied ideology to the most recent innovations in chemistry. Those political sentiments were not however shared by Watt who was increasingly alarmed by the rise of republicanism. If France was in turmoil, Britain was in alarm. While Priestley may have once been prophetic, in England's republican tide in the 1790s, Watt wrote, "The Rabble of this country are the mine of Gunpowder that will one day blow it up & violent will be the explosion."[18] Yet the attractions of chemistry remained understandable enough. In 1774 Joseph Priestley had discovered the

---

14 BCL, JWP, MS 3219. Watt to James, jun., July 24, 1790.

15 See Roy Porter, *Quacks: Fakers & Charlatans in Medicine* (Stroud: Tempus, 2000), 168–169; 176ff.; Michael Neve, "Orthodoxy and Fringe: Medicine in Late Georgian England," in W.F. Bynum and Roy Porter, eds., *Medical Fringe & Medical Orthodoxy 1750–1850* (London and Sydney: Croom Helm, 1987), 40–55, esp. 48–51; Roy Porter, "Plutus or Hygeia? Thomas Beddoes and the Crisis of Medical Ethics in Britain at the Turn of the Nineteenth Century," in R. Baker, Dorothy and Roy Porter, eds., *The Codification of Medical Morality* (Dordrecht: Kluwer, 1993), 73–91.

16 BCL, JWP 4/23/45. J. Ewart to Watt, October 5, 1794.

17 BCL, JWP W/9/53. Beddoes to Watt, April 5, 1796.

18 BCL, JWP copy books. Watt to Joseph Black, July 17, 1793. See Robert G.W. Anderson and Jean Jones, eds., *The Correspondence of Joseph Black* (London: Ashgate, 2012), II, 1208.

anti-septic properties of nitrous air and happily predicted "a very great medicinal use."[19] Shortly thereafter, Priestley, following the notion of the medical effects of atmospheres, had asked Matthew Boulton in Birmingham to capture some phials of air not, he cautioned, from a place in the country but "actually breathed by the different manufacturers in this kingdom" especially "in those places where you expect the air to be the worst, on account of bad fumes or a number of people working together."[20] It is surely notable that the fears of poisonous airs were ubiquitous. But this evolved alongside the anti-septic properties Beddoes would discuss in 1787 with the chemist Guyton de Morveau in Dijon, and simultaneously in a rapidly increasing number of industrial sites.[21]

Throughout the 1790s iconoclastic physicians were drawn to newly-discovered airs in the battle both against disease and a recalcitrant medical establishment. Consequently, the democrat Erasmus Darwin would employ pure oxygen in his country practice. Such was the case of a Mr. Welhouse who breathed seven gallons a day for two or three weeks where the much-praised foxglove proved to no avail. But temporary relief induced inflated claims. His was a matter of hydrothorax [excessive pleural fluid] for which Darwin contemplated a trial of hydrocarbonate [CO] in small quantities "as his case is desperate without some new aid."[22] Despair could easily lead physicians like Darwin or surgeons like George Goodwin of Hampstead and Edward Coleman of Norwich to trials of anything which might promise, including electric shocks or seltzer waters.[23] All were elements of desperate innovation. Pneumatic medicine was, sometimes literally, simply a breath of fresh air in sick-rooms whether of the poor which John Farriar encountered at Manchester Infirmary, of the trials of the royal physician James Lind at the Court at Windsor or, for that matter, even on a voyage to China with John Ewart.[24]

---

19 Quoted in Levere, "Dr. Thomas Beddoes," 10.

20 BCL, Matthew Boulton Papers, 249/212, n.d.; 249/213, Priestley to Boulton, November 25, 1777; reprinted in Robert E. Schofield, ed., *A Scientific Autobiography of Joseph Priestley, 1733–1804* (Cambridge, Mass., and London: M.I.T. Press, 1866), 161–163.

21 Larry Stewart, "His Majesty's Subjects: From Laboratory to Human Experiment in Pneumatic Chemistry," *Notes & Records of the Royal Society* 63 (September, 2009): 231–245, esp. 232–233.

22 BCL, JWP C1/25, Erasmus Darwin to James Watt, July 9, 1795.

23 Dorothy Porter and Roy Porter, *Patient's Progress: Doctors and Doctoring in Eighteenth-Century England* (Cambridge: Polity Press, 1989), 171–172; Beddoes, *Considerations*, Pt. I, p. 12.

24 BCL, JWP W/9/2. Beddoes to Watt, n.d.; W/9/46, James Lind to Watt, August 7, 1796; See also Jan Golinski, *Science as Public Culture: Chemistry and Enlightenment in Britain, 1760–1820* (Cambridge and London: Cambridge University Press, 1992), pp. 58, 116, 161, 163, 172; and Stewart, "His Majesty's Subjects," 231–232.

The seductions of experiment were compelling, even among those not entirely convinced by their own experience. Thus the Glasgow physician, Dr. Robert Cleghorn, recalled his own trials on four patients, *all* of whom died: "Two of them were hopeless before the air was tried, but none of my patients experienced any kinds of the sensations said to happen so uniformly with you. The heat-pain-oppression &c. all continued totally unaltered." Similarly, "The application of carbonick acid air relieved one case of Cancer very remarkably, but failed in all the rest." Diagnosis was as deficient as were claims of improvement. Ewart, in 1794, proposed the surface application of the so-called 'carbonic acid air' in cases of breast cancer.[25] Likewise, Thomas Creaser, surgeon also at Bath, informed Watt that he went "on with Pneumatic Medicine in a way to ascertain if possible the general laws of its action on the living system" although, in the end, he could actually only report on one case – that of asthma – ostensibly cured by hydrogen.[26] Even so, Cleghorn, as chemist echoing both Ewart and Beddoes, claimed further tests were essential:

> Those who oppose these trials, & still more those who ridicule them, deserve the reprobation of every lover of mankind–but it was always so & always will be while the profession of physick is follow'd as a Craft. Improvements must come from such men as yourself, or some of your Coadjutors whose circumstances are somewhat different from those of our profession in general.[27]

Beddoes and Watt found it possible to accumulate many such reports, and turned them to advantage in promoting pneumatic medicine.

The net of innovators was cast as widely as possible. The result was hundreds of claims of cures of the hitherto incurable. Beddoes sought out physicians and surgeons, chemists and apothecaries, possibly inclined to test airs in the endless intractable diseases that sadly defeated the medics. Much then rested on the adventurous. Hydrogen did not only raise balloons. Beddoes frequently consulted with Darwin on its curative effects. And he reported on the experience of Dr. Robert Thornton who had studied chemistry at Guy's Hospital and whose large London practice gave him many opportunities – like that of the emergency application of azotic [nitrogenous] air in a desperate case of croup. Beddoes filled volumes with such cases and made a plea for

---

25 John Ewart, *The History of Two Cases of Ulcerated Cancer of the Mamma; One of Which Has Been Cured, the Other Much Relieved by A New Method of Applying Carbonic Acid Air* (Bath: Crutwell, and London: Dilly, 1794).

26 BCL, JWP W/9/1. Thomas Creaser to Watt, October 23, 1798.

27 BCL, JWP W/9/56. Cleghorn to Watt, May 12, 1796.

the continued trials of modified gases in order to explore the effects of prolonged respiration. It was not only a question of types of airs, uncertain in their chemistry, but of their concentrations and their continued, if inconsistent, application.

Here was opportunity. Chemists too sought out the physiology of airs and acids. Hence, both Jan Ingenhousz and Joseph Priestley provided instructive examples from which Beddoes drew highly optimistic conclusions.[28] Limits were ill-defined, at least in aerial trials and for the many subjected to them. Beddoes would report on the uses of nitrous oxide in cases of palsy, and published the history of seventeen-year old Elizabeth Lambert of whom he had heard from John Alderson, physician to the Hull Infirmary. She had been treated with oxygen for chlorosis, a disease named for the green pallor in anaemic young women. Beddoes also revealed that in the case of Ann Smith, eighteen years old and victim of epileptic fits, her treatment by Dr. Richard Pearson of the Birmingham General Hospital offered no sustainable cure. The failure suggested "some organic affection of the brain."[29] The range of ailments was as wide as the expanding list of airs. As pneumatic trials were ventured, vast was the confusion over how, and in what circumstances, in what concentrations, gases might prove beneficial. Positive accounts fuelled further enthusiasm. These were sustained when those in perfect health reported high spirits upon inspiration of the so-called vital air, still more with the endlessly promising nitrous oxide.[30] How then, the chemists demanded, could such vapours fail to aid the sick?

Consuming Airs

It was the devastating, rampant, consumption that demanded the most urgent attention. Many medics were seduced by proposals to attack the apparent contagion and fevers that overwhelmed their consumptive patients.[31] There were more than sufficient claims of success to justify experiments. George Birkbeck,

28 Beddoes, *Considerations*, Part I (1796), pp. 29, 44, 176–177; Golinski, *Science as Public Culture*, 160.

29 BCL, JWP 6/33/70. Beddoes to Watt, December 23, [1800?]; Beddoes and Watt, *Considerations*, Part III, Second Edition (1796), 55–58; Golinski, *Science as Public Culture*, 160–161.

30 Beddoes and Watt, *Considerations*, Part III, Second Edition (1796), 66–67; cf. Mike Jay, *The Atmosphere of Heaven. The Unnatural Experiments of Dr Beddoes and His Sons of Genius* (New Haven and London: Yale University Press, 2009), Chapter 9.

31 Levere, "Dr. Thomas Beddoes", 4–5; Neve, "Young Humphry Davy", 13–15.

while still a medical student in Edinburgh, witnessed the rapid improvement of a patient to whom Beddoes had given hydrogen in a suspected case of super-oxygenated blood. The notion was that by altering the blood oxygen level symptoms might be alleviated – revealed by the florid complexions of some consumptive patients. This idea was proposed to Beddoes by Dr. Thomas Garnett, Glasgow public lecturer in chemistry, and later of the Royal Institution, that relief of consumption might result from decreasing the levels of atmospheric oxygen to which patients were seemingly subjected.[32] Variability of gases and their concentrations presented immense obstacles. Dosage could not easily be determined nor could adverse reactions – as in one of Beddoes' exceptionally delicate consumptive patients who kept fainting on application of hydrogen even when diluted by atmospheric air.[33] What appeared promising in such a deadly disease also suggested hope in many other common afflictions. Among his correspondents was Thomas Creaser in Bath who tried oxygen on a woman with herpes on the face who had also been continually cursed by difficulties of breathing and with oedema of the limbs. But the oxygen was discontinued after some weeks for fear that it was losing its effect. At this point Beddoes also began his exploration of the ingestion of nitrous acid or even oxygenated muriate of potash (potassium chloride) in venereal disease.[34] This treatment gained much attention in the naval hospital in Plymouth and as far afield as the East India Company in Bombay.[35] Many a site was transformed into a laboratory.

Beddoes soon faced the resistance of patients who feared the prospect of being transformed into *subjects* of experiment. Not all would do so willingly. Moreover, many trials remained private and often went ignored or unreported. Such, for example, was the effort of the chemist Jan Ingenhousz at Calne when he assisted a consumptive journeyman tailor, aged twenty-two, to find relief

---

32 Beddoes, *Considerations*, Part I (1796), pp. 119–120, 125–126. Watt shortly wrote to Garnett about a programme of types of airs that ought to be tried via the lungs to access the vascular and nervous systems. BCL, JWP, MS 3219/118. Watt to Garnett, Feburary 9?, 1797.

33 BCL, JWP W/9/20. Beddoes to Watt, October 4, 1797.

34 BCL, JWP W/9/24. Thomas Creaser to Beddoes, copy to Watt [1797?]; W/9/20. Beddoes to Watt, October 4, 1797; W/9/23. Beddoes printed circular, September 5, 1797.

35 Thomas Beddoes, M.D., *Reports Principally Concerning the Effects of the Nitrous Acid in the Venereal Disease, By the Surgeons of the Royal Hospital at Plymouth, and By Other Practitioners* (Bristol: N. Biggs; London: J. Johnson, St. Paul's Church-yard, 1797); Beddoes, *A Collection of Testimonies Respecting the Treatment of the Venereal Disease by Nitrous Acid* (London: J. Johnson, 1799), 2ff.; Mike Jay, *The Atmosphere of Heaven. The Unnatural Experiments of Dr Beddoes and His Sons of Genius* (New Haven and London: Yale University Press, 2009), 143.

through breathing the fumes of a lime kiln. Lime kilns then had some reputation for curative powers. But, however much he tried, Ingenhousz was unable to procure any further pulmonary patients willing to repeat the effort.[36] Recruiting for airs remained problematic. Were patients then inevitably subjects?

There was little alternative but to turn to the generally healthy in testing factitious airs. At least, one hoped, these subjects might give credible reports of reactions driven by neither desperation nor self-delusion. The effects of airs could nonetheless be alarming. In the many cases of inspiration of hydrogen, the result was invariably a debility of pulse, blueness of the lips and short dizziness, even though some felt the moment highly agreeable. And there was obviously the problem of reliability. Beddoes, for example, reported on one consumptive patient who loved the experience of inhalation – to the point of asphyxiation. However appreciative, or spirits uplifted, the patients soon might find themselves incapable of walking. Among Beddoes' early supporters was also the Bristol actor William Tobin who, for two weeks during 1794, administered oxygen to himself in an effort to cure impaired eyesight caused by an apparent amaurosis [a disease of the optic nerve], all to no effect. But Tobin, like many of Beddoes' enthusiasts, happily graduated to nitrous oxide the virtues of which he famously commended: "I threw myself into several theatrical attitudes, and traversed the laboratory with a quick step; my mind was elevated to a most sublime height."[37]

Such self-experiments provided the most immediate proof of apparent benefit. In the eyes of medical reformers, nothing should be permitted to inhibit pneumatic adventures. This meant many risks. Thus, in 1796, John Seward of the Worcester Infirmary consulted Watt on devices for the preparation of oxygen and hydrocarbonate informing Watt he intended, "trying again the effects of hydrocarbonate on myself." Throughout the next month Seward was breathing airs and dealing, perhaps not too surprisingly, with a persistent cough. His symptoms were meticulously recorded after ingesting hydrocarbonate diluted with common air: "It affected my head very much that is I felt a remarkable tinnitus aurium & pain in the forehead, which was greatly increased by stooping or using any exertions. My heart also I could perceive by my hand as I breathed beat more feebly." Seward hoped to control the reactions by mixing the gas with oxygen.[38] Beddoes faced this same problem with Mr. Greene, the M.P., who came to Bristol to try the new azotic (nitrogen) gas for a cure in a

---

36 Beddoes, *Considerations*, Part I (1796), Addendum, 93.

37 Beddoes, *Considerations*, Part I (1796), pp. 41–44; See also Golinski, *Science as Public Culture*, 167–168; Jay, *Atmosphere of Heaven*, 177, 183.

38 BCL, JWP W/9/42. Seward to Watt, August 24, 1796; W/9/38. Seward to Watt, September 28, 1796.

case of palsy. Beddoes welcomed the opportunity.[39] In the pneumatic campaign a young Scarborough woman with a haemorrhage, treated by Watt, improved although her father complained of the lack of access to a breathing apparatus. Among the variously afflicted any distinction between patient and subject largely disappeared. A Dr. Alexander was certain he had fully cured phthisis and calmed pulmonary irritations with fixed air. Many medics provided Watt and Beddoes with case histories of their own patients. It is not surprising that Beddoes was hopeful.[40] If published histories provided credible testimony, Beddoes and Watt privately discussed their own self-experiments, although the surviving details are surprisingly scant. There were as many motives as there were self-experiments. Among the most optimistic practitioners, even they had to admit that no one could be certain of the consequence of inhalations of otherwise "unrespirable airs" beyond a marginal or momentary effect, however dramatic. In such circumstances, human testimony was as inescapable as it was unreliable.[41]

The assessment of new airs was widely reflected in the simultaneous pursuits of many medical men. For example, the chemical lecturer and physician, Thomas Garnett of Anderson's Institute in Glasgow, and later demonstrator of London's Royal Institution, described his self-experiments in the spa town of Harrowgate. His efforts were focused on remedies for super-oxygenation with *kali sulphuratum* of potassium (commonplace in gastric fevers, oespohagitus):

> Before twenty hours had elapsed, I found the sens [sic] of tightness in the thorax considerably lessened, some degree of expectoration came on, and the cough was much relieved. In three days, by pursuing this method, my countenance became considerably paler, and I found myself perfectly free from any complaint. Since that time I have prescribed the kali sulphuratum in several cases of florid consumption, with considerable relief; and in some other cases there were evident marks of superoxygenation.[42]

39 BCL, JWP 6/33/69. Beddoes to Watt, January 20, 1801; JWP 6/33/38, Beddoes to Watt, January 8, 1801.

40 BCL, JWP, W/9/19. Beddoes to Watt, October 24, 1797.

41 Thomas Beddoes, *Considerations on the Medicinal Use, and On the Production of Factitious Airs*: Part I. By Thomas Beddoes, M.D. Part II. By James Watt, Engineer. Edition the Third. Corrected and Enlarged (Bristol: Printed by Bulgin and Rosser, For J. Johnson, in St. Paul's Church-Yard, London, 1796), 177.

42 Beddoes, *Considerations on the Medicinal Use*, 119–121. On Garnett see Berman, *Social Change and Scientific Organization. The Royal Institution, 1788–1844* (Ithica: Cornell University Press, 1978), 2–23; Robert Fox, 'Garnett, Thomas (1766–1802)', *Oxford Dictionary*

Going first, or self-experimentation, required persistence and gaseous preparations demanded some chemical skill. But auto-experimentation was not the only route to pneumatic testimony.

The expansion of hospitals and infirmaries during the early industrial revolution afforded many opportunities for experimental adventures.[43] Thus, for example, William Cruickshank, surgeon and chemist of the Royal Artillery at Woolwich, recommended pursuing oxygenated muriatic [hydrochloric] acid, even citric and nitric acids, in the treatment of lues or primary syphilis. Nitrous acid, delivered by tube into the back of the mouth beyond the salivary glands provided what he was convinced was "an experimentum crusis."[44] In some hospitals, desperate patients could only accept the unpleasantness and, despite occasional resistance, experimenters thus found an ample pool of subjects on which to try new remedies.

In the circumstances, new therapies had ultimately to confront the absence of an acceptable theory of the curative powers of gases. By 1793 James Carmichael Smyth, of the Middlesex Hospital and physician to George III, worried in a letter to Matthew Boulton about pressing far too fine a theory of gases than the chemistry could support. This, at least, was his view of Beddoes' campaign.[45] Even with increasing analytical precision, chemistry remained too novel and unclear to account for any explanation of medical benefit. However, despite his radical reputation, Beddoes still attracted the attention of prominent society doctors like Smyth. Similarly, George Smith Gibbes of Bath, known for his eudiometry of spa waters, had hoped not only to attend Beddoes's new chemical lectures but to undertake an experiment which might shed light on the possibility of the direct application to the lungs of mercury "as it were dissolved in a Gas."[46] The vast range of physicians and surgeons in touch with

of National Biography, Oxford University Press, 2004 [http://www.oxforddnb.com/view/articles/57592].

43 See Bill Bynum "Reflections on the History of Human experimentation" in Stuart F. Spicker et al. eds., *The Use of Human Beings in Research: With Special Reference to Clinical Trials* (Dordrecht & Boston: Kluwer Academic, 1988), 36.

44 Thomas Beddoes, *Reports Principally Concerning the Effects of the Nitrous Acid in the Venereal Disease, By the Surgeons of the Royal Hospital at Plymouth, and By Other Practitioners* (Bristol: N. Biggs, and London: J. Johnson, 1797), 90; See K.D. Watson, 'Cruickshank, William (d. 1810/11)', *Oxford Dictionary of National Biography*, Oxford University Press, 2004 [http://www.oxforddnb.com/view/article/57592].

45 See BCL, Matthew Boulton Papers 220/22. Robinson Boulton to Beddoes, November 19, 1799; 254/228. James Carmichael Smyth to Boulton, October 25, 1793.

46 BCL, JWP W/9/13. George S. Gibbs to Watt, January 10, 1798. On tests of spa waters see Simon Schaffer, "Measuring Virtue: Eudiometry, Enlightenment and Pneumatic Medicine,"

Beddoes and Watt were often highly engaged in chemical researches in their own laboratories. Gibbes' proposal for further trials clearly followed notions put by James Watt. Gibbes proposed that various substances "chemically divided and obtained in the state of solution in air of some congenial species... might have their full effect." Moreover, pneumatic physicians were inundated with endless trials. A sense of desperation obviously drove the search for therapeutic invention and obscured the risk. There were surely enough reasons. Watt's loss in 1794 of his young daughter Jessy, to the plague of consumption, took him through grief to the grail of pneumatic medicine. This was a tragedy that stalked him, and ultimately took his son Gregory as well. Watt was driven to provide Beddoes an extensive list of substances that might be employed and, most significantly, the manner in which to do so. As a chemist of some note, at least during his lifetime if not so much now, his argument was compelling.[47] He suggested numerous chemical remedies, besides oxygen, calx of zinc, inflammable air, fixed air, azotic air and various compounds of nitre. He proposed "It would be desirable that a list were made out of all substances, which are known to be soluble in air of any kind, or are themselves reducible to vapour or steam, that experiments may be made upon their sanative effects in cases of diseased lungs. This list will prove more numerous than may appear at first glance."[48] The immediate idea was to overcome diseases thought to have arisen from super-oxygenation. But, in the summer of 1794, to test such a notion could not simply rest on trials with patients. Driven by their failing health, their crisis could hardly provide credible testimony.

Laboratory Practice

Laboratories compounded the risk of airs. Pneumatic effects were difficult to predict and, consequently, proved more dangerous than Watt at first expected. Manufacture was difficult, the effects sometimes worse. In his own workshop he tried numerous methods in making a "heavy inflammable air, or carbonated

in Andrew Cunningham and Roger French, eds., *The Medical Enlightenment of the Eighteenth Century* (Cambridge and New York: Cambridge University Press, 1990), 281–318.

47 See David Philip Miller, *James Watt, Chemist: Understanding the Origins of the Steam Age* (London and Vermont: Pickering & Chatto, 2009).

48 Beddoes, *Considerations on the Medicinal Use, and on the Production of Factitious Airs.* Part I. By Thomas Beddoes, M.D. Part II. By James Watt, Engineer. Edition the Third, 107–112, esp. 111–112.

hydrogene, being principally a solution of charcoal in inflammable air." Typical of the chemical practice Watt tried it on himself, breathing through a tube until it made him giddy and unable to stand, finally followed by nausea. He attempted faithfully to record many reactions, notably of his assistants. He told Beddoes in September, 1794, that:

> Another young person, merely from smelling…it as it issued from the bellows, fell upon the floor insensible, and wondered where he was when he awaked. None of us experienced any disagreeable effects in consequence of the vertigo, &c. only in going to bed six hours afterwards, I felt some small remains of the vertigo. Several other persons have inhaled it since; and all were affected in the same manner. I have no doubt, from what I have observed, that if inhaled in a pure state, this air would speedily bring on fainting and death; when given as a medicine, it ought therefore to be much diluted with common air, I should think with 12 times its bulk.[49]

Caution was obviously required. But how much? To the point where trials would be delayed or even rejected? And who would decide? Laboratory life was obviously hazardous. And Watt certainly felt airs affected him dramatically. He tried dephlogisticated airs, variously prepared, as well as nitrous airs, and a mixture of azotic and fixed air. When his assistants breathed the mixture without any ill effects, Watt then tried them on himself. Auto-experiment was part of the general arsenal of the chemist.[50] But some unidentified airs proved highly dangerous merely in their presence. He tried using red-hot charcoal of beef, producing what he called "pyrofarcate" and "pyr-hydro-farcate" which, he reported, "first made me sick, though I did not inspire any purposely, and not above one third of the quantity mentioned was let loose in my laboratory, and 3 doors and a chimney were open; we were, however obliged to leave the place for some time." His son Gregory, a knowledgeable chemist, was made distinctly giddy. These were the hazards of unknown airs. Even with great care, some were so pervasive there was no escape. In one case, Watt was so worried that he resolved to go no deeper for fear of creating a miasma and spreading disease. There was no chemical theory here. As he put it, "One may

49 Beddoes, *Considerations on the Medicinal Use*, 113–114.

50 On the significance of self-experiment, cf. Simon Schaffer, "Self-Evidence," Critical Enquiry 18 (1992): 327–362; and Stuart Walter Strickland, "The Ideology of Self-Knowledge and the Practice of Self-Experimentation," *Eighteenth-Century Studies*, 31 (Summer, 1998): 453–471.

discover, by accident, the air which causes typhus, or some worse disorder, and suffer for it."[51]

Making Airs Easy

Others did not adopt Watt's periodic alarm. His demands for chemical instruments continued, mirrored most immediately, in the laboratory practices of Wedgwood, Keir and Priestley. Their proto-democratic sympathies had insisted on instrumental simplicity. Here was an experimental cosmology that elevated public curiosity and, in medical therapies, induced urgent engagement. But we should not, of course, assume that pneumatic chemistry was always the offspring of a rising republicanism. Disturbances of mob and republic were among Watt's personal terrors. But it did become increasingly apparent that there was, at least in the British circumstance, a strong doctrine of instrumental accessibility that served ample polemical purposes, especially among those averse to French innovations in chemical theory. Even so, there remained a broadly utilitarian agenda soon promoted by the great chemist, and former Beddoes' assistant, Humphry Davy after the turn of the century – thus, for example, his image of the enlightened landowner suitably equipped for chemical trials with "a small portable apparatus; a few phials, a few acids, a lamp and a crucible are all that are necessary."[52] The simplicity of devices, it seems, was crucial to the conquest of privilege as much as of nature. When Thomas Beddoes tried gases in the treatment of patients at Bristol, he claimed that he had simply adapted the gazometers of Lavoisier and of Martinus Van Marum for uses in sick rooms.[53] Not all, however, were inexpensive. Democratic rhetoric did not always match the demands of the enlightened rich. Watt's design of a small furnace and

51 Beddoes, Considerations on the Medicinal Use, 117–118; see also F.F. Cartwright, "The Association of Thomas Beddoes, M.D. with James Watt, F.R.S.," *Notes & Records of the Royal Society of London*, 22 (September, 1967): 131–143, esp. 135; David Miller and T.H. Levere, "'Inhale it and See'? The Collaboration between Thomas Beddoes and James Watt in Pneumatic Medicine," *Ambix* 55 (2008): 1–24.

52 Humphry Davy, *Elements of Agricultural Chemistry, in a Course of Lectures for the Board of Agriculture* (London, 1813), quoted in Morris Berman, "The Early Years of the Royal Institution 1799–1810: A Re-Evaluation," *Science Studies* 2 (July, 1972): 205–240, esp. 230; Berman, *Social Change and Scientific Organization: The Royal Institution, 1799–1814* (Ithaca: Cornell University Press, 1978).

53 Thomas Beddoes, *A Letter to Erasmus Darwin, M.D.* ON *A New Method of Treating Pulmonary Consumption, and Some other Diseases hitherto found Incurable* (Bristol: Bulgin and Rosser; London: J. Johnson [1793]), 41–42.

breathing apparatus, was arranged to be had by private practitioners through Chippendale's in Fleet Street, London.[54] For pneumatic medicine, access was fundamental. For many medics by then, chemical apparatus, just like new static electrical machines, had evolved into highly popular therapeutics.[55]

Gases created were gases used. This resulted, in large measure, from the evolution of James Watt's own breathing apparatus which was made widely available with explanations and directions published with Beddoes. While the device could be purchased from Watt at the Soho foundry, this was hardly convenient. The intent to market through Thomas Chippendale, did not always work to advantage however. The experimentalist Tiberius Cavallo, himself a champion of the diffusion of inexpensive instruments, complained to Watt that Chippendale would not sell him pieces of the portable apparatus, which had been broken or had been lost, finding the "abovementioned Chippendall[sic]… as kind as the Keeper of Hell."[56] Even so, the dispersal of devices for the ready breathing of factitious airs was not confined to Watt.

The promise of pneumatic chemistry at least gave practitioners and patients hope. Amidst London's grey pollution the physician and botanist Robert Thornton, seeing endless cases of pulmonary disease, took up the cause. After consulting with the chemist Adair Crawford at Woolwich, Thornton treated Frederick, Duke of York with super oxygenation as well as the celebrated optician and instrument maker, George Adams, junior.[57] Thornton drew on his earlier Cambridge thesis that animal heat actually arose from the decomposition of airs; hence oxygen was effective in the release of caloric or "the matter of heat." Fevers or consumption focused on the chemical reactions of new gases.[58]

---

54 Beddoes, *Considerations on the Medicinal Use*, 220–222.

55 Cf. Paola Bertucci, "Therapeutic Attractions: Early Applications of Electricity to the Art of Healing," in Harry Whitaker, C.U.M. Smith and Stanley Finger, eds., *Brain, Mind and Medicine. Essays in Eighteenth Century Neuroscience* (New York: Springer Science, 2007), 271–283.

56 Beddoes and Watt, *Considerations*, Part II. Third edition (London, 1796), 181–222; BCL, JWP W/9/11. Cavallo to Watt, January 15, 1798; Golinski, *Science as Public Culture*, pp. 122–123, 127–128.

57 BCL, JWP 4/23/15 (3)?, Thornton to Watt, August 20, 1795. On Adams, see John R. Millburn, *Adams of Fleet Street: Instrument Makers to King George III* (Aldershot, Burlington: Ashgate, 2000).

58 Thomas Beddoes, *Letters from Dr. Withering, of Birmingham, Dr. Ewart, of Bath, Dr. Thornton, of London, and Dr. Biggs, Late of the Isle of Santa-Cruz; Together with Some Other Papers, Supplementary to Two Publications on Asthma, Consumption, Fever, and other Diseases* (Bristol: Printed by Bulgin and Rosser; London: J. Johnson and H. Murray, 1794), 22ff.; R.J. Thornton to Beddoes, December 7, 1793; Martin Kemp, 'Thornton, Robert John (1768–1837)', *Oxford Dictionary of National Biography*, Oxford University Press, 2004 [http://www.oxforddnb.com/view/article/27361].

And Thornton claimed to have endured phthisis himself. Crawford likewise attacked his own evident consumption with hydrocarbonate, declaring "that it transfused over his body at the time of soothing tranquillity, such as opium is known to produce, but with slight vertigo." He gained some relief but survived no more than three months after this experiment in 1795.[59] In his treatments, Thornton was well aware of the cautions of John Barr, the Birmingham surgeon, and of Carmichael in the administration of airs. The self-experiments of others were to be carefully considered, insofar as "one quart of Hydrocarbonate to 100 pints of Atmospheric produced Giddiness in Dr. Crawford, & he would not inhale the whole."[60] New substances also produced doubt as much as hope.

Among the chemists there were many skeptics. Thus, the London physician George Pearson, F.R.S. and lecturer on chemistry, argued in 1794 that "In cases which bid defiance to cure by medicines already known its [clearly?] lawful to make a Cautious trial of new substances." Yet, in the midst of the rage for airs, he sensibly expressed reservations about the likelihood of the vast benefits Beddoes was enthusiastically promoting. Despite his admiration for the effort, Pearson complained about Beddoes' unrestrained enthusiasm: "I am afraid his genius has rendered him *wild*," he wrote to the aspiring chemist Tom Wedgwood.[61] Uncertainty and variability were at least as large obstacles as medical resistance.

Thornton enthused about the pneumatic trials of Beddoes and Watt. He commented on them fully, describing the "happy discovery" of hydro-carbonate following from Watt's work, whom he thought "one of the first chemists of the age."[62] Thornton had an extensive medical practice in which he adopted

---

59 Beddoes and Watt, *Considerations on the Medicinal Use*, 137–138. Thornton to Beddoes, August 1, 1795. Crawford was a member of the Chapter Coffee House Society which had extensive connections amongst the chemists. See Trevor Levere and Gerard L'E Turner, *Discussing Chemistry and Steam: The Minutes of a Coffee House Philosophical Society 1780–1787* (Oxford and London: Oxford University Press, 2002).

60 BCL, James Watt Papers, 4/65/5. Thornton to Watt, July 13, 1795.

61 Keele University, Wedgwood Correspondence, George Pearson to Tom Wedgwood?, March 19, 1794. See also Noel G. Coley, "George Pearson MD, FRS (1751–1828): 'The Greatest Chemist in England,' *Notes & Records of the Royal Society*, 57 (2003): 161–175; E.L. Scott, 'Pearson, George (bap. 1751, d. 1828)', *Oxford Dictionary of National Biography*, Oxford University Press, 2004 [http://www.oxforddnb.com/view/article/21713]; Susan C. Lawrence, *Charitable Knowledge. Hospital Pupils and Practitioners in Eighteenth-Century London* (London and New York: Cambridge University Press, 1996).

62 Robert John Thornton, *The Philosophy of Medicine: or, Medical Extracts on the Nature of Health and Disease, Including the Laws of the Animal Economy, and the Doctrines of Pneumatic Medicine* (London :C. Whittingham, H.D. Symonds, J. Johnson, et al., 1799–1800), vol. III, 16.

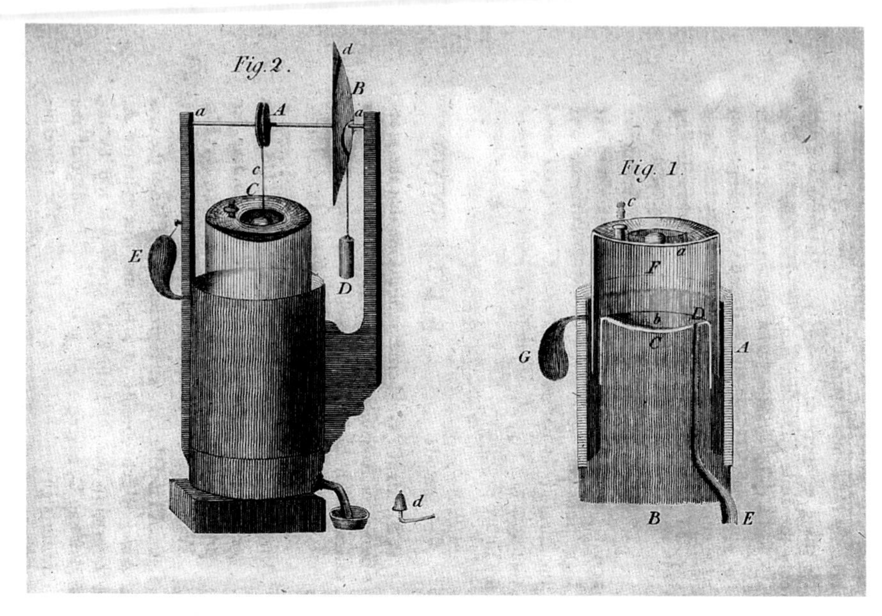

FIGURE 5.2 *Apparatus for producing and containing factitious airs. Engraving from Considerations on the medicinal use of factitious airs and on the manner of obtaining them in large quantities by Thomas Beddoes, 1794* (*Wellcome Library, London*). HTTP://WELLCOMEIMAGES.ORG/ INDEXPLUS/IMAGE/L0000133.HTML

the latest pneumatic devices, seemingly inventing his own even before Watt. Thornton, in 1795, in correspondence with Watt, claimed his device "was constructed a long time back, which I believe you saw when in Russel Street." The multiplication of pneumatic practitioners depended upon the production of a useable device. Thus, Thornton asked Watt to "turn your mind towards the contrivance of a graduated glass apparatus, & tin japanned reservoir, & having them in abundance, & of a moderate price, neat, simple, & elegant, they would ensure the sale of the pneumatic apparatus, & intend the benefit of your discoveries as well as those of Dr. Beddoes..."[63] The prospect of a pneumatic conquest induced euphoria among some physicians.

---

63 BCL, JWP 4/65/5. Thornton to Watt, July 13, 1795. On the spread of pneumatic apparatus see Trevor H. Levere, "Measuring Gases and Measuring Goodness," in Frederick L. Holmes and Trevor H. Levere, eds., *Instruments and Experimentation in the History of Chemistry* (Cambridge, Mass., and London: MIT Press, 2002),105–135; David Philip Miller and Trevor H. Levere, "'Inhale it and See?' The Collaboration between Thomas Beddoes and James Watt in Pneumatic Medicine," *Ambix* 55 (March, 2008): 1–24; also Leslie Tomory, "Let it Burn: Distinguishing Inflammable Airs 1766–1790," *Ambix* 56 (November, 2009): 253–272.

Despite Chippendale's best efforts, the spread of Watt's portable apparatus continued fuelled by the fashion for pneumatic chemistry. The result was a large number of reports from those surgeons, apothecaries and physicians who used it or something similar. David Dundas, royal surgeon, commented that he soon "expected…to find a pneumatic apparatus in every apothecary's house." Not only would this allow more experimental airs to be administered than ever, but a portable device required less skill in chemical arts. Health was to be engineered as much as steam. For Beddoes to set up an apparatus amidst the suburbs of Bristol was no great difficulty and the exercise was repeated throughout many provincial hospitals, infirmaries and laboratories.[64] Not surprisingly John Barr, who worked with Watt on scrofulous cases, encouraged further trials. Beddoes believed the contrivance "extremely convenient for the chemists as well as the practitioners of pneumatic medicine."[65] The Manchester chemist, William Henry, planned to use a Watt contrivance in a public course on chemistry for gentlemen – thus demonstrating once again the way instruments crossed apparent boundaries. According to one report, even Guyton de Morveau and Antoine François de Foucroy, who had already been identified as enemies of the British establishment – by setting on foot in Paris a subscription for the benefit of the victims of the Birmingham Church & King mob – were both interested in Beddoes' medical practice.[66]

The anxiety of patients was magnified by their chronic exposure to polluted airs – especially in the new industrial centres. In the sites of manufactures of Britain and France, Beddoes took note of: "An immense list of artisans of different name, whose labours are carried on amid the floating particles of earth and metals, might be subjoined to the needle-grinders" who were notoriously afflicted. Taking close notice of similar reports from France, Beddoes lamented the difficulties of the needle manufacturers, flax-dressers, plaster and marble workmen, carpet-makers, and especially the young girls in the silk industries of the Cevennes. Many became victims to pulmonary complaints no matter how much they were forewarned.[67] It was, perhaps, their lot. And the promises of chemists were not always comforting.

---

64 Levere, "Dr. Thomas Beddoes," 5–7; Beddoes and Watt, *Considerations*. Second edition (1796), 8.

65 BCL, JWP, 4/23/25. Beddoes to Watt, August 21, 1794.

66 Beddoes and Watt, *Considerations*. Third edition (1796), 68; BCL, JWP W/6/14. James Watt, jun. to Watt, August 14, 1791; JWP 6/35/32. William Henry to Watt, December 15, 1798; JWP W/9/29. Beddoes to Watt, May 16, 1797.

67 Beddoes, *Hygeia: or Essays Moral and Medical, On the Causes Affecting the Personal State of our Middling and Affluent Classes*, vol. II (Bristol: J. Mills; London: R. Phillips, 1802), Essay seventh, 27–33; cf. Jonathan Simon, *Chemistry, Pharmacy and Revolution in France 1777–1809* (Aldershot & Birmingham, Vt.: Ashgate, 2005), 93–128.

The rage in gas therapy, especially following Lavoisier and Fourcroy, had magnified the urgency of trying the airs. As Beddoes put it, "considering what an apprenticeship mankind have served to the art of making experiments, we may hope to be able to bring our certain results sooner than could be done by the antients [sic], or in those half-barbarous ages, when controversy ran high concerning antimony, mercury, and bark. And the longer we are likely to be before we arrive at the end proposed, the sooner we ought to start, and the most briskly to advance."[68] This experimental campaign, he told Tom Wedgwood, was most likely to produce results. Christoph Girtanner meanwhile was publishing cases in Gottingen, and there was likewise much interest in Vienna "but I think want a good apparatus." Beddoes was receiving reports daily, notably from one practitioner whose "experience goes to 2000 patients & upwards." This was likely Thornton. Pneumatic chemistry required human trials. Despite some resistance, there was no shortage of subjects and even desperate volunteers.

New instruments made possible respiration of factitious airs, both in domestic settings and in the medic's surgery. Such was the case of paralysis in the Newcastle hospital where new airs offered new hope, and induced the physicians to order the apparatus. All manner of new devices attracted 18th century medics who had no effective alternatives. Apparatus, like Voltaic piles or Galvanic instruments, as Iliffe, Steigerwald, and Bertucci here reveal, became quite ubiquitous by the end of the century.[69] Beddoes also became one of those seduced. Innovation attracted him. By 1792, having heard of the latest on Alessandro Volta' experiments, he became convinced he could apply electricity "to create a new system of medicine."[70] He noted the many similar attempts of the surgeon Henry Sully of Somerset to restore the damaged optic nerves of patients. By the turn of the 19th century, new technologies were clearly driving the practice of medicine.[71]

---

68 Beddoes, *Contributions to Physical and Medical Knowledge, Principally from the West of England*, 371–372.

69 Cf. Paola Bertucci, "The Electrical Body of Knowledge: Medical Electricity and Experimental Philosophy in the Mid-Eighteenth Century," in Paola Bertucci and Giuliano Pancaldi, eds., *Electric Bodies: Episodes in the History of Medical Electricity* (Bologna: Centro Internazionale per la Storia delle Universita e della Scienza, 2001), 43–68.

70 Cornwall Record Office, Davies-Gilbert Correspondence, 41/54. Beddoes to Davies Gilbert, October 8, 1792.

71 BCL, JWP W/9/52. Beddoes to Watt, June 20, 1796; JWP W/9/55. Erasmus Darwin to Watt, June 2, 1796; JWP 6/20/6. Henry Sully to Watt, February 20, 1805; See also Roy Porter, *Mind-Forg'd Manacles: A History of Madness in England From the Restoration to the Regency* (Cambridge, Mass.: Harvard University Press, 1987), 221–222.

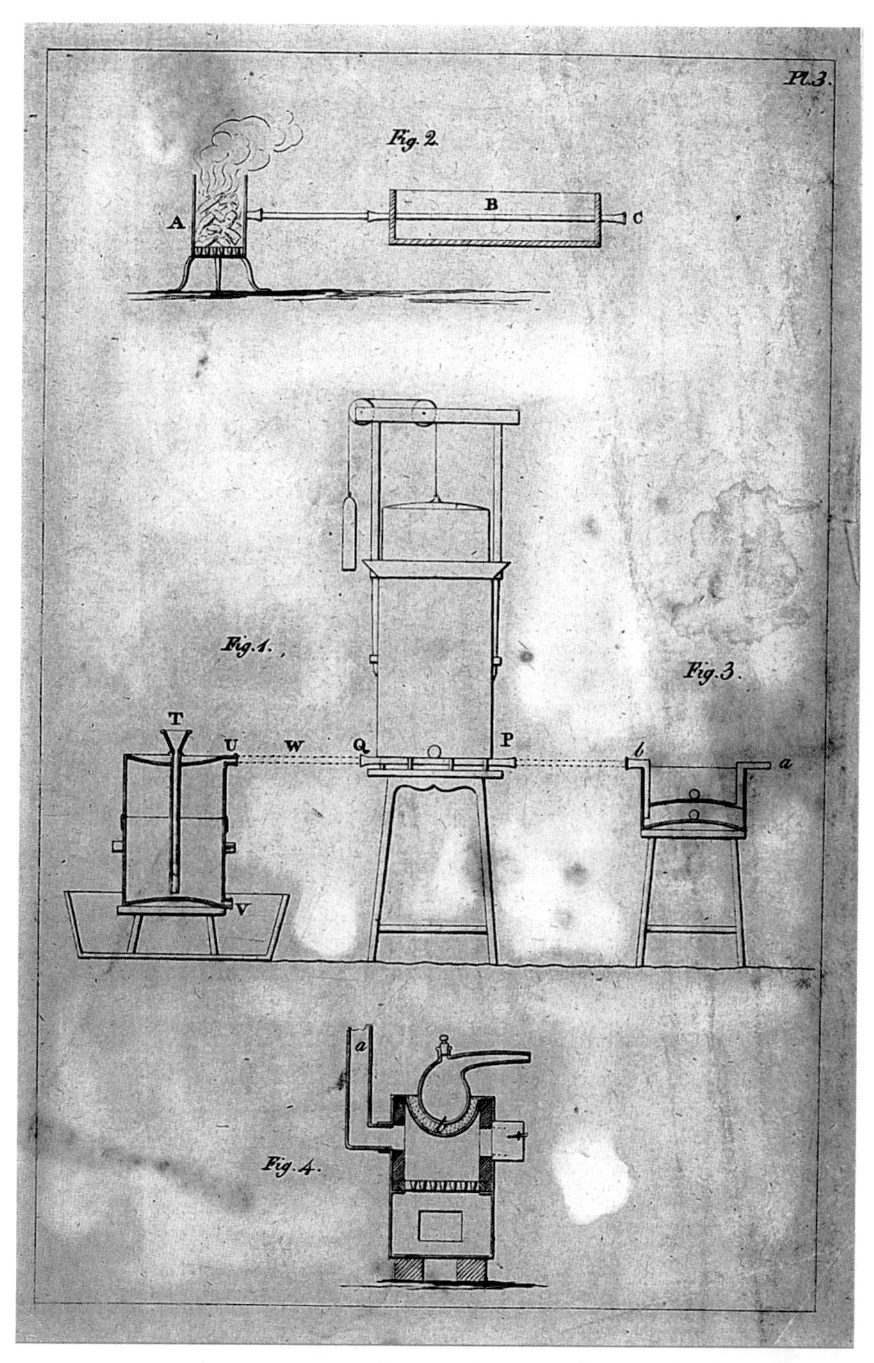

FIGURE 5.3 *Illustration for equipment.* From Considerations on the medicinal use of factitious airs and on the manner of obtaining them in large quantities by *Thomas Beddoes, 1794.* (*Wellcome Library, London*).
HTTP://WELLCOMEIMAGES.ORG/INDEXPLUS/IMAGE/L0000132.HTML

Between the spread of a simplified apparatus, and the blurring of the distinction between subject and patient, emerged a plethora of cases. Trials sometimes avoided even the most superficial supervision. It was surely part of the appeal that even menial servants might mange Watt's device, even to the point that "one maid servant has proved equal to the task" – although the acids which splashed occasioned difficulty. But, reported Richard Pearson, such problems must not stand in the way of promise, especially in those manufacturing towns overrun by vast numbers of consumptive cases.[72] It was easy enough for chemical initiates to make mistakes, as in the case of an old woman who was accidentally "stupefied into sound sleep by hydrocarbonate" given instead of oxygen. Expertise was frequently wanting, but democratic rhetoric precluded supervision. On the other hand, Beddoes was always on the lookout for an assistant "who has the genius of experiment."[73] This was Humphry Davy's opportunity.[74]

Finding Subjects

The uses of airs were obviously highly contested. But the issue was not simply one of risk. In the late 18th century, therapeutic innovation frequently followed ideology. In a disease like the ubiquitous consumption, no prophylactic preserved masters before their servants, nor factory proprietors more than their mechanics. Disease was the great leveller. In illness or contagion, the democratic Beddoes argued that his nemesis Edmund Burke had no sound reason to imagine the "numerous gangs of our manufactures" might envy their social betters.[75] Furthermore, for Beddoes as for Priestley, no social privilege could yet be asserted above a democratic, therapeutic, practice. Yet apparent differences in exposure to disease, notably in the mills and in urban crowding, clearly increased the dangers to the "two-legged cattle stalled in our manufactories." Beddoes' remarkably prescient essay *Hygeia*, on the effects of manufactures, tells us much about the emerging awareness of environmental illness brought on by irritating, chemical, and polluting effects in a rapidly industrializing Britain. As Beddoes pronounced, "If we raise our view from the

72 Beddoes and Watt, *Considerations*. Third edition (1796), 55–56, 76.

73 BCL, JWP W/9/43. Beddoes to Watt, August 31, ?; JWP W/9/32. Beddoes to Watt, n.d.

74 See Fullmer, *Young Humphry Davy*, chapter 8.

75 Thomas Beddoes, *Hygeia: or Essays Moral and Medical, On the Causes Affecting the Personal State of our Middling and Affluent Classes*, vol. 1 (Bristol: J. Mills, and London: R. Phillips, 1802), Essay Second, 77–81.

lower to the higher classes, we shall here perceive that it is upon the lilies of the land, that neither toil nor spin, that the blight of consumption principally falls."[76] Pneumatic medicine undermined social privilege. For this reason, if for no other, Beddoes' vision clearly had many critics. As Beddoes put it, he "must expect to be decried at home as a silly projector, and by others as a rapacious empiric."[77]

The alarm of pneumatic patients, trapped between disease and therapy, was the measure of their fears. Many were afflicted with ill-defined and, mostly, incurable disease. To many observers, consumption was particularly to be feared and was constantly spreading.[78] Undoubtedly even the most knowledgeable physicians, surgeons and apothecaries were defeated by what Darwin called the "giant malady."[79] There appeared no place to hide from miasma. Did industrializing Britain breed an advancing affliction? Could distinctions be drawn between narrow alleys and the freer air of rural England? Even this difference was too tidy for the chemically enlightened. As Beddoes put it, "it may be remarked, that neither confinement in close rooms, unwholesome fumes, or an unfavourable posture, contribute any more than an effeminate education, to the production of consumptions. Again among the peasantry of Warwickshire and Staffordshire, I understand that consumptions are extremely frequent; not less so than among the Birmingham manufacturers."[80] Citing cases in Birmingham where elastic fluids, notably hydro-carbonate, had seemed effective, Beddoes told Tom Wedgwood that a unique ward in the Birmingham hospital was finally to be appropriated to pneumatic medicine in the spring of 1795.[81] In the chemist's airs there was a wisp of hope.

Patients' Progress

Many cases were obviously critical, and the diagnosis was never straightforward. Watt was directly consulted by patients about their asthma and the

76 Beddoes, *Hygeia*, vol. I, 79; Hygeia, vol. II, 24–25, 34.

77 Beddoes, *A Letter to Erasmus Darwin*, 3.

78 Thomas Beddoes, *Hygeia: or Essays Moral and Medical, on the Causes Affecting the Personal State of our Middling and Affluent Classes*, vol. II (Bristol: J. Mills, St. and London: R. Phillips, 1802), Essay the Seventh, 19.

79 Beddoes, *Letters*, 22. Thornton to Beddoes, December 7, 1793.

80 Beddoes, *Observations on the Nature and Cure of Calculus*, 68–169: Cf. Stock, *Memoirs of Thomas Beddoes*, 230–231.

81 Keele University, Wedgwood-Mosley Collection, WM 35, Beddoes to Thomas Wedgwood, March 17, 1795.

methods for the production of airs to alleviate their difficulties.[82] Watt's apparatus not only attracted these cries for help. It also raised fears among physicians of being overly optimistic. James Watt's commitment to chemistry has only recently come to be appreciated.[83] But it was this which led him to medicine. He was much involved with John Barr, the Birmingham surgeon, in the treatment of phthisis. Watt was in the thick of the action. His own household became a pulmonary laboratory. One afflicted was Rebecca Stanley, a thirty-five year old servant to Watt attended by Barr. According to Barr, she remained in extreme difficulty even after Watt attempted to assist her breathing by a variety of treatments, including the ubiquitous laudanum, but also with significant inspirations of hydrocarbonate.[84] A year later, Barr was involved in another Watt case. This one involved Richard Newbury, forty-six, a labourer employed by Watt. His drinking did him no good. Newbury suffered from haemoptysis, characterized by expectoration of the blood, likely of tubercular origin. Watt had prepared and prescribed inhalations of hydro-carbonate but, in frustration, called in Barr and the Birmingham physician, John Carmichael. According to Carmichael, "Newberry himself uniformly expressed much thankfulness for the benefit he invariably received from breathing hydrocarbonate. Had the inhalation of the modified air been repeated more frequently, [he speculated] would it not alone have been adequate to the complete removal of the pain of his side, and consequent cure?"[85] Watt apparently had few qualms in advancing factitious airs with his own servants. He was nonetheless very cautious as to their chemical preparation and regarding the concentrations, specifically in the inhalation of oxygen and hydrocarbonate, in what he clearly understood as an experimental process.[86] Even so, as he told James Lind, trials might yet be worthwhile "especially if you have in mind to amuse the patient with something strange & scientific."[87] If nothing else, it

---

82 BCL, JWP C/1/15. Watt to James Lind, March 18, 1795; W/9/49. Robert Holny? to Watt, August ?.

83 See David Philip Miller, *James Watt, Chemist: Understanding the Origins of the Steam Age* (London and Vermont: Pickering & Chatto, 2009).

84 Thomas Beddoes and James Watt, *Considerations of the Medicinal Use and Production of Factitious Airs*. Part III. Second edition, corrected, and enlarged (London: Printed J. Johnson, 1796), 132. John Barr to Beddoes, August 5, 1795; On Barr, see Irvine Loudon, "The Nature of Provincial Medical Practice in Eighteenth-Century England," *Medical History* 29 (1985): 1–32, 7.

85 Beddoes and Watt, *Medical Cases and Speculations*, 97–99, 117–118. John Carmichael to Beddoes, August 15, 1796.

86 Beddoes and Watt, *Considerations of the Medicinal Use*, 164–166.

87 BCL, JWP C1/15. Watt to James Lind, August 10, 1796.

might divert a patient's mind. Any reservations, such as they were, were ultimately overcome. What has remained obscure, however, is that many patients were as much those of James Watt the engineer as of the local surgeons or physicians. Watt was directly supervising experimental therapies including cases of scrofulous phthisis, notably where consumptive patients "are not susceptible to the bad effects of [the inhalation] of lead."[88] In the laboratories and the infirmaries of the late 18th century, the range of inspirations was spreading rapidly and Watt's portable apparatus was largely responsible. His medical communications, reproduced with many others by Beddoes, ensured medicine addressed the afflictions of mechanics.

By the end of the century, pneumatic practices were substantial. Included in Beddoes' own lists were Mr. Tobin, probably the actor William Tobin, and likewise one of the sisters of Horace Walpole on whom Mr. Hill, the surgeon, combined both vital air and electrical shocks in an effort to cure her obscured vision in one eye. Similarly, the notable surgeon Jonathan Wathen of St. Thomas's Hospital tried vital air in an otherwise incurable case of partial blindness in a young boy.[89] It is likely that Mrs Keir, wife of the chemist James Keir, might have been one of those Beddoes treated for fever, as he also tried azote, prepared by Humphry Davy, on Mr. Green the M.P. for some pulmonary difficulty.[90]

Significantly, not all subjects were ill. Beddoes was fascinated by the experimental attempts at breathing pure hydrogen:

> I have remarked them in a number of healthy persons who were curious to try how long they could breathe this air. The frequency and debility of pulse, blueness of the lips and coloured parts of the skin, were always observable in a minute, or a minute and an half. Besides, dizziness was felt, and the eyes have grown dim; in animals, the transparent cornea has appeared sunk and shrivelled. Several individuals agree in describing the incipient insensibility as highly agreeable. One consumptive person loved to indulge in it; for this purpose, contrary to my judgment, he used to

88 BCL, JWP, W/9/53. Beddoes to Watt, April 5, 1796.

89 BCL, JWP 4/65/7. Beddoes to Watt, December 12, 1794; JWP 4/27/26. Beddoes to Watt, January 21, 1803; JWP 4/27/70. Beddoes to Watt, n.d. (prob. 1803); On Tobin see Golinski, *Science as Public Culture*, 167–168; On the effects of nitrous oxide see Humphry Davy, Researches, Chemical and Philosophical; Chiefly Concerning Nitrous Oxide, or Dephlogisticated Nitrous Air, and Its Respiration (London: J. Johnson, 1800), 497–559.

90 BCL, JWP 4/23/4. Beddoes to Watt, January 5, 1796; JWP 6/33/69. Beddoes to Watt, January 20, 1801.

> inspire a cubic foot of hydrogene at a time. This quantity most commonly produced little change in his feelings. Sometimes it brought on almost complete asphyxia. During this process, I have felt the pulse nearly obliterated. Afterwards, as he recovered, it was sensibly fuller, and stronger than before the inspiration.[91]

The useful effect of gases was obviously highly speculative and necessarily required much medical research and many subjects. The means lay in the encouragement of chemists and medics to report their experiences, whether in the laboratory, the infirmary or the hospital. Chemical politics and Beddoes' experience provided not a little too much enthusiasm. He held that *"The more widely any species of knowledge is disseminated, the more rapidly may we expect that it will make advances."* Here he clearly echoed Priestley's earlier proto-democratic manifesto that "…progress might be quickened, if studious and modest persons, instead of confining themselves to the discoveries of others, could be brought to entertain the idea, that is was possible to make discoveries themselves."[92] This, of course, became a mantra among the enlightened. Beddoes later argued, in his Bristol public lectures, that, "by multiplying the number of minds in activity, we multiply the chances of fortunate combinations."[93] At that revolutionary moment, the politics of airs were reflected in the suppression of reform.[94] But, before the benefits of airs could be established a credible, or less than haphazard, method of experimentation had also to be achieved. This was the proposal put forth by James Watt.

The bodies of the public were the most compelling sites for experimental tests. Little could be achieved without instruments which traversed the space between laboratory and infirmary. In England, pneumatic chemists routinely tried the new gases. The design and application of instruments for physicians, surgeons and in self-medication was reflected among the electro-therapists. The London instrument maker, Edward Nairne patented a 'medical electrical

---

91 Beddoes, *Considerations on the Medicinal Use*, 41–42.

92 Priestley, *The History and Present State of Electricity*, 546. Italics in original.

93 Thomas Beddoes, *A Lecture Introductory To a Course of Popular Instruction On the Constitution and Management of the Human Body* (Bristol: Printed by N. Biggs, For Joseph Cottle, and London: J. Johnson, 1797), 23.

94 See E.P. Thompson, *The making of the English working class* (New York: Vintage, 1963), esp. 114–160; Kenneth R. Johnson, *Unusual Suspects. Pitt's Reign of Alarm & the lost Generation of the 1790s* (Oxford: Oxford University Press, 2013), 98–109.

machine' marketed simultaneously for both experimental and medical purposes.[95] Barriers constructed out of theory and of philosophical sophistication proved quite unnecessary. Some experiences were not only dramatic, as in the balloons over the Champ de Mars – even in parlour games or the workshop they could both delight and confound the participants. In Bristol, the radical Beddoes would put Humphry Davy to work exploring the effects of the nitrous acid bath. Davy's self-experiment was an experience he quickly determined not to repeat. In August, 1800, Davy reported:

> On the 22d I bathed both in the morning and at night, remaining in the bath nearly an hour and three quarters each time. On this day slight eruptions began to appear on parts of the skin: On the 23d I bathed twice; my gums, examined in the morning, appeared redder than usual about the bases of the teeth; towards the evening they became a little painful: On the 24th the redness and soreness of the gums were increased, I thought the secretion of saliva greater than usual; my breath had no foetor. At this time I was obliged to desist from the experiment, in consequence of violent diarrhoea, which had been slightly felt during the two preceding days, and which was probably occasioned by causes unconnected with the acid bathing.[96]

All was seemingly not lost. The willingness to undertake trials nevertheless had its occasional benefits. Hope continually conquered caution. It was quite possible to secure lightning in a jar, or find relief in laudanum, as both were ubiquitous in the late 18th century, especially amongst self-administering physicians and their desperate patients. In the private medical journal of James Watt, jun., under Beddoes' care throughout 1797–99, there were recounted many attempts at the nitrous acid bath and the constant application of laudanum.[97] And Samuel Taylor Coleridge, friend to Beddoes, the Watts and the

95 Simon Schaffer, "Measuring Virtue: Eudiometry, Enlightenment and Pneumatic Medicine," in Andrew Cunningham and Roger French, eds., *The Medical Enlightenment of the Eighteenth Century* (Cambridge and New York: Cambridge University Press, 1990), 281–318; Paola Bertucci, "A Philosophical Business: Edward Nairne and the Patent Electrical Machine (1782)," *History of Technology* 23 (2001): 41–58.

96 Thomas Beddoes, *Communications Respecting the External and Internal Use of Nitrous Acid; Demonstrating its Efficacy in Every Form of Venereal Disease, and Extending its Use to Other Complaints: with Original Facts, and a Preliminary Discourse* (London: Printed for J. Johnson, 1800), xxxvii.

97 BCL, Murihead I /6/6. James Watt, jun. Medical Journal, 1797–99.

Wedgwoods, later reported that "...the practice of taking Opium is dreadfully spread. – Throughout Lankashire & Yorkshire it is the common Dram of the lower orders of People – sold in the small Town of Thorpe the Druggist informed me, that he commonly sold on market days two or three Pound of Opium, & a Gallon of Laudanum – all among the labouring Classes."[98] The reference to down market medication reflected the willingness to put faith in fashion. Nitrous oxide could likewise lift the spirits even if cures were ultimately elusive. William Tobin described the agitation it induced, but it was difficult to distinguish between high times and the healthful promise of "Dr. Beddoes Breath." Suitably Mrs. Beddoes partook of the delights of nitrous oxide so that she "could walk much better up Clifton Hill [and] frequently seemed to be ascending like a balloon."[99] No hot air was thus needed, nor even a Sadler.

Mundane practices were widespread and, sadly, of little practical benefit. But there was experimental achievement nonetheless. The deeper we delve into the worlds of experimenters, demonstrators, and the adventurous volunteers, the distinctions between the library and the bench, between scholar and mechanic seem as tidy, historiographical, contrivances now reflecting class and intellectual self-image, of both philosophers and historians. As James Delbourgo recently reminded us, many of the most famous demonstrations by renowned experimentalists were often repeated as parlour games. And as Paola Bertucci has revealed in the previous chapter, enthusiasts provided amusement, and crowds were drawn to both the air and electrical men. Diversions such as the electrical shocks from wires hidden beneath a rug were among those tried by William Watson in London, reported to the Royal Society, and noted both by Priestley in Birmingham and Ebenezer Kinnersley in America.[100] Reservations closely held by the proper gentlemen of the Royal

98 Quoted in Neil Vickers, *Coleridge and the Doctors 1795–1806* (Oxford: Clarendon Press, 2004), 103.

99 Quoted in Golinski, *Science as Public Culture*, 167–168.

100 William Watson, *A Sequel to the Experiments and Observations Tending to Illustrate the Nature and Properties of Electricity*. Second Edition (London, 1746), 22, 69–70; Delbourgo, *A Most Amazing Scene of Wonders. Electricity and Enlightenment in Early America* (Cambridge, Mass. and London: Harvard University Press, 2006), 122. Among the works of Paola Bertucci see especially "The Electrical Body of Knowledge: Medical Electricity and Experimental Philosophy in the Mid-Eighteenth Century," in P. Bertucci and G. Pancaldi, eds., *Electric Bodies. Episodes in the history of medical electricity* (Bologna: CIS, Dipartimento di Filosofia, 2001), 43–58; Bertucci, "Revealing sparks: John Wesley and the Religious

Society, or later even in the Royal Institution, seem now like fingers in a very public dike, by the late 18th century leaking profusely, impossible to police, the outcome hotly contested. In any case, as Mrs. Beddoes reported, neither new airs nor high times were very hard to find. Innovation was also a mirror held up to the incurable.

Utility of Electrical Healing," *British Journal for the History of Science* 39 (September, 2006): 341–362; and "Therapeutic Attractions: Early Applications of Electricity to the Art of Healing," in Harry Whitaker, C.U.M. Smith, and Stanley Finger, eds., *Brain, Mind and Medicine: Essays in Eighteenth-Century Neuroscience* (New York: Springer, 2007), 271–283.

CHAPTER 6

Food Fights: Human Experiments in Late Nineteenth-Century Nutrition Physiology

Elizabeth Neswald[1]

A History of Experimental Agency

From November 1889 to January 1890, Max Rubner, Germany's leading nutrition physiologist, conducted a series of experiments on the metabolism of alcohol. In each experiment, a Mr. Bretschneider, presumably a laboratory technician, spent four hours in a respiration chamber. Atmospheric air was pumped into this airtight box the size of a small room, thermometers measured the chamber temperature and barometers its humidity, while the air in the chamber was pumped out through a series of chemical filters that enabled the carbon dioxide content of the experimental subject's exhalations to be extracted and weighed as an indirect measure of metabolism. Rubner varied the temperature and humidity in the chamber, measured the average carbon dioxide and water vapor produced per hour under these different conditions both with and without the consumption of alcohol, and recorded the experimental subject's perception of temperature. He discarded the first in the series, crossing it out with an 'X' across the page. In the section reserved for notes, Rubner remarked, "During the experiment, the experimental subject consumed 440 ccm. of 'Nordhäuser' [a schnaps], that is, he imbibed as he wished from a measured half-liter. As a result, by the end of the experiment, he was drunk. In the following experiments, he will receive a measured quarter-liter."[2]

At first glance an amusing anecdote from the laboratory records of an experimenting physiologist, the episode described in these few lines encapsulates one of the central problems that nineteenth-century nutrition physiologists, and presumably many other researchers in the life sciences, encountered in their trials on living, and especially human subjects: A methodologically "good" experiment required that the subjects and their bodies functioned in particular ways, but neither the subjects nor their bodies were consistently willing or

1 The author would like to acknowledge the generous financial support for this project from the Canadian Institutes of Health Research.

2 Max Rubner, "Alkoholversuche", Max Rubner Papers, Archive of the Max Planck Society, Abt. III, Rep. 8, 123/7. All translations from the German by EN.

 | DOI 10.1163/9789004286719_008

able to perform in the ways most convenient for the physiologists. Subjects had agency and bodies resisted the physiologists' interventions. A "successful" experiment, in contrast, was the result of a compromise between the methodological needs of the experimenter and the physical and psychological needs of his or her subject. Without cooperation and collaboration between experimenter and subject, or, at least, a relationship that included mutual consideration and concessions, most nutrition physiological experiments would have had little chance of completion, let alone success. Experiments in nineteenth-century nutrition physiology thus challenge assumptions about clear hierarchies and power differentials between experimenters and their subjects. They highlight both the strategies of and limits on controlling living beings and the role not only of voluntary consent, but also of active cooperation and participation to enable success.

The status of living subjects in scientific and medical research has long been viewed as problematic.[3] What it means to experiment on beings capable of feeling and suffering, is a question that has long occupied scientists, philosophers, ethicists, and the public. The use of poor medical patients, children, institutionalized populations, and others not provided with sufficient knowledge, freedom or the capacity to consent to these interventions, raised questions about the value of an individual human life, choice, and dignity in relation to the potential benefits these efforts could provide for the population, or, more cynically, to the career of a particular experimenter.[4] Such concerns became even more pressing from the second half of the nineteenth century, as the experimentalization of physiology and the laboratory revolution in medicine led to an increase in vivisection and the testing of various procedures and therapies on animals and humans. In the twentieth-century, after the horrors of medical experiments on concentration camp prisoners under National Socialism and the Tuskegee syphilis study in the United States, among many other examples, the ethics of human experimentation and consent moved to the foreground.[5]

---

3 Anita Guerrini, *Experimenting with Humans and Animals: From Galen to Animal Rights* (Baltimore: Johns Hopkins University Press, 2003); Nicolaas Rupke (ed.), *Vivisection in Historical Perspective* (London: Routledge, 1990).

4 Barbara Elkeles, *Der moralische Diskurs über das medizinische Menschenexperiment im 19. Jahrhundert* (Stuttgart: Urban & Fischer, 1996); Jordan Goodman, Anthony McElligott, and Lara Marks (eds), *Useful Bodies. Humans in the Service of Medical Science in the Twentieth Century* (Baltimore: Johns Hopkins University Press, 2003); Allen M. Hornblum, Judith L. Newman and Gregory J. Dober (eds), *Against their will: the secret history of medical experimentation on children in cold war America* (New York: Palgrave MacMillan, 2013).

5 Harriet A. Washington, *Medical Apartheid. The Dark History of Medical Experimentation on Black Americans from Colonial Times to the Present* (New York et al.: Doubleday, 2006); George

In light of the suffering caused by human experimentation, it is both unsurprising and appropriate that most historical research on the topic has focused on the ethically-sensitive question of informed consent and its disregard, and on scientific and medical abuses of power and status. This perspective has led historians of human experimentation, however, to focus on a limited range of scientific fields, in which these abuses appeared most egregious. Studies of medical experiments in early bacteriology, immunology, and vaccination development, of experiments undertaken on individuals denied the information, ability or status to voluntarily consent, and of manipulative psychological experiments clearly demonstrate how social relationships of power, authority, and dependency have been – and continue to be – abused.

Nutrition research is not without its own history of coercive experimentation, most often undertaken on institutional populations. While interest in the relationship between diet, health and disease has been a part of medical practice since the antique, and the phenomena of metabolism have been explored since the seventeenth century at the latest, it is noteworthy that the field of nutrition science, when presenting its own history, most often traces its origins not to the tradition of dietetics, but to the work of Benjamin Thomson, Count Rumford. An American adventurer at the court of the Bavarian king, Rumford envisioned the workhouse as a new method of administering the poor and tested various means of economically feeding its residents. He developed a barley and legume-based gruel, supplemented with bits of vegetables and meat for flavor and dry bread for texture.[6] Supposedly palatable, nutritious, and inexpensive – the latter was accurate at least – this gruel was distributed to the destitute poor of Munich, provided they worked in exchange. A few decades later, the French Academy of Science recruited physiologist François Magendie to study whether bone gelatin, a waste product, was a nutritious replacement for meat and meat broth in the diet of hospital patients, again a population of the poor. While Magendie conducted precise experiments on gelatin feeding in dogs, the hospital physicians conducted their own, less structured trials on hospital populations, observing the effects in varying amounts in the hospital diet.[7] To their credit, most eventually rejected this

---

J. Annas and Michael A. Gordin (eds), *The Nazi doctors and the Nuremberg Code: human rights in human experimentation* (New York/Oxford: Oxford University Press, 1992); James Jones, *Bad Blood: The Tuskegee Syphilis Experiment* (New York: Free Press, 1981).

6 Benjamin Thomson, Count Rumford, "An Account of an Establishment for the Poor at Munich," in Benjamin Thompson, *The Complete Works of Count Rumford*, Vol. IV (Boston, 1875), 229–326.

7 François Magendie et al, "Rapport fait à l'Académie des Sciences au nom de la Commission dite de la gélatine", in *Compte Rendu des Séances de l'Académie des Science* (1841): 237–295;

attempt by gelatin producers and state interests to see the sick poor nourished on a substance rejected by Magendie's dogs.

Prisons, as well, were sites for observational experimentation. In the second half of the nineteenth-century, prison physicians attempted to reduce endemic and cyclical prison diseases including scurvy, rickets, edema, and night blindness through changes in diet.[8] They noticed increases in scurvy rates, for example, when potatoes were removed from the prison diet during years of high prices, and declines, when they were reintroduced. They discussed improvements in night blindness, when inmates were given cod liver oil, and tested other substances, to see whether they could produce beneficial effects. Although these observations on the relationship between specific foods and diseases became part of the knowledge that led to the recognition of the role of micronutrients in preventing deficiency diseases, such interventions had primarily therapeutic and only secondarily experimental intentions. Prisoners were treated, whether effectively or ineffectively, with the aim of reducing the incidence of prison disease.

Humanitarian concerns doubtlessly motivated the nutrition experimentation of many nineteenth-century prison physicians, who observed the effects of the dietary restrictions imposed by administers and lobbied governments to improve the diets of prisoners. In the twentieth century, institutions became sites for active, invasive experimentation. In the second decade of the twentieth century, Joseph Goldberger conducted a multi-year experiment on prisoners in Mississippi, giving them a diet designed to induce pellagra, a disease caused by niacin deficiency and common in the American South due to the heavily corn-based diet of the poor population.[9] Although the prisoners volunteered, with a pardon as their reward, one can hardly consider them to have

---

Frederic L. Holmes, *Claude Bernard and Animal Chemistry: The Emergence of a Scientist* (Cambridge, MA: Harvard University Press, 1974), 6–15.

8 Abraham Adolf Baer, "Die Morbidität und Mortalität in den Straf- und Gefangenenanstalten in ihrem Zusammenhange mit der Beköstigung der Gefangenen (mit besonderer Berücksichtigung der Salubritätsverhältnisse und des Kostregimens in dem Strafgefängniss bei Berlin-Plötzensee)," *Deutsche Vierteljahresschrift für öffentliche Gesundheitspflege* 8 (1876): 185–208; J. Michel, "Bericht über das Vorkommen von Nachtblindheit im Arbeitshause Rebdorf," *Bayerisches ärztliches Intelligenzblatt* 1882: 335–338; Heinz Goerke, "Anstaltsernährung im 19. Jahrhundert," in Edith Heischkelt-Artelt (ed.), *Ernährung und Ernährungslehre im 19. Jahrhundert. Vorträge e. Symposiums am 5. und 6. Jan 1973 in Frankfurt am Main* (Göttingen: Vandenhoeck & Ruprecht, 1976), 303–317.

9 Joseph Goldberger, "The Experimental Production of Pellagra in Human Subjects by Means of Diet," in *Goldberger on Pellagra*, ed. Milton Terris, (Baton Rouge: Louisiana State University Press, 1964), 54–94.

been sufficiently informed of the progress of the disease they would later develop. The codification of experimental ethics in the Nuremberg Code changed little. Heinrich Kraut, who conducted nutrition experiments on concentration camp prisoners and forced laborers in Nazi Germany to explore caloric minimums under different degrees of physical activity, continued this work after the war under the umbrella of "developmental aid", moving his research site away from Europe to the African continent.[10] From 1942–1952 Canadian nutrition researchers conducted studies on malnourished Aboriginal populations and on Aboriginal children in residential schools.[11] Diverse experiments tested the effect of adding dietary supplements to the nutritionally and calorically inadequate diets of some children, while using others as a control group. Others manipulated welfare food provision systems to steer the foods that families could obtain. In neither case did researchers obtain the consent of the experimental subjects or provide necessary information.

As these examples show, the history of human experiments in nutrition physiology can be told as a history of coercive experiments undertaken by sovereign and government researchers on individuals and groups with limited ability to truly consent to their participation. Without a doubt, the relationship between an experimenter and an experimental subject is characterized by structural asymmetry. The experimenter determines the question and the experimental arrangement, while the subject is expected to be passive and obedient. Nonetheless, as Susan Lederer hints in her seminal study on the history of human experimentation, experimenter-subject interaction was often more complex, requiring the active cooperation of informed and knowledgeable subjects.[12]

This essay takes Lederer's observation as a starting point to explore the complex dynamics of human experimentation in nutrition physiology from the

---

10 Jennifer K. Alexander, "An Efficiency of Scarcity: Using Food to Increase the Productivity of Soviet Prisoners of War in the Mines of the Third Reich," *History and Technology* 22 (2006), 391–406; Susanne Heim, *Kalorien, Kautschuk, Karrieren. Pflanzenzüchtung und landwirtschaftliche Forschung in Kaiser-Wilhelm-Instituten 1933–1945* (Göttingen: Wallstein Verlag, 2003), 102–120; Ulrike Thoms, "Das Max-Planck-Institute für Ernährungsphysiologie und die Nachkriegskarriere von Heinrich Kraut," in Theo Plesser and Hans-Ulrich Thamer (eds), *Arbeit, Leistung und Ernährung. Vom Kaiser-Wilhelm-Institut für Arbeitsphysiologie in Berlin zum Max-Planck-Institut für molekulare Physiologie und Leibniz Institut für Arbeitsforschung in Dortmund* (Stuttgart: Franz Steiner Verlag, 2012), 295–356.

11 Ian Mosby, "Administering Colonial Science: Nutrition Research and Human Biomedical Experimentation in Aboriginal Communities and Residential Schools, 1942–1952," *Social History* 46 (2013): 145–172.

12 Susan Lederer, *Subjected to Science. Human Experimentation in America before the Second World War* (Baltimore/London: Johns Hopkins University Press, 1995), 14–15.

perspective of the history of experiment, rather than from biomedical ethics. The "experimental turn" in science studies and the history of science in the past decades has given new prominence to the objects of experiment and experimentation in the generation of scientific knowledge. "Epistemic things", "things that talk" and hybrid "quasi-objects" populate the landscape of the history of science, and these studies have shown that the status of these things – whether microorganisms and stuffed animals or gas discharge tubes and quarks – is highly problematic.[13]

Andrew Pickering describes this problematic relationship of experimenters to their scientific things as a dialectic of resistance and accommodation.[14] Using examples from modern physics and beginning with the assumption that practices and action precede theories and observation, he ascribes a material agency to the things of science: "They are objects that do things in the world."[15] Scientific practice is, in Pickering's words, a "dance" between the experimenter, who acts with intention and manipulates an apparatus in order to generate and study an effect, the apparatus that produces or does not produce the desired effect, and the effect, which is or is not produced. Difficulties in conducting the experiment as first conceived are experienced as resistance. In order to overcome this resistance, the experimenter must develop new strategies, make compromises, modify his experiments, change his expectations for the results, redefine his goals, or make alterations in the technical apparatus.[16] Scientists find themselves in a feedback relationship with both their apparatus and their objects. What changes and modifications may be necessary cannot be known before the experiment begins. They emerge temporally during and through the process of experiment and are a product of experimental practice.

Although Pickering aimed to describe experimental practices in physics and machine sciences, his dialectic of resistance and accommodation and of the ways in which experimenter, apparatus, and experimental object interact, is a useful way to approach the dynamics of experimenter-experimental subject

13 Hans-Jörg Rheinberger, *Toward a History of Experimental Things. Synthesizing Proteins in the Test Tube* (Stanford: Stanford University Press, 1997); Lorraine Daston (ed.), *Things that Talk. Object Lessons from Art and Science* New York: Zone Books, 2004; Bruno Latour, *We have never been modern* (Cambridge, MA: Harvard University Press, 1994); Falk Műller, *Gasentladungsforschung im 19. Jahrhundert* (Berlin: Verlag für Geschichte der Naturwissenschaft und Technik, 2004); Andrew Pickering, *Constructing Quarks. A Sociological History of Particle Physics* (Chicago: University of Chicago Press, 1984).

14 Andrew Pickering, *The Mangle of Practice. Time, Agency, and Science* (Chicago: University of Chicago Press, 1995).

15 Pickering, *Mangle*, 9.

16 Pickering, *Mangle*, 22.

relationships in nutrition experiments. Nutrition physiology emerged in the 1860s as part of the general trend in nineteenth-century physiology toward quantitative, experimental methods. Aiming for the status of an exact science, physiologists looked to physics and chemistry for their models, bringing precision instruments and apparatus into their laboratories. They manipulated experimental objects and conditions, isolated variables, analyzed and quantified complex physiological phenomena, and structured their experiments according to the methods of physics, chemistry, and engineering.[17] The organic objects and living subjects of their experiments occupied the same position in the experiments as the magnets, chemical compounds, and gas discharge tubes of the physicists and chemists, and experiments were designed with the expectation that these living objects would be similarly passive and malleable, that they would "function" as required.

There is, of course, nothing particularly insightful in ascribing agency to the human subjects of these experiments, but within the structure of these experiments and during the process of experimentation, their agency became problematic in unexpected ways. These experimental subjects were both conscious, intentionally acting agents and bodies that neither the subject nor the experimenter could fully control. Bodies as well as subjects cooperated and resisted, and both needed to be accommodated for an experiment to succeed.

I focus on human experiments conducted in the formative decades of nutrition physiology, from around the 1870s through the early twentieth century, when only the macronutrients protein, fat, and carbohydrate were considered to have dietary significance. In the 1880s, after Max Rubner proved that the energy conservation law applied to living organisms and proposed that caloric content was the crucial factor in nutrition, protein and energy became the main assessment categories.[18] Nutrition experiments were designed as intake-output experiments, in which the food entering the body was compared to the solid, liquid, and gaseous components leaving it.[19] Dogs provided the model animal for two basic kinds of experiments, on digestion and respiration, which were then modified for the study of human beings, as techniques and apparatus were developed for use on these more cumbersome and demanding subjects.[20]

17 For a statement of physiological methods, see Carl Ludwig, *Lehrbuch der Physiologie des Menschen. Erster Band* (Heidelberg: Winter, 1852), 11–12.

18 Max Rubner, "Die Quelle der tierischen Wärme," *Zeitschrift fur Biologie* 30 (1894): 73–142; Wilbur Olin Atwater and Francis Gano Benedict, *Experiments on the metabolism of matter and energy in the human body* (Washington D.C.: US Government Printing Office, 1899).

19 Frederic L. Holmes, "The Intake-Output Method in Physiology," *HSPS* 17 (1987): 245–270.

20 Carl Voit, "Die Gesetze der Zersetzungen der stickstoffhaltigen Stoffe im Thierkörper," *Zeitschrift für Biologie* 1 (1865): 69–107; 109–168; 281–314, on p. 92.

In basic digestion trials, after a brief period of fasting, the experimental subject consumed a specified amount of one foodstuff or foodstuff combination. Control samples of the foods were taken prior to weighing, chemically analyzed to determine their components, and burned in a chemical calorimeter to determine their energy value. The subject ate the food and collected its own urine and feces, which were then chemically analyzed and compared with the contents of the consumed food, as projected from the sample. Differences between the two were then evaluated for their significance. Respiration experiments analyzed carbon dioxide production, among other variables, as a measure of metabolism or energy balance. Some experiments were done in a respiration chamber, such as the one used in Rubner's alcohol experiments, or a respiration calorimeter, which was thermally isolated, so that the subject's heat production could be measured as well. Other trials, especially after the turn of the century, used a respiration apparatus, since they were less expensive to build and easier to use than the unwieldy chambers. This apparatus required that the subject breathe through a mask, so that the exhaled carbon dioxide could be captured for analysis. Respiration and digestion experiments could then be combined to study total intake-output balance. Experimenters varied the amounts and kinds of food consumed, the length of experiments, temperature, and humidity, and had the subjects perform various kinds of physical activity, from riding stationary bicycles and lifting weights to singing, typing, doing housework, or playing the piano, to study the effects of these variables on metabolism and calorie needs. The body of the experimental subject was thus conceived as an apparatus for the transformation of chemical compounds and energy. The internal processes of assimilation were not studied. This body was a black box.

As much as the structure of the trials turned bodies into transformation machines, the experimenters depended on the voluntary participation and cooperation of their subjects. Eating is a voluntary activity, and for reasons of modesty and cultural convention, subjects collected their own urine and feces. Physiologists were also aware that any coercion and pressure would skew their results. Appetite, digestion, urination, defecation, and metabolism are affected by anxiety, and, since most experiments aimed to determine normal metabolism and digestion, they required subjects in a normal physical and psychological state, as calm and comfortable as possible. Although physiologists did occasionally conduct experiments with subjects unable to consent, such as an inmate of a psychiatric institution or a patient in catatonic state, their ability to do so successfully was limited by the lack of active cooperation. Thus, while the catatonic patient could be catheterized and her metabolism measured in a respiration chamber, it could not be compared with her waking metabolism.[21] Studies conducted on the diets of residents in a psychiatric institution were compromised by the patients' refusal to allow the collection of their urine.[22]

The problem of agency clearly confronted physiologists at every stage of their experiments. The dance of resistance and accommodation permeated the experimental process, from the pre-experimental stage of subject selection, through the set-up and conduction of the experiment, and post-experimentally into the interpretation of the results. Physiologists struggled to control these unruly objects, to coax them into performing as the experiment required, and to deliver the desired results. They tried to shape their subjects in ways that made them conform as much as possible, adapted their experimental arrangements to meet the demands of their subjects, and modified their experimental procedures to accommodate active and unintentional resistance.

Selecting the Subject

The first strategy of controlling subject agency and potential resistance lay in the selection of the subjects. Nutrition physiologists were very aware of the need to carefully select the individuals used in their experiments. How to find a suitable subject and the difficulties caused by subject resistance and non-compliance during the experiment were recurrent themes in experimental reports and physiological handbooks. Writing in 1911, and able to look back on a history of over fifty years of experimentation in the field, Wilhelm Caspari and Nathan Zuntz began their chapter on metabolism methods in the *Handbuch der physiologischen Methodik* by emphasizing the crucial importance of selecting an appropriate subject.[23] Some selection criteria were basic and pragmatic: Unless the experimental question required a particular age range or health condition, experimental subjects needed to be foremost healthy and adult, that is, in a state of normal equilibrium, neither growing nor in a phase of decline. In addition, they recommended using male rather than female subjects, since due to the anatomical differences, the urine of men could be collected with more precision. Menstruation was also a ground to

21 Robert Tigerstedt, "Das Minimum des Stoffwechsels beim Menschen," *Nordiskt Medicinskt Arkiv* (Festband) 37 (1897): 1–19, on pp. 7–11.

22 Otto Folin and Philip A. Shaffer, "On Phosphate Metabolism," *American Journal of Physiology* 7 (1902): 135–151, on p. 136.

23 Wilhelm Caspari and Nathan Zuntz, "Stoffwechsel," in Robert Tigerstedt (ed.), *Handbuch der physiologischen Methodik, Vol. 1*, (Leipzig: Herzel, 1911), 1–70, on p. 2–3.

avoid female subjects, both because its potential effect on metabolism was unknown and because it could lead to interruptions in the experimental regimen.[24] The healthy, adult male within the normal spectrum for height and weight became nutrition physiology's "standard subject".

If pragmatic considerations, research question, and the need for an at least minimally standardized subject shaped this first selection filter, the necessity of voluntary participation and cooperation placed additional pressures on the selection process. Successful nutrition and metabolism experiments required specific psychological characteristics and behavior from the subjects. Physiologists clearly perceived human subjects as actors with their own interests and preferences and realized that they needed to be recruited as cooperative collaborators. Digestion experiments, in particular, were lengthy and inconvenient, with most conducted over a period of days, and some over weeks or months. Subjects could rarely be interred in laboratories for the duration and most needed to continue with their daily routines and working lives. They could leave or withdraw from the experiment at any time and deviate from the protocol without the experimenter's knowledge. Even basal metabolism studies, which might require only an hour of preparation and measurement, demanded that the subject fast beforehand for a twelve-hour period. Unless the subject spent the night in the laboratory or hospital, his word was the only assurance of compliance.

Given their inability to completely control the actions of the subject, when he was outside the laboratory, the experimenter's vigilance could not compensate for a lack of cooperation and discipline on the part of the subject. As Caspari and Zuntz noted, "one cannot forget that such metabolism experiments place great demands on the conscientiousness, will power, and energy of the experimental individual."[25] Money, they emphasized, was thus not incentive enough to endure the discomfort and inconvenience of the experiment, and one could not expect that a "rented" subject would strictly abide by the occasionally draconian dietary regimen or exercise the necessary care in collecting his excrement. Subjects needed to be reliable, disciplined and focused, capable of self-control and self-denial, able to endure discomfort for the sake of the greater good. In short, an ideal experimental subject embodied late-nineteenth century moral and cultural values. Considering the importance of "good character" for the experiment, researchers turned to similar means for finding suitable subjects as they did for finding any other kind of

24 Caspari and Zuntz, "Stoffwechsel," 3.

25 Caspari and Zuntz, "Stoffwechsel," 3.

service. The workers studied by Swedish physiologists Hultgren and Landergren in their survey of Swedish working men's diets, for example, came with character references.[26] Researchers often turned to subjects they personally knew, such as the employees of their institutes, and experimented repeatedly on the same subject, once an individual had shown himself to be experimentally compliant, dependable, and malleable.

Conducting repeated experiments with the same subject had the additional benefit that the subject not only became familiar with the experimental routine, but could also be taught to play his part with a minimum of additional effort. Experimental subjects were trained to perform to expectations. In many cases they needed to develop skills in order to function "normally" in their interaction with the physiological apparatus. Volker Hess has described this phenomenon in early body temperature measurement, which required that the patient hold his arm tightly to his side for about half an hour, and "depended on patients who practiced and learned such sequences of controlled movements."[27] Metabolism measurements done with a respiration apparatus, rather than in a respiration chamber, made similar demands on the subjects, who needed to learn the skill of mask breathing, so that their respiration and metabolism remained normal despite the extra effort demanded by overcoming the resistance of the tubes or compensating for their plugged noses.[28] Not every reliable subject was therefore a suitable subject or suitable for every kind of experiment.

Reliability and trustworthiness was, however, often difficult to establish with certainty in advance. While experimenters could try to judiciously select their subjects for conscientiousness or their willingness to submit to the experimental protocol, they could not know what their subjects were thinking or how they perceived their role in the process. Subjects often entered the laboratory with their own interests or expectations of the experimenter's interests. They had diverse reasons for their participation, and they had their own interpretations both of what the physiologists wanted from them and of what was

---

26 E. Hultgren and E. Landergren, *Untersuchung über die Ernährung schwedischer Arbeiter bei frei gewählter Kost* (Stockholm: Samson & Wallin, 1891), 4.

27 Volker Hess, "Standardizing Body Temperature: Quantification in Hospitals and Daily Life, 1850–1900," in Gérard Jorland, Annick Opinel, and George Weisz (eds), *Body Counts. Medical Quantification in Historical and Sociological Perspective*, Montreal & Kingston/London/Ithaca: McGill-Queen's University Press 2002, 109–126, on pp. 111–112.

28 Francis Gano Benedict, "The Influence of the Preceding diet on the Respiratory Quotient after Active Digestion has Ceased," *American Journal of Physiology* 17 (1911): 383–405, on p. 383.

in their own best interest. They could over or underreport their own excretions, according to what they thought would satisfy the experimenter, or, in the case of verbal protocols, tell the experimenter what they thought he wanted to hear. When the Yale biochemist Russell Chittenden attempted to prove that contemporary protein and calorie norms were too high and that practicing "physiological economy" would lead to both greater health and more disposable income, he tested his diet on a group of soldiers for several months.[29] A semi-voluntary but nonetheless disciplined group accustomed to obeying orders, the soldiers submitted to the diet and the accompanying exercise program, as well as allowing regular controls of their excretions and physical condition. According to Chittenden, they all reported feeling stronger and healthier than ever before on a diet with about half the protein and considerably fewer calories than described in the dietary norms and provided to them under normal circumstances. According to a contemporary, however, by the end of the experiment they were "restless, nervous, so eager to get back to regular rations that they would say *anything* about their feelings which would tend to bring the experiment to a close...".[30] Anomalies in some results suggest that several of the soldiers did not adhere to the experimental dietary regimen, but supplemented their meager meals elsewhere, and several deserted.[31]

Chittenden's experiment on the soldiers is a fairly straightforward example of the different perceptions of interests and the different goals pursued by experimenter and subject. Interactions could be more complex. The nutrition laboratory was a site in which various alternative nutrition activists and practitioners tried to engage science to further their own agendas. Nutrition physiologists were both intrigued by and wary of vegetarians, professional fasters and hunger artists such as Giovanni Succi, Francesco Cetti, and Agostino Lavanzin, and dietary reformers such as Horace Fletcher and Mikkel Hindhede, who presented their bodies and diets for experimental review or climbed into respiration chambers seeking confirmation of their unique metabolism. Individuals with unusual dietary practices were popular experimental subjects, since the study of normalcy also demanded a study of deviation and variation as reference points. Chittenden's interest in "physiological economy" began when he accepted the offer by the American dietary reformer, Horace Fletcher, of his body as a subject for nutrition and metabolism studies. Known

29 Russell Chittenden, *Physiological Economy in Nutrition. With special reference to the minimal proteid requirements of the healthy man* (New York: Frederick A. Stokes, 1904).

30 Hutchinson Woods, *Health and Common Sense* (London: Cassell and Co., 1909), 19.

31 Chittenden, *Physiological Economy*, 133.

as the "Great Masticator", Fletcher promoted a dietary practice based on extensive chewing. He claimed that thorough chewing allowed the body to better absorb the nutrients of its food, thus reducing the amount of food it needed to consume and the amount of work the body needed to perform for digestion, leaving more energy available for physical activity.[32] If Chittenden was convinced by the performance of Fletcher's body in the apparatus and in the gymnasium, other nutrition researchers were skeptical of his claim to exist and maintain stable body weight on a diet so far below the established norms, and they opened their own laboratory doors to test it. Conducting his own studies on Fletcher, Francis Gano Benedict, director of the Carnegie Nutrition Laboratory, noted that Fletcher ate very little during his time in the respiration calorimeter, but also that "no other subjects in the same kind of experiment have been quite so inactive". In addition, Fletcher's diet, which he had been allowed to choose himself in conformity with his purported nutrition regimen, was so carbohydrate-heavy that the water retention it facilitated could easily have kept his weight stable for the duration of the three day experiment.[33] Consciously or unconsciously, Fletcher had acquired enough experimental experience to manipulate the results in his favor. For Fletcher, this attention from the scientific community was a source of confirmation and a form of advertising. He not only offered his body, but also funded these investigations, including that of Chittenden, and complained bitterly, when these experiments, in which he had starred as the subject, used the standard forms of identification by initials or subject number, rather than his full name.[34] For Fletcher, the knowledge gained from experiments on his body belonged to him and not to the experimenting physiologists, since his body was the site of that knowledge production. In consequence, he claimed, his interpretation of his subjective experience had more credibility than that of the physiologists who could merely quantitatively evaluate data from an outsider's distance.

---

32 Horace Fletcher, *The A.B.-Z. of our Own Nutrition* (New York: F.A. Stokes, 1903); Horace Fletcher, *The New Glutton or Epicure* (London/New York: F.A. Stokes, 1903); Margaret L. Barnett, "Fletcherism. The chew-chew fad of the Edwardian era," in David F. Smith (ed.), *Nutrition in Britain. Science, Scientists and Politics in the Twentieth Century* (London/New York: Routledge, 1997), 6–28.

33 Francis Gano Benedict, "The Nutritive Requirements of the Body," *American Journal of Physiology* 16 (1906): 409–437, on pp. 432–434.

34 Horace Fletcher to Irving Fisher, 26 February 1908, p. 5, Horace Fletcher Papers, Houghton library, Harvard University, MS Am 791 (12); Fletcher to Nathan Zuntz, 25 March 1912, HFP, MS Am 791 (48).

Hunger artists also offered their bodies for physiological study and were welcomed into the laboratories.[35] While few other subjects, including self-experimenting physiologists, were willing to fast for longer than a few days, late nineteenth-century hunger artists were professional showmen and performers, trained and able to fast for periods of up to a month. In exchange for medical care, occasionally considerable financial compensation, and publicity, they allowed nutrition researchers to study on their bodies the physiological effects of prolonged starvation and the physical changes it caused. Engaging in a professional exchange, hunger artists were largely cooperative subjects, submitting to the frequent blood analyses, urine collection, temperature measurements and the collection of other physiological data. Although they presented their bodies as significant sites for the production of knowledge, like Fletcher, they frequently came with their own agendas.[36] Money was certainly one motivation, since they when they were not demonstrating their fasting skills to physiologists and laboratory assistants, they were presenting them to paying publics at fairgrounds and similar locations. Since they were frequently accused of clandestinely eating during their public performances, they used the physiological studies as references, citing both the chemical analyses and the controlled laboratory situation as proof of their fast and the physiologists as credible and respectable witnesses of their feats. Like Fletcher, they insisted that the knowledge produced through their bodies was theirs, and not the physiologists' to interpret.

Although some experimental co-operations between nutrition physiologists and vegetarians, dietary reformers, and hunger artists went smoothly, the diverging interests often created difficulties, if not directly in the laboratory, than in the public arena afterwards.[37] Subjects with so much agency and

35 Curt Lehmann et al., "Untersuchungen an zwei hungernden Menschen," *Archiv für pathologische Anatomie und Physiologie und klinischen Medicin* 13. Folge – 131 Supp. (1893): 1–228; Francis Gano Benedict, *A Study of Prolonged Fasting* (Washington DC: Carnegie Institute of Washington, 1915). Benedict summarizes previous studies with hunger artists on pp. 13–18.

36 Agustí Nieto-Galen, "Mr Giovanni Succi Meets Dr Luigi Luciani in Florence: Hunger Artists and Experimental Physiology in the Late Nineteenth Century," *Social History of Medicine* 28 (2015), 64–81.

37 For one of the more bitter battles see: Anon, "Torture in Starving Test. Levanzin Complains of Treatment at Carnegie Nutrition Laboratory," *New York Times*, 21. May, 1912. Benedict to Robert Woodward, 3 May 1915, Carnegie Institution of Washington Archive, Nutrition Laboratory Box 1, File 15; Agostino Levanzin to Woodward, 12 June 1912, CIW Archive, Box 1, File 15; Woodward to Benedict, 14 June 1912, CIW Archive, Nutrition Laboratory, Box 1, File 14.

purpose were difficult to manage and impossible to effectively control. Interested in avoiding both unintentional and intentional deception, and unable to exert more than limited control over their subjects in these largely non-coercive situations, nutrition physiologists relied heavily on subjects whose interests and background most closely conformed to their own. As Caspari and Zuntz explained, "one does best to select an educated and knowledgeable individual for the experiments, one who understands what is being investigated and is interested in it himself."[38] A knowledgeable subject who identified with the goals of the experiment would be more willing to endure inconvenience and discomfort and subsume his own will and needs to those of the experimenter. Nutrition physiologists recruited subjects from their own circles, frequently using medical students, other physiologists, and themselves, as well as technicians from their laboratories, although, as Rubner's alcohol experiments show, laboratory personnel could also exhibit undesired degrees of agency. Through personal familiarity and their familiarity with the experimental procedures, goals, and apparatus, colleagues were considered more trustworthy and reliable than random individuals. They could be entrusted with fulfilling their role in the experiment with the goals of the experimenter in mind. Benedict emphasized this in 1911, when he described the subjects of one of his experimental series: "All of the subjects were engaged in scientific work in the laboratory, and were thoroughly familiar with the methods of experimenting and the object of the research. Their intelligent co-operation was thus assured."[39] Not a hierarchical relationship with an ignorant subject, cooperative collaboration with subjects who were not only fully informed but also familiar with experimental protocols and methodology offered the best prerequisite for success.

Physiologists also used their own bodies as sites for the production of nutritional knowledge.[40] Enduring inconvenience, discomfort, and indigestion, as well as the occasional health benefit, they were assured that they would stick to the experimental protocol and conduct the experiment as designed. They could also report their physical sensations while subsisting on the experimental diet or using the apparatus, to provide additional information about the effect of the diet or the experience of the apparatus that could not be derived from quantitative data alone. What functioned in calculations of nutrient

38 Caspari and Zuntz, "Stoffwechsel," 3.

39 Benedict, "Influence of the Preceding Diet," 391.

40 Lawrence K. Altman, *Who goes First? The Story of Self-Experimentation in Medicine* (New York: Random House, 1986); Elsie M. Widdowson, "Self-Experimentation in Nutrition Research," *Nutrition Research* 6 (1993): 1–17.

content and consumption needs, did not always function in practice. Bodies had agency of their own.

The Agency and Resistance of Experimental Bodies

While the subject's primary active role in the experiment was to consume the experimental food and collect its excretions, between intake and output stood a body, and neither experimenter nor subject could completely control this body and its reactions. Dealing with these bodies and their limitations forced the experimenters to develop a variety of strategies designed to both overcome and accommodate resistance. They used this resistance to generate new knowledge about physiological needs and functions, as well as to refine their experimental methods.

Experimenting with human subjects required changes in the apparatus. Although simple digestion experiments needed no further apparatus than the digesting body, experiments in respiration chambers required a number of modifications. The basic apparatus and equipment of respiration experiments has been described earlier. A respiration chamber was largely an airtight box fed by a ventilation system, with numerous attached measurement devises. This was all the experimenter needed to conduct the experiment, but the human subjects of these experiments had a variety of additional needs. Chamber respiration experiments lasted hours, and in the early phase, twenty-four hours with sleeping and waking phases were considered necessary. In addition, canine chambers were small, since it was considered desirable to limit the movement of the animals, so that they would not expend energy in motion, which could not be measured by the instruments. By contrast human beings could not be kept immobile in virtual darkness for even a few hours, so human experimental respiration chambers were the size of a small room. Size and ventilation were interconnected. The chamber had to be large enough that the experimental subject was not disturbed by ventilation drafts.[41] Large chambers also made temperature and humidity control more complex, and windows added a source of error. A result of this complexity was that, while many researchers built smaller apparatus for animals, few had the skills or funds to build human respiration chambers and conduct experiments on this scale.

Human respiration chamber experiments also had to make concessions to the very social nature of this body. For a twenty-four hour period, a human

41 Max Pettenkofer, "Ueber die Respiration," in *Annalen der Chemie und Pharmacie*, Supp 2, pt. 1 (1862): 1–52, on pp. 7–8.

subject required a number of things for his or her comfort, and the experimenter had to adjust his experiment to accommodate them. In addition to enough space to move, the subject needed material objects such as a table, chair, and bed, sheets, blankets, clothing, and a nightshirt, lighting, and objects to keep him occupied – books, perhaps for a student, tools for an artisan or technician. All of these objects entered the chamber and became part of the experimental set-up, but they all could potentially affect the results.[42] Paper and textiles absorb water, so they needed to be weighed before they were brought into the chamber and again, when they left it. Metals absorb heat and might affect the temperature measurements. Their heat capacity had to be determined before they were brought into the apparatus. The light source and devices for communication between experimenter and subject could also affect the measurement and needed to be evaluated. The needs of the subject forced the experimenter to develop techniques to integrate these changes into their measurements and experiments. Although these changes added layers of complication and uncertainty to the experiment and introduced more potential for error, accommodation was, in this case, a largely technical problem and it had a technical solution in the addition of tedious measurement routines.

More difficult to manage was the resistance offered by bodies. Nutrition experiments depended upon the functioning of the digestive machine, and they frequently pushed it to its limits. A robust digestion was mandatory, but even willing subjects were sometimes unable to perform as the experimenter desired. Part of this physiological resistance arose from the experimental conditions of specific research questions, especially in digestion experiments. In order to conduct a coherent experiment, it is necessary to reduce the variables as far as possible and, ideally, to test only one variable at a time. By testing the variables individually, the experimenter can isolate the effects of each one on a particular process and test for correlations between multiple variables. While this procedure is suitable for physics and chemistry, the models for nineteenth-century physiology, translated into digestion experiments it meant feeding the experimental subject one food or food combination in an amount sufficient to cover daily needs for a period of days.

Early nutrition physiologists, who had studied dogs, faired very well with this method, since the dogs eagerly consumed any amount of meat provided for weeks on end and were similarly willing to subsist on starch and fat balls, brown bread, and other simple diets for extended periods. Agricultural researchers designed livestock feeding studies in a similar manner, testing one kind of fodder

42 Max Pettenkofer and Carl Voit, "Untersuchungen über den Stoffverbrauch des normalen Menschen," *Zeitschrift für Biologie* 2 (1866): 459–573, on pp. 473–474.

at a time. When physiologists attempted to apply this approach in human experimentation, both subjects and bodies rebelled. Physiologists were already aware that some subjects experienced abdominal distress when fed large quantities of milk. When studying the digestion of cheese with its high fat content, Rubner was unable to find an experimental subject able to consume sufficient quantities to cover his daily needs for even one day and was forced to alter the experiment to include milk.[43] A pure meat diet presented similar problems. The self-experimenting physiologist Johannes Ranke described in 1862 his attempt to live on lean meat alone.[44] Already on the first day of the experiment, his body resisted, and he was unable to fully consume the nearly two kilograms of meat that would have fulfilled his daily nutritional requirements of protein and carbon according to the contemporary dietary norms. Feelings of disgust and repugnance combined with indigestion and fatigue. With continuing experimentation, he complained of insomnia, sweating, abdominal discomfort, nausea, headaches, constipation, severe thirst, and retching, when he drank. Losing weight, he concluded that the human body was simply physiologically incapable of consuming enough pure meat to maintain itself.

Even experiments with a potato diet met with swift resistance. The subject in this experiment, a Bavarian soldier, had been carefully selected for his ability to consume large quantities of potatoes, since this was his normal diet. Although Rubner had originally planned to test this food in the simplest form possible, as boiled potatoes, he quickly realized that the subject would be unable to consume enough of this bland food, and instead altered the experimental procedure to include salt and butter, fried the potatoes, and made potato salad. In order to consume the three kilograms of potatoes his daily needs required, the experimental subject ate for most of the day and excreted both day and night.[45] Similar difficulties arose with most foodstuffs, and few experiments could be conducted for more than four or five days, while some only lasted a day, the minimum period determined for a metabolic cycle. Rubner named this inability of the human subject to consume sufficient amounts of a single foodstuff over the course of a longer experiment one of the most significant problems facing experimenters in nutrition physiology, attributing it primarily to a lack of will. With palpable frustration, he complained,

---

43 Rubner, "Ausnűtzung," 136.

44 Johannes Ranke, "Kohlenstoff- und Stickstoff-Ausscheidungen des ruhenden Menschen," *Archiv für Anatomie und Physiologie* (1862): 311–380.

45 Rubner, "Ausnűtzung," 146–149.

> Most people are so pampered that they cannot tolerate consuming only one foodstuff prepared in the same manner over a longer period. They become revolted by the monotonous fare and even experience intestinal disturbances and diarrhea. One would think that one could easily live for several days from rice, grain, bread, potatoes, milk, eggs, meat etc. alone, since entire peoples live almost exclusively from one of these foods. But this is misleading. Most people who are accustomed to our varied fare consume such foods with aversion after only a few meals, so that the experiment can only be continued for few days. Rarely does one encounter an individual accustomed to a simple and frugal fare, who can enjoy large quantities of the food for a somewhat longer duration.[46]

Although Rubner seems to find fault with the experimental subjects, who did not have the necessary will-power to adhere to the experimental regimen, most physiologists drew different conclusions. They interpreted this inability to consume large amounts of a single foodstuff as proof that the varied mixed-diet was physiologically necessary and that variety stimulated the appetite and assisted digestion. With these difficulties in mind, however, nutrition physiologists were particularly fascinated by individuals who showed a tolerance for such extreme diets, such as the Arctic explorer Vilhjalmur Stefansson, who claimed to be able to live on a meat diet for a year,[47] and the Inuit, who they assumed also subsisted on meat.

These unruly bodies created additional difficulties for the kinds of experiments that nutrition physiologists aimed to conduct. Unlike a machine, the body could not be turned on at the beginning of the experiment and turned off, when it was over. And unlike many objects of nineteenth-century physics and chemistry, it did not have a static or latent state, out of which it could be activated at a specific point in time. Every nutrition experiment began in the middle of ongoing physiological processes, and the effect of these processes on the experiment and its results could not be entirely predicted. Nutrition physiologists needed to develop techniques and strategies for controlling and compensating for these processes as much as possible. Selecting primarily healthy, adult, male subjects within a normal weight and height range, was one way to exclude the possible effects of illness or physical anomalies before the experiment began, but it could not compensate for the particularities of individual digestion.

---

46 Rubner, "Ausnűtzung," 120.

47 Graham Lusk to Max Rubner, 7 January 1927, Max Rubner Papers, Archive of the Max Planck Society, Abt. III, Rep. 8, Nr. 59/8/21.

Experiments in early nutrition physiology were primarily concerned with establishing intake-output relationships. They studied the various outputs of the body, compared them to the inputs and established relationships between the two. The outputs – urine and feces – had to be collected as carefully and completely as possible, and numerous techniques were developed to enable this, both in animals and in humans. What physiologists could not do, however, was to look inside the black box of the body to observe intake becoming output, and they could thus not verify the intake-output relationship with certainty. Unlike built machines, bodies and digestive systems do not function with predictable, clockwork-like regularity. An experimenter could not be certain that the feces he was analyzing was a result of the food that the subject had consumed during the experiment, that it did not include food the subject had consumed on previous days, and that there was no more food from the experiment still inside the body.

Nutrition physiologists devised a variety of techniques in their attempts to solve this problem. One way of increasing control of the process was to try to make the body as much like the static object of inorganic science as possible, that is, to approximate turning it off, or, at least, turning off the digestive process. To do this, experimenters had their subjects fast for a day before experiments and used enemas to try to empty the intestines of pre-experimental waste. Although these procedures increased the likelihood that output would reflect experimental input, it still contained much uncertainty, and these methods were not suitable for all experiments or experimental subjects.

In the 1860s, Carl Voit, one of the first physiologists to focus on nutrition research, and a crucial figure in the development of its experimental methods, introduced a technique for determining what excrement in the continuing digestive process was a product of the experimental food intake.[48] The technique, which was universally adopted by nutrition physiologists, exploited the phenomenon that digestion is incomplete and that different kinds of foods lead to excrement that differs in color and consistency. Experimenting on the metabolism of dogs, Voit established that the excrement from a pure bread diet was brown, that of a meat diet black, and from bone feeding white and crumbly. If a dog was given bones to eat twelve to twenty-four hours before and after a meat feeding experiment, the excrement from the bone feedings would form a white parenthesis around the dark excrement from the meat, thus providing a clear visual marker of the experimental output. The experimenter

48 Carl Voit, *Beiträge zum Kreislauf des Stickstoffs im thierischen Organismus* (Augsburg, 1857), 19.

could then carefully separate the feces from the experiment out of the continuum of ongoing digestive processes. The feces separation technique externalized hidden internal processes.

Finding the best marker substance and separation techniques for human subjects was indeed so crucial for successful nutrition experimentation that Rubner, at the time a pupil of Voit's, devoted the first five pages of his dissertation to his exploration of this question.[49] The ideal marker substance was one that produced feces with a unique and characteristic color and consistency and was completely indigestible itself, to reduce the possibility that nutrients from the marker material would contaminate the experiment feces. Substances such as cranberries, which had indigestible skins, were not sufficiently precise, since they did not pass evenly through the intestines, and other indigestible substances could cause abdominal distress. As an alternative, physiologists turned to specific foodstuffs. Rubner discussed the characteristic excrement from several foods and their potential use as a marker material before concluding that for most experiments milk provided the best results. Milk excrement had, he wrote, a light color, a form reminiscent of an ear of corn, and the consistency of soap.[50] Marker material and digestion had to be carefully calibrated. Unfortunately, Rubner did not leave a record of the experiments he conducted to establish that milk best fulfilled its function as a marker material if two liters were given to the subject sixteen to twenty-four hours before the first and twenty-one hours after the last experimental food intake. In order to facilitate separation, the experimental subject was instructed to evacuate his bowels on a long tray, which he drew out behind himself, so that the straight line of feces visualized the order in which the marker material and food material were consumed and digested.[51] This feces separation technique never became as exact a method as the physiologists had hoped, although they explored various other marker substances and coloring agents such as silica and lampblack. Nonetheless, the use of marker substances and the "preservation of intestinal order" collection technique provided them with a means to overcome the body's resistance and to trace the experimental intake through the unobservable realm of the digestive system. What the subjects thought of these techniques is unknown.

---

49 Rubner, "Ausnűtzung," 115–120.

50 Rubner, "Austnűtzung," 119.

51 In addition to Rubner, "Ausnűtzung," 120, see: Traugott Cramer, "Die Ernährungsweise der sog. Vegetarier, vom physiologischen Standpunkte aus betrachtet," *Zeitschrift für physiologische Chemie* 6 (1882): 346–384, on pp. 354–355.

Conclusion

In many ways, the history of human experiments in early nutrition physiology can be read as comedy. Physiologists struggled to get their unruly objects to eat and excrete, to remain still in the respiration chamber during basal metabolism studies, behave normally in the artificial environment of the nutrition laboratory, and become the passive, obediently functioning and performing objects that the experimental models required. This asymmetrical model, which, as Pickering has elsewhere shown, does not function without friction with inorganic objects, required even more adjustment in the case of organic subjects, whose personal agency affected every part of the experimental set-up.

It is important to note that the particular experimental model used in late nineteenth-century nutrition physiology was a product of changes in experimental approaches in the nineteenth century in general, and of changes in approaches to the study of living bodies in the second half of the nineteenth century, in particular. Over the course of a few decades, physiology had separated itself from the study of anatomy, which looked largely at static structures, to define itself as the study of biological processes, while medical research similarly moved from observations in hospital settings to manipulation in the laboratory.[52] These changes introduced new methods and new kinds of subjects into physiological and medical research. Described as the "laboratory revolution in medicine" by Cunningham and Williams and as the "experimentalization of life" by Hans-Jörg Rheinberger and Michael Hagener,[53] they brought the experimental approaches of nineteenth-century inorganic sciences to the study of living creatures. Researchers in the life sciences sought to realize in their experiments the quantification, precision measurement, instruments and apparatus, single variable experiments, isolation of phenomena, experimental systems, and use of standard objects that characterized the exact sciences of the non-living world. These methods and devices were obviously compelling. These experimental models shaped physiologists' approaches to life phenomena, the kinds of questions they could ask, and the ways they tried to interpret their results. All required modification.

---

52 Andrew Cunningham and Perry Williams, "Introduction," in Andrew Cunningham and Perry Williams (eds), *The Laboratory Revolution in Medicine* (Cambridge: CUP, 1992), 1–13, on p. 3.

53 Cunningham and Williams, *Laboratory Revolution*; Hans-Jörg Rheinberger and Michael Hagener (eds), *Die Experimentalisierung des Lebens. Experimentalsystem in den biologischen Wissenschaften 1850–1950* (Berlin: Akademie Verlag, 1993).

While anatomical studies were conducted on dead bodies – passive objects composed of organic materials –, much physiological and medical experimentation required living beings. Nineteenth-century physiologists were, of course, not the first to experiment on living humans and animals, but in the later part of the century, this experimentation became increasingly systematic and routine, a necessary component of research, rather than an occasional or opportunistic one. In particular, when experimenting on human beings, physiologists faced different limitations than physicists and chemists. Leaving aside the ethical and social limits on the experimental manipulation of human beings, which most, but clearly not all, physiologists and medical researchers internalized, both the subjects and the physiological systems of their bodies imposed limits on what any experimenter could do. Human subjects were actors and bodies were complex systems that resisted control.

This chapter has looked two different kinds of subject agency in nutrition experiments, the agency of intentionally acting human beings, and the agency of bodies. Subjects made their needs known, chose to comply with experimental protocols or secretly break them, and claimed ownership of the knowledge generated by their bodies. Physiologists needed to sooth their anxieties and attend to their comfort, modify the conditions of the experiment and teach them the skills they, the subjects, needed to co-conduct it. While the structure of the experiment remained hierarchical, the structure of the relationship required cooperation and concessions from both parties, if the experiment was to succeed to the physiologist's satisfaction. Subjects and experimenters entered into a voluntary relationship, with both negotiating the terms of their engagement and not always agreeing. Managing subject agency was such an important problem for experimental physiologists that a very significant proportion of their subjects were recruited from the experimenter's immediate environment. Laboratory assistants, medical students, and colleagues were not only familiar with experimental protocols. They also identified with the goal of the experiment or, at least, understood the goals of experimentation in general, and were thus more likely to share the interests of the experimenter and take an interest in their own contribution to the experiment's success. More than cooperators, such subjects could become collaborators, reading instruments in the chamber, contributing detailed verbal protocols, compiling written reports of their participation, and finding themselves named and thanked in the acknowledgements, along with the support staff and assistants.[54]

---

54 Wilbur Olin Atwater, and Francis Gano Benedict, *Experiments on the metabolism of matter and energy in the human body* (Washington, DC: US Government Printing Office, 1903), 3; Francis Gano Benedict and Edward P. Cathcart, *Muscular work. A metabolic study with*

Bodies also exhibited agency of a kind that is perhaps closer to Pickering's description of "material agency", in that it was not a result of intentional action, but of manipulations undertaken on a physiological system. Even when experimental subjects were cooperative, bodies often resisted the experimental conditions, whether these were the bodies of humans or of animals.[55] In the human experiments described here, the bodies resisted the single variable experiments that best suited the experimenters and their methods. Consuming large quantities of single foodstuffs pushed them to their physiological limits, leading to digestive disorders, loss of appetite, and an end to the experiment. Physiologists were forced to develop new techniques to harness the particularities of digestive processes for experimental purposes, again relying on their subjects to violate cultural norms of modesty and hygiene to assist them.

The relationship of authority and submission between experimenter and experimental subject is certainly a fundamental one in the history of human experimentation. It is not the only possible relationship, however, and one that only functions in particular kinds of experiments. In others, cooperation and collaboration, communication of knowledge and identification with experimental goals, voluntary participation and concessions, resistance and accommodation, characterize a different kind of relationship, with both experimenter and subject acting as agents. A history of human experimentation must thus consider not only the different kinds of experiments that were conducted, but also the varied and often complex relationships that emerged.

---

special reference to the efficiency of the human body as a machine (Washington DC: Carnegie Institution of Washington, 1913), 111.

55 For examples of resistance and accommodation of the animal body in nutrition experiments see Theodor Bischoff and Carl Voit, *Die Gesetze der Ernährung des Fleischfressers durch neue Untersuchungen festestellt*, (Leipzig: C.F. Winter, 1860); Caspari and Zuntz, "Stoffwechsel".

CHAPTER 7

Experimenting with Radium Therapy: In the Laboratory & the Clinic

Katherine Zwicker

In 1932 a leading American cancer specialist, physician Francis Carter Wood, addressed the New York Academy of Medicine about the state of cancer therapy and research.[1] For the treatment of cancer, he said, surgery and radiation were the only effective options. Though considered effective, Wood spoke of the need for research aimed at improving the practices of both surgery and radiation therapy. He noted a particular need for animal experimentation conducted by scientists in laboratories since, as he stated, "human patients bearing cancer cannot be experimented upon. They must be given the best treatment known to science."[2] As Wood stood before his colleagues, he painted a picture in which cancer patients and the treatment they received benefitted from scientific experimentation, but were separate from it. On the contrary, the development of radium therapy as a cancer treatment in the United States from the mid-1910s to the mid-1920s reveals that physicians' efforts to provide their patients with "the best treatment known to science" meant that patients were not, in fact, separate from scientific experimentation. Rather, as physicians worked with scientists – mostly physicists – to create technologically sophisticated cancer therapies, they blurred the practices of clinical therapy and scientific experimentation. As a result, patients became objects of study. They served as experimental subjects that helped advance a process through which physicians and physicists collaborated in an effort to establish radium therapy as an effective and scientific cancer treatment. The collaborative work of physicians and physicists also fuelled the development of biomedical science.

While this paper draws attention to the dual role of cancer patients as both patients and subjects, its primary focus is on the relationship established between physicians and physicists. Physicians who sought to exploit the therapeutic value they believed existed in radium enlisted the help of physicists to

1 Francis Carter Wood, "The Wesley M. Carter Lecture: Fundamental Research in Cancer," *Bulletin of the New York Academy of Medicine*, vol. 8, no. 11 (November 1932): 653–657.

2 Wood, "Fundamental Research in Cancer," 659.

 | DOI 10.1163/9789004286719_009

develop radium therapy in the tradition of scientific medicine. The proliferation of scientific medicine in the nineteenth century had, as historian William Bynum argues, a transformative effect on the medical profession. For Bynum, the development of research-based medical knowledge was integral to the modernization and professionalization of medicine.[3] Bacteriology, for instance, informed public health measures that sufficiently reduced the occurrence of infectious disease. Contributions from chemistry, such as anaesthesia and antiseptics, had a profound impact on surgery.[4] These and other advances in medical research and practice bolstered the reputation of physicians, whether they were personally engaged in scientific research or not. Hoping to further the development of scientific medicine and medical professionalization, many physicians pushed for reform within medical education in the late nineteenth century. They sought to incorporate more training in pre-clinical or biomedical sciences in a concerted effort to make clinical practices more scientifically informed.[5]

The transformative effects of scientific medicine were also evident in the evolution of hospitals. Hospitals were once considered the treatment centers for the poorest members of society – those who could not afford to pay for a visit from a doctor.[6] By the twentieth century, the stigma associated with hospital-based health care was rapidly disappearing. Hospitals had become the locus of medical education and were often the only site at which patients could access sophisticated technologies. In his examination of medical technologies, historian Joel Howell suggests that the greater use of technology within hospitals was a key means through which physicians self-consciously

3 Bynum's study is not exclusive to the United States, though his argument about the role of science in transforming medicine is the same for the three regions he studies: the United States, the United Kingdom, and Germany. See, William F. Bynum, *Science and the Practice of Medicine in the Nineteenth Century* (New York: Cambridge University Press, 1994); see also, Harry M. Marks, *The Progress of Experiment: Science and Therapeutic Reform in the United States, 1900–1990* (New York: Cambridge University Press, 1997). Like Bynum, Marks attributes medicine's modernity to its reliance on the physical and biological sciences.

4 Bynum, *Science and the Practice of Medicine*, 91–141.

5 The terms "biomedical" or "biomedicine" did not come into being until the twentieth century. The development of biomedical science is examined thoroughly in Robert E. Kohler, *From Medical Chemistry to Biochemistry: The Making of a Biomedical Discipline* (New York: Cambridge University Press, 1982). On the role of biomedical science in medical education see, Kohler and George Weisz, *Divide and Conquer: A Comparative History of Medical Specialization* (New York: Oxford University Press, 2006).

6 Bynum, *Science and the Practice of Medicine*, 25–54.

branded hospital-based medicine as scientific.[7] As this paper and its focus on radium therapy will illustrate, the hospital was very much a site at which the links between science, medicine, and technology were strengthened. The development of radium as a new therapeutic technology necessitated the assistance of scientists whose participation in clinical medicine compelled the creation of hospital-based laboratories.

Physicians adopted the concept of rational medicine to characterize their commitment to scientific medicine or, more explicitly, their efforts to ground clinical practice in the knowledge derived from laboratory experimentation.[8] The term, which was used interchangeably with scientific or experimental medicine, was peppered throughout the medical literature of the time. For radium therapy pioneers, the ideal of rational medicine required the pursuit of laboratory experiments related to cancer and the therapeutic action of radium upon it. Cancer pathologist James Ewing demonstrated his adherence to this ideal in a 1922 paper in which he argued that the rationalization of radium therapy depended on "a thorough study of the mechanism by which physical agents affect neoplastic disease."[9]

Historian and radiation oncologist Charles Hayter argues that, rather than representing the rise of rational medicine in which laboratory research preceded clinical application, radium therapy developed in the tradition of empirical medicine. That is, physicians used radium as a therapeutic tool and assessed its merits based on the progress of their patients' recoveries. According to Hayter, scientific research that investigated radium's effects on tissues and disease came later.[10] Physicians were quick to apply radium in medicine soon after it had been discovered and research did, indeed, follow. The rapid development of radium therapy was an important stimulus for scientific research

7 Joel D. Howell, *Technology in the Hospital: Transforming Patient Care in the Early Twentieth Century* (Baltimore, MD: The Johns Hopkins University Press, 1995).

8 Historian Harry Marks provides a clear articulation of rational medicine as being physicians' aim to, first, obtain knowledge in the laboratory and, only then, apply it at the bedside. See, Marks, *The Progress of Experiment*.

9 James Ewing, "The Mode of Radiation Upon Carcinoma," *The American Journal of Roentgenology*, vol. 9, no. 6 (June 1922): 336.

10 Hayter's study examines radium therapy as a response to cancer in Canada. While focusing on Canada, his discussion on the discovery of radium, early application of radium in medicine, and related research encompasses the United States as well. Thus, his study is relevant here. See, Charles Hayter, *Radium and the Response to Cancer in Canada, 1900–1940* (Montreal, QC: McGill-Queens University Press, 2005), 21–25; and Hayter, "The Clinic as Laboratory: The Case of Radiation Therapy, 1896–1920," *Bulletin of the History of Medicine*, vol. 72, no. 4 (1998): 663–688.

and helped trigger the emergence of new fields of biomedical research. However, the clinical development of radium therapy and scientific research associated with it, were not entirely separate endeavors. By the time that radium was reasonably accessible in the 1910s, the clinical use and scientific study of radium therapy were blended. They were blended in practice, in the collaboration that took place between physicians and physicists, and in institutional space as physicists worked alongside physicians in hospitals.

Hospital-based scientists conducted animal experimentation and other laboratory work in parallel to the therapeutic use of radium, yet scientists and physicians greatly valued knowledge obtained from therapy delivered to humans. This contributed to the dual role of patient-subjects as physicians delivered therapy to patients that was carefully planned to both treat the patient and yield desired knowledge. Patients' subject status was masked, though, by an emerging narrative that conveyed the positive effects of science and technology in medicine. The discovery and application of radiation in medicine benefitted from and helped to establish that narrative. Professional physicians increasingly framed radiation-based therapeutics as a form of advanced medicine delivered by skilled professionals whose credentials were enhanced and interconnected with the work of scientists.

This is a history in which physicians' and physicists' paths of professionalization continually crossed. Physicians introduced elements of experimentation into their clinical practice as a means of refining cancer therapy. Furthermore, they worked with physicists to achieve a scientific understanding of radium's physiological effects and to carefully design and control the instruments they used for measuring and delivering doses of radium to patients. Physicists readily participated in the clinical use of radium therapy and built research laboratories within hospitals to support this endeavor. Their involvement in medicine stemmed from a burgeoning field of physical research on radioactivity and the atom. The collaborative work and intersecting professional identities of physicians and physicists – both of which reflected each group's commitment to scientific medicine – contributed significantly to the reframing of patients as subjects even though neither group explicitly described patients as such.

Attracting the Attention of Americans: Radium & Cancer in the Early Twentieth Century

In 1898, French physicists Marie and Pierre Curie discovered a naturally occurring radioactive element they called radium. This came just a few years after Wilhelm Roentgen's discovery of X-rays and Henri Becquerel's discovery that uranium emits radiation. Together, these breakthroughs were lauded within the scientific community and opened many avenues for scientific research in what became known as the "new physics."[11] The discovery of radioactivity also helped usher in a new era of clinical practice and biomedical research. While physicists were most interested in understanding the properties of radiation and the atom, physicians were keen to develop radiation as a diagnostic and therapeutic tool. X-rays were known to reveal images of the internal structures in the body and radium seemed an effective means to destroy unwanted tissue. This understanding of radium was derived from the experiences of scientists who had worked with radium and suffered burns or loss of hair due to radiation exposure. In 1903, Francis H. York of Boston and Robert Abbe of New York were the first North American doctors to obtain and use radium therapeutically.[12] Their interest in radium, like physicians elsewhere, was rooted in its physically destructive power.

Physicians were most hopeful that radium could be effectively deployed as a means of removing cancerous tumors. The prospect of using radium as a cancer therapy was particularly appealing since cancer had long been considered a scourge on society. The ominous reputation of cancer was informed by the extreme suffering of those it afflicted, its potential to strike individuals indiscriminately, and the perception that deaths caused by cancer were increasing.[13] The heightened fear of cancer was, in part, a reflection of the late-nineteenth-century decline in deaths caused by infectious disease. As the medical community gained control over infectious disease, cancer loomed as a problem for which they had insufficient solutions.[14] Indeed, the prognosis for most types of cancer was very bleak. Surgery was the only treatment option and was only marginally successful, not to mention, mutilating. Given these circumstances, individuals often chose not to seek medical attention when they developed tumors they suspected to be cancerous. They believed that since surgery would likely not result in a cure, it was better to avoid going through a traumatic treatment experience.

---

11 R.A. Millikan, "The Significance of Radium," *Science*, vol. 54, no. 1383 (July 1921): 1–8; and Daniel J. Kevles, *The Physicists: The History of a Scientific Community in Modern America* (Cambridge, MA: Harvard University Press, 1995), 75–90.

12 Hayter, *Radium and the Response to Cancer*, 12.

13 For an excellent examination of the history of cancer culture in the United States, see James Patterson, *The Dread Disease: Cancer and Modern American Culture* (Cambridge, Mass.: Harvard University Press, 1987). See also, Siddhartha Mukherjee, *The Emperor of All Maladies: A Biography of Cancer* (New York: Scribner, 2010).

14 Charles Hayter, "Cancer: "The Worst Scourge of Civilized Mankind," *Canadian Bulletin of Medical History*, vol. 20, no. 2 (2003): 251–263; and Wood, "Fundamental Research in Cancer," 654–655.

Amidst this culture of hopelessness, radium therapy seemed a possible, if not favorable, alternative to the surgical removal of tumors.[15] Accordingly, the development of radium therapy helped strengthen the medical profession's and the public's resolve to fight cancer. Renewed efforts to control cancer were, perhaps, best symbolized by the 1913 formation of the American Society for the Control of Cancer (ASCC), which led a public education campaign that aimed to overturn Americans' resistance to seeking cancer care.[16] Representing the medical community, the ASCC – later renamed the American Cancer Society – publicized the importance of early diagnosis as well as advances in medical science and technology as key to the successful treatment of cancer.[17]

While cancer was the primary focus of most physicians and physicists interested in developing radium as a therapeutic agent, it was not the only illness for which radium offered promising applications. According to physician-scientist, Dr. John Gofman, cancer "was only part of the story."[18] Gofman, whose career as a biomedical scientist spanned the middle decades of the twentieth century, reviewed the earliest applications of radium in medicine and reported that for "every disease you can think of, there is a paper."[19] Historian Claudia Clark drew the same conclusion arguing that, within the first few years of the twentieth century, X-rays and radium were being tested for the treatment of a variety of ailments.[20] Aside from cancerous tumors, radium was often used to perform tonsillectomies and to treat dermatological conditions.[21]

---

15 Hayter, *Radium and the Response to Cancer*, 13.

16 For a comprehensive history of the American Society for the Control of Cancer – now the American Cancer Society – see Walter S. Ross, *Crusade: The Official History of the American Cancer Society* (New York: Arbor House, 1987); and David Cantor, "Introduction: Cancer Control and Prevention in the Twentieth Century," *Bulletin of the History of Medicine*, vol. 81, no. 1 (Spring 2007): 1–38.

17 For an examination of the controversies that developed amidst public education campaigns, see, David Cantor, "Uncertain Enthusiasm: The American Cancer Society, Public Education, and the Problems of the Movie, 1921–1960," *Bulletin of the History of Medicine*, vol. 81, no. 1 (Spring 2007): 39–69.

18 Oral History of John W. Gofman, M.D./Ph.D., interview conducted December 20, 1994, as part of the Office of Human Radiation Experiments' Oral History Project, *Human Radiation Studies: Remembering the Early Years*. Conducted by Loretta Hefner and Karoline Gourley in San Francisco, CA, n.p.

19 Oral History of John Gofman, 1994, n.p.

20 Claudia Clark, *Radium Girls: Women and Industrial Health Reform, 1910–1935* (Chapel Hill, NC: University of North Carolina Press, 1997), 43.

21 Though not examined here, radium water gained increasing popularity within the general population and medical community throughout the 1910s and 1920s. It was used in internal medicine or as a health tonic and gained a reputation amongst the general public

Medical Motivation for Securing a Supply

Optimism about radium in medicine abounded, but there were obstacles to overcome in terms of obtaining radium and establishing a scientific understanding of its therapeutic powers. First, radium is not an element that occurs in nature in great abundance and the cost of mining ore from which radium could be extracted was extremely high. For American physicians it was particularly difficult to obtain radium. From the time it was discovered in 1898 until the early 1910s, there were no American refineries processing radium. Thus, throughout that period Americans had to buy their radium from refineries in Paris, Vienna, and Germany and they did so at an extraordinary cost of $120,000 to $180,000/gram. As of 1914, there were just two grams of radium in the Unites States, an amount that the Secretary of the Interior reported could only adequately supply a few surgeons.[22]

Throughout the 1910s, the federal government, specifically the Secretary of the Interior and the Bureau of Mines, became very interested in securing a domestic supply of radium.[23] The government's actions aimed to address both the expense of radium and the disruption in the European supply caused by the outbreak of World War I. Secretary of the Interior Franklin Lane crafted a national policy that outlined the government's intentions to survey public lands for radium-rich ore, to extract the ore, and to establish refineries to process it. The Bureau of Mines oversaw much of this work.[24] Individuals from industry who were interested in radium's fluorescent properties welcomed the

---

as a magic cure. See, Clark, *Radium Girls*, 43–64; and Aimee Slaughter, "Harnessing the Modern Miracle: Physicists, Physicians, and the Making of American Radium Therapy," (Ph.D. Dissertation, University of Minnesota, Minneapolis, MN, forthcoming).

22 N.A. "The Radium Resources of the United States," *Science*, vol. 39, no. 993 (Jan. 1914): 60; and N.A. "The Production of Radium in Colorado," *Science*, vol. 42, no. 1075 (Aug. 1915): 184–185.

23 The federal government's involvement in radium mining marks an early-twentieth-century example of its efforts to advance cancer care. Though the government did not formally launch a "war on cancer" until 1971, it made other strides to develop or improve the availability of new cancer therapies involving radium and, later, radioisotopes. See, David Cantor, "Radium and the Origins of the National Cancer Institute," in *Biomedicine in the Twentieth Century: Practice, Policies, and Politics*, ed. Caroline Hannaway (Washington, D.C.: IOS Press, 2008), 95–146; and Angela N.H. Creager, "The Industrialization of Radioisotopes by the U.S. Atomic Energy Commission" in *The Science-Industry Nexus: History, Policy, Implications*, eds. Karl Grandin, Nina Wormbs, Sven Widmalm (Sagamore Beach, MA: Science History Publications, 2004), 141–167.

24 N.A. "Radium Resources," 60–61; and N.A. "Radium in Colorado," 184–185.

government's involvement in radium mining. So too, did the medical and scientific communities. Physicians and scientists were very vocal about the need for radium in medical practice and scientific research. In fact, two physicians – Drs. Howard Kelly and James Douglas – successfully lobbied the United States Congress to stimulate radium production. They negotiated an agreement with the federal government that committed the Bureau of Mines to mining ore. In return, Kelly and Douglas endowed a refinery that was incorporated as the National Radium Institute in 1913.[25]

Underlying the negotiations between Kelly, Douglas, and the federal government was a widespread notion that radium, while hard to obtain, was invaluable to medicine. The medical or humanitarian value of radium publicized by the medical community was endorsed in frequent statements and reports released by the Secretary of the Interior and the Bureau of Mines.[26] Due to the initiatives of the medical community, federal government, and – increasingly – private companies throughout the 1910s, the United States' radium industry became the most productive in the world. Commercial success helped drive the industry, but the dominant rhetoric espoused by private companies continued to focus on the medical uses of radium and the necessity of ensuring a stable supply. For instance, when American radium companies faced stiff competition following the discovery of radium-rich ores in the Belgian Congo, the American government opted not to impose tariffs on imported radium. As the Eastern Manager of an American company, the Radium Company of Colorado, reported, "The world needs radium to combat cancer and it was recognized that the lower the price the more widespread would be the use of this valuable agent."[27]

The narrative crafted by physicians, scientists, government and radium mining industry representatives exuded confidence in radium as a cure for cancer. As historian Matthew Lavine has persuasively argued, Americans quickly developed a fascination or obsession with the curative powers of radiation in the early decades of the century.[28] The confidence in radiation was necessary

25 N.A. "Radium Resources," 60–61; and N.A. "Radium in Colorado," 184–185.

26 Statements from the Secretary of the Interior as well as reports produced by the Bureau of Mines on the extraction and refinement of radium-containing ores were published regularly in the journal *Science* throughout the 1910s and 1920s. See, for example, N.A. "Radium Resources," 60–61; N.A. "Radium in Colorado," 184–185; and H.E. Bishop, "The Present Situation in the Radium Industry," *Science*, vol. 57, no. 1473 (March 1923): 341–345.

27 Bishop, "Radium Industry," 342.

28 Levine argues that Americans' obsession with radiation began to subside and turn to disillusionment in the 1930s and 1940s even though radiation-based therapeutic practices were far more refined and often less damaging to patients. Matthew Lavine, *The First*

to inspire a federal commitment to developing and expanding the domestic radium mining industry. Relatedly, however, physicians presented radium therapy as a more developed practice than it actually was by the early 1910s. The public education campaigns presented by the ASCC after its creation in 1913 helped reinforce the perception that new scientific and technological cancer treatments such as radium-based ones were well-developed and more effective than the treatment options previously available. The messaging both around the radium mining industry and from the ASCC helped obscure the extent to which physicians and scientists were still in very early phases of developing radium-based cancer therapies. Patients became subjects in experimental therapeutic uses of radium therapy and not just because radium therapy was new and poorly understood. They also became subjects in cases where physicians and physicists collaborated to deliver radium therapy and also measure the effects of dosing, equipment and other variables involved in therapeutic procedures.

Scientific Investigation of Radium Therapy

Physicians and physicists blended experiment into clinical practice because, aside from the supply problem, the other major obstacle that loomed over the application of radium in medicine was the lack of knowledge regarding its physiological effects. The discovery of radium had prompted a series of scientific experiments that began yielding knowledge in the earliest years of the twentieth century. By 1910, physicists had defined the basic physical properties of radium. They knew that it was the sixth member of the uranium series, meaning that, as uranium undergoes radiological disintegration or decay, it transforms into other elements, the sixth being radium. They also determined that radium emits three forms of radiation – alpha particles, beta particles, and gamma rays. As physicists investigated the different forms of radioactivity, some of them dedicated themselves to research on the medical application of radium. They sought to determine how and in which forms radioactivity could be used to treat diseased tissues on the surface of or located deep within the body. Physicists recognized that the pursuit of such knowledge was valuable to society and benefitted the scientific endeavours of all physicists.[29]

Atomic Age: Scientists, Radiations, and the American Public, 1895–1945 (New York: Palgrave Macmillan, 2013), 25–88 and 145–182.

29 For instance, at a celebration held in honour of Marie Curie and her discovery of radium, renowned University of Chicago physicist R.A. Millikan highlighted the knowledge

Radium companies encouraged the scientific study of radium as a therapeutic agent – they helped involve non-medical actors in the effort to realize radium's medical potential. For instance, the Radium Chemical Company, Inc., a subsidiary of the Standard Chemical Company, established a biological research laboratory to investigate medical applications of radium. The new company also founded the journal *Radium* in 1913 to publish related research in biology, chemistry, physics, and medicine.[30] Other American companies joined the Radium Chemical Company in supporting research, developing instruments, and training physicians in the use of radium in medicine. As the Eastern Manager of the Radium Company of Colorado explained, many radium companies maintained medical departments which carried out such research. "These departments," he said, "publish periodicals containing the latest articles on the therapeutic use of radium and act in a general consulting capacity to doctors using or contemplating the use of radium."[31] Company-sponsored periodicals published articles with the intent of educating the medical community on the therapeutic value of radium. They disseminated information about improvements in medical instruments, treatment methods, and other issues being studied by the companies' scientists. These periodicals also served as a form of publicity, helping to stimulate demand for radium. Indeed, radium companies supported medical research to ensure a market for their product.

Research conducted by scientists employed at industrial laboratories or, as radium became more accessible, in university and hospital laboratories, made use of seeds, embryos, bacteria, and animals as experimental bodies. Experiments with flies, mice, guinea pigs, rabbits, dogs, and many more different species were common as a means of determining the effects of radium's radioactive rays on living cells. These experiments were well documented by Chicago dermatologist and medical professor Frank Edward Simpson who presented a thorough discussion of the literature published during the 1910s in his 1922 textbook, *Radium Therapy*. Aiming to help orient his fellow practitioners

---

physicists had produced that contributed to the clinical use of radium. He focused on physicists' contributions to medicine while standing before a room of eminent scientists who gathered at the National Museum in Washington, D.C. in the early 1920s. See, R.A. Millikan, "The Significance of Radium," *Science*, vol. 54, no. 1383 (July 1921): 4.

30 American companies followed an example set earlier by a French refinery, Armet De Lisle, which subsidized a radium clinic and published a free journal called *Le Radium* from 1904–1914. R.E. Rowland, *Radium in Humans: A Review of U.S. Studies* (Argonne, IL: Argonne National Laboratory, 1994), 3; and Robison, "Telecurie Therapy," 1212–1213.

31 Bishop, "Radium Industry," 342.

in the use of radium as a clinical agent, he reviewed the use of countless animal species to study the effects of radium on various parts of the body including the skin, reproductive and vital organs, the nervous system, the lymphatic system, and cancerous cells.[32] Simpson also took account of the literature on the clinical use of radium. In an environment in which laboratory research proceeded alongside the clinical use of radium therapy, Simpson's text demonstrated an explicit attempt to translate knowledge gained in the laboratory to the clinic.

Physicians and scientists have long debated the substitution of experimental bodies for human patients. Those who engaged in radium research in the early twentieth century did as well. They questioned whether data obtained using animal subjects could be accurately applied in human contexts, but they also identified a few advantages in using animals as subjects.[33] For instance, when they questioned the long-term effects of radium therapy, the use of animals with shorter life spans seemed a reasonable means to carry out studies lasting only days or weeks. Specifically, the use of short-lived animals as subjects provided scientists and physicians an opportunity to examine the lifetime progression of harmful effects such as the inflammation and fibrosis of connective tissues exposed to radium. Scientists and physicians also turned to the study of animals to investigate whether radium therapy served only as a means of arresting a tumor's growth or might be considered a cure for cancer – then defined as no recurrence for five to ten years. Generally, the results of these studies were used to make projections of how radium therapy might affect humans over the span of many years.[34]

In addition to certain scientific advantages, animal experimentation made good sense for ethical reasons, especially since physicians considered the therapeutic use of radium to be a work in progress. They regarded the treatment of patients with radium therapy to be *therapy*, but recognized that radium posed health risks that were potentially significant and required further investigation.[35] The development of radium therapy occurred decades prior to the

---

32 Frank Edward Simpson, *Radium Therapy* (St. Louise, MO: C.V. Mosby Company, 1922), 73–89.

33 For a historical survey of debate regarding experimental bodies, see Ilana Löwy, "The Experimental Body," in Roger Cooter and John Pickstone, eds. *Companion to Medicine in the Twentieth Century* (New York: Routledge, 2003), 435–449.

34 Bishop, "Radium Industry," 343.

35 The hazards of radium became especially apparent in the 1920s due to the illness and fatalities that emerged amongst dialpainters (young women employed to apply radioluminous paint to watch faces). See, Clark, *Radium Girls*. For an account of the role played by physicists in regulating the hazards of radiation see, Vivien Hamilton, "Incongruent

United States government's implementation of comprehensive regulatory structures to govern scientific and medical experimentation involving humans.[36] The lack of government regulation did not, however, mean that physicians used patients as experimental subjects without any regard for ethical standards. Historian Susan Lederer persuasively argues that, prior to World War II, clinical experimentation was informally regulated within the medical community. The American Medical Association (AMA) was at the centre of a medical ethics debate that resulted in the creation of a code of ethics for animal experimentation in 1909 and a proposal for a similar set of guidelines for human experimentation in 1916.[37] Although the AMA opted not to impose an official code of research ethics on clinical investigators who used humans as experimental subjects, the debate that unfolded on the issue allowed physicians to reassert their personal responsibility to patient well-being. The physicians who developed and implemented the practice of radium therapy expressed their commitment to patient well-being by collaborating with scientists, participating in or consuming the knowledge generated by laboratory experimentation, and adopting an experimental approach to study and improve their clinical work with patients.

Radium Therapy at Memorial Hospital: A Clinical Practice & Scientific Enterprise

Physicians' desire to collaborate with physicists, apply knowledge obtained through laboratory experimentation, and adopt an experimental approach to their clinical work was particularly well illustrated by those working at Memorial Hospital for the Treatment of Cancer and Allied Diseases in New York. Memorial Hospital, now known as the Memorial Sloan-Kettering Cancer Center, was established in 1884 and was the United States' first hospital specifically devoted to the treatment of cancer. Soon after the hospital opened,

Bedfellows": Physics and Medicine in the formation of North American and British Radiology, 1896 – 1930," (Ph.D. Dissertation, University of Toronto, Toronto, ON, 2012).

36 For a thorough account of the development of bioethics and a regulatory framework for medical research see, David J. Rothman, *Strangers at the Bedside: A History of How Law and Bioethics Transformed Medical Decision Making* (New York: Aldine de Gruyter, 1991).

37 The AMA did not establish a formal code of ethics for research involving humans until the 1940s. See, Susan E. Lederer, *Subjected to Science: Human Experimentation in America Before the Second World War* (Baltimore, MD: Johns Hopkins University Press, 1995), 73–74.

though, its focus on cancer began to wane. As examined above, the treatment options available to individuals with cancer in the late nineteenth century inspired little confidence and many Americans resisted treatment. Amidst such a climate, Memorial Hospital began to function more like a general hospital, that is, until the development of radium therapy played a pivotal role in re-establishing the hospital's initial purpose.[38] As one of hospital's physicians proclaimed, radium therapy helped launch a new era in the study of cancer within the hospital and beyond.[39] Memorial Hospital's reinvestment in cancer coincided with the formation of the American Society for the Control of Cancer which helped publicize both the importance of seeking treatment and the promise of radium therapy.[40] To realize that promise, Memorial's physicians set out to transform radium therapy from a work in progress to a successful and scientifically tested cancer therapy. This goal resulted in the careful study of patients, not only by their physicians, but by scientists as well.

Memorial Hospital began acquiring radium in the spring of 1914. By 1915 it had obtained sufficient supplies for physicians to employ radium therapy on a regular basis.[41] Of those who did, the hospital's Chief of Surgery Henry H. Janeway and Benjamin S. Barringer, a urologist, were particularly notable for the experimental approach they brought to their clinical work. For Janeway and Barringer, radium therapy was a treatment option that could be employed, but also needed to be meticulously studied and refined.

Over the course of 1915 and 1916, Janeway and Barringer used radium therapy to treat hundreds of patients, most of whom were terminal. After delivering treatments, they tended to their patients carefully. They, of course, sought to manage their patients' health, but also aimed to better understand and improve their therapeutic practices. Together, these physicians compiled data from their treatment of 424 patients with malignant tumors and published

---

38 James B. Murphy, "James Ewing, 1866–1943: A Biographical Memoir," in *National Academy of Sciences Biographical Memoir* (Washington, D.C.: National Academy of Sciences, 1951), 48–49.

39 James Ewing was a pathologist at Memorial Hospital and was very committed to both the study of cancer and the development of radium therapy. See, Ewing, "Radiation Upon Carcinoma," 336.

40 Ross, *The American Cancer Society*, 15–27.

41 The purchase of radium from various radium companies was a gradual process due to the cost of radium. Memorial Hospital also benefitted from a donation of radium made by Dr. James Douglas who, along with Dr. Howard Kelly, endowed the National Radium Institute. See Henry H. Janeway, Benjamin S. Barringer, and Gioacchino Failla, *Radium Therapy in Cancer at the Memorial Hospital: First Report, 1915–1916* (New York: Paul B. Hoeber, 1917), 52; and Murphy, "Ewing Memoir," 48.

their results in a comprehensive report in 1917. Janeway explained their motives as stemming from their concern that the available literature describing the use of radium therapy lacked the specific information practitioners required to reproduce successful treatment practices. He claimed: "it has become more and more apparent that the successful use of radium in cancer, requires careful consideration of each particular type of the disease in each organ as separate problems in which different methods must be devised and different results expected."[42] The specificity he desired guided his own and his colleague's approaches to radium therapy and culminated in a collection of case studies that, when taken together, constituted a clinical experiment meant to test the efficacy of their methods. Put another way, human patients were an integral part of the process through which physicians came to understand the physiological action of radium therapy on cancerous tumors. Furthermore, the clinical use of radium for the treatment of patients with cancer encouraged the involvement of scientists who shared the goal of improving radium therapy by expanding the scientific research associated with it.

Janeway's and Barringer's report described the treatment of various types of cancer in more than twenty regions of the body including tissue, bone, and organs. They presented statics to quantify their assessment of success or failure and categorized their statistical analyses according to the type or location of their patients' tumors. Janeway's and Barringer's use of radium therapy for the treatment of 424 patients could be characterized as being just marginally successful given that of those 424, only 120 patients experienced a clinically complete retrogression at the site of their tumor. A year after treatment, 21 of those 120 patients were free from recurrence. Janeway and Barringer deemed another 134 patients "improved," but 162 patients experienced no improvement. The remaining 8 patients were either still undergoing treatment or their tumors had been found to be benign.[43] To Janeway and Barringer, these results were encouraging, especially since most of their patients had advanced cancers that were considered inoperable. Based on the number of patients who experienced at least some improvement in their condition, if not a complete retrogression at the site of their tumor, Janeway concluded that their use of radium therapy was sufficiently promising to urge the testing of the procedure on patients with earlier-stage cancers.[44] In other words, radium therapy did not seem so risky as to limit its use to terminal patients.

---

42 Janeway, Barringer, and Failla, *Radium Therapy in Cancer*, Preface and 46.

43 Janeway, Barringer, and Failla, *Radium Therapy in Cancer*, 223.

44 Janeway, Barringer, and Failla, *Radium Therapy in Cancer*, 226.

Case analysis was common at this time, but Janeway and Barringer aimed to accomplish much more than a statistical assessment of the clinical results they achieved during the 1915 to 1917 period. In fact, the evaluation of instruments and refinement of radiation dosages were important priorities throughout the course of their work. Through these variables, they sought to better understand the clinical results they had observed in relation to the applicators they used to deliver radium therapy. Janeway and Barringer tested numerous radium applicators of different structure and design. For internal tumors they used glass, aluminum, or platinum iridium tubes, and for external tumors, lead or silver plaques.[45] Overall, they attributed improved clinical results to the use of applicators if appropriately selected for tumors of varying size and location. Their adoption and modification of these devices, while partially achieved through a trial-and-error approach to clinical therapy, was also the result of laboratory experimentation and consultation with scientists. Here, the role of physicists emerges as an important factor in the technical development of radium therapy. Indeed, the technical considerations of radium therapy prompted the start of close collaboration between Memorial Hospital's physicians and a small, but growing scientific community that formed within the hospital.

Bringing a Scientist to the Hospital

In 1915, when Henry Janeway began treating cancer patients with radium and collecting data related to this work, he hired Gioacchino Failla, an engineer and physicist-in-training, to provide technical and scientific expertise. When hired, Failla had already completed one degree in electrical engineering and was in the midst of a Masters degree in Physics. After completing his Masters in 1917 he took a two-year leave of absence from his position at Memorial to serve in World War I as an assistant to the scientific attaché of the American Embassy in Rome. He subsequently pursued a doctoral degree during which he spent a year at the Sorbonne training under Marie Curie. As the discoverer of radium, Curie was already a renowned expert in radium research.[46]

Although he was hired for his technical and scientific expertise, Failla's contributions to clinical practice were significant. One of the primary responsibilities he assumed at the hospital was the construction and operation of a plant for the extraction of radon. Radon is a daughter product of radium. It is a gas

45 Janeway, Barringer, and Failla, *Radium Therapy in Cancer*, 69–78.

46 Hymer L. Friedell, L.D. Marinelli, and Titus C. Evans, "Gioacchino Failla, 1891–1961," *Radiation Research*, vol. 16, no. 5 (1962): 620.

that, at that time, was also known as radium emanation.[47] Failla collaborated with the hospital's physicians who used both radium salt and radon as their radium source, helping to develop appropriate instruments for handling them and delivering therapy. The selection of either radium salt or radon as a source of therapeutic radiation was one of the many variables that physicians tried to standardize or rationalize during the early years of radium therapy at Memorial Hospital. Janeway believed that radon contained in glass tubes was the most effective means for delivering radium therapy. However, two decades later a leading cancer physician, Max Cutler, reported that physicians were more likely to use radium rather than radon. Cutler cited the technical difficulties of collecting and measuring radon as one reason for this choice. As the medical and scientific communities studied and debated such issues, physicians continued to employ both. Their choice of one form or the other often depended on whether they had sufficient scientific and technical support to produce and prepare radon.[48]

While managing the hospital's radon plant and assisting his clinical colleagues on scientific and technical matters, Failla helped write the radium report. He provided a thorough explanation of the physical properties of radium as well as the basic principles of radium therapy. His contribution to the text represented his own and his clinical collaborators' hopes to strengthen the clinical use of radium through laboratory experimentation and scientific knowledge. Failla's early years of employment at Memorial Hospital proved to play a significant role in shaping the direction of his career as a physicist. He became increasingly convinced that the study of radium in medicine was a worthy pursuit.

Following his wartime service and doctoral studies in France, Failla returned to Memorial Hospital to continue his research on the clinical application of radium and gained approval to expand his research laboratory at the hospital. The first scientist he hired was physicist Edith H. Quimby. Together, Failla and Quimby conducted research that was closely intertwined with the clinical work of Memorial's physicians and was guided by the same goals. Like their clinical colleagues, Failla and Quimby strove to improve the methods, instruments and, ultimately, the results of radium therapy as a cancer treatment. They were motivated by the problem of inconsistent therapeutic practices and

47 Eric J. Hall, "From Beans to Genes: Back to the Future," *Radiation Research*, vol. 129, no. 3 (March 1992): 235.

48 Janeway, Barringer, and Failla, *Radium Therapy in Cancer*, preface; and Cutler, "Radium Treatment," 644–645.

clinical results that Janeway had identified in the 1917 radium report. According to Failla, Quimby, and Archie Dean, one of the physicians with whom they collaborated, variation in clinical practice was a problem that continued to plague the use of radium as a therapeutic agent. Failla, Quimby, and Dean described this pressing issue in a 1922 paper, stating that, "a perusal of the literature on the subject of radiation therapy reveals the difficulty of comparing doses of radiation used by different radiologists."[49] Thus, they focused their research on problems such as the quantification of radiation dose and the standardization of the methods used to deliver radium therapy to patients.

Throughout the early 1920s, Failla and Quimby published a steady stream of papers including, "Dosage in Radium Therapy," "The Effect of the Size of Radium Applicators on Skin Doses," and "The Economics of Dosimetry in Radiotherapy."[50] These papers reported the results from a number of experiments, all of which aimed to achieve a greater understanding of the variables involved in the delivery of radium therapy. Some of their experimentation took place within the laboratory and was conducted with animal subjects. However, their results also included the study of cancer patients being treated with radium therapy and it was their observation of patients to which they attributed the greatest authority. Quimby signalled this clearly in a paper she published to present the results of her effort to chart the erythema dose, or the reddening of the skin caused by radium. In it, she discussed the veracity of the different erythema doses she determined for each applicator at different distances and stressed the important role played by patients and physicians in her research. She stated, "These doses for the applicators have been determined clinically by different physicians, and have been used in the hospital on many patients, so that the identity of their effects is firmly established."[51] Quimby considered the evaluation of patients undergoing radium therapy to be invaluable

49 Gioacchino Failla, Edith H. Quimby, and Archie Dean, "Some Problems of Radiation Therapy," *The American Journal of Roentgenology, Radium Therapy and Nuclear Medicine*, vol. 9, no. 8 (August 1922): 479.

50 Gioacchino Failla, "Dosage in Radium Therapy," *The American Journal of Roentgenology*, vol. 8, no. 11 (1921): 674–685; Edith H. Quimby, "The Effect of the Size of Radium Applicators on Skin Doses," *The American Journal of Roentgenology*, vol. 9, no. 10 (October 1922): 671–683; and Gioacchino Failla and Edith H. Quimby, "The Economics of Dosimetry in Radiotherapy," *The American Journal of Roentgenology*, vol. 10, no. 12 (1923): 944–967. See also, Gioacchino Failla, "The Absorption of Radium Radiation by Tissues," *The American Journal of Roentgenology*, vol. 8, no. 5 (1920): 213–226; and Gioacchino Failla, "Ionization Measurements," *The American Journal of Roentgenology*, vol. 10, no. 1 (1923): 48–56.

51 Quimby, "Effect of the Size of Radium Applicators," 682.

to the pursuit of establishing a scientific understanding of the therapeutic practice that those patients and others received. She did so, even as the medical community aspired to ground their clinical practices in laboratory research with animals and as laboratory research on radium expanded. That Quimby identified the clinical use of radium on patients as an important factor in determining the effects of various doses delivered by particular applicators implies that the clinical use of radium was part of the scientific process for verifying knowledge. Although her clinical colleagues who treated patients with radium therapy did so with the intention of benefitting patients, patients also served as experimental subjects under analysis. Patients' role as subjects was not obvious to them, largely because radium therapy was promoted as a scientific and technologically advanced therapy.

Building an Interdisciplinary Laboratory in the Hospital

Radium therapy research involving experimental subjects other than humans accelerated throughout the 1920s. As laboratories devoted to the study of radium therapy expanded, they became increasingly interdisciplinary. At Memorial Hospital, collaboration between physicians and physicists had begun in the 1910s. By the 1920s, the scientists participating in the clinical use of radium and the laboratory investigation of it included chemists, biologists, and biochemists. The interdisciplinary research that developed around the application of radiation in medicine drew from many disciplines and fueled the development of hybrid research, or research that did not fit neatly into existing disciplinary boundaries.

Failla's Memorial Hospital laboratory illuminates the development of the interdisciplinary research communities and hybrid expertise that formed around the practice of radium therapy. When Failla began to build a research laboratory at the beginning of the 1920s, he did so with the conviction that the study of radium in medicine was a field of research to which he, as a physicist, could make important contributions. Furthermore, he believed that such work should encompass more than just the expertise that he and the hospital's physicians could offer.[52] As Failla expanded his laboratory staff he hired physicists, chemists, and biologists, all of whom worked closely with the hospital's physicians. Failla's commitment to crossing disciplinary boundaries was very

52 Edith H. Quimby, "Gioacchino Failla (1891–1961) and the Development of Radiation Biophysics," *The Journal of Nuclear Medicine*, vol. 6, no. 5 (1965): 377–378.

evident by the late 1920s. By then he thought it was not enough to employ biologists in his laboratory; he made it clear to the physicists he employed that they must develop a good understanding of biology as well.[53]

The interdisciplinary collaboration and hybrid expertise that Failla explicitly sought to facilitate was integral to the development of radium therapy. It eased the translation of knowledge between physicians and variously trained scientists by helping to nurture biomedical science that was closely related to clinical practice. His colleagues credited his role as a pioneer in the fields of biophysics – specifically radiation biophysics – as well as radiobiology.[54] The former applied the principles and practices of physics and chemistry to the study of biological processes or functions. The latter encompassed the study of the biological effects of radiation.[55] In the 1920s, biophysics and radiobiology were both emerging fields of research.

Therapeutic Experimentation: Hidden in the Shadow of Laboratory Research

The development of biomedical science in support of radium therapy reflected the growing importance of laboratory research in medicine. Physicians and scientists alike welcomed the advancement of radium-related research. In fact, they thought it necessary. In a 1922 paper that drew attention to the inconsistent methods used to deliver radium therapy, Gioacchino Failla, Edith Quimby, and Archie Dean attributed the problem to "the more rapid empirical development of the *art* as compared with the *science* of radiation therapy."[56] Here, the two scientists and one physician made a distinction between the practice of medicine and the pursuit of scientific research. Their depiction of an uncomplicated dichotomy between the art and science of radium therapy was misleading, though. It suggested that clinical therapy and experimentation fell on opposite sides of a line dividing medicine and science.

Such a distinction between the therapeutic use and scientific investigation of radium did not exist at Memorial Hospital during the mid-1910s to

53 Quimby, "Gioacchino Failla," 378.

54 Quimby, "Gioacchino Failla," 377.

55 Quimby, "Gioacchino Failla," 377. For an excellent discussion of the history of radiobiology, see, Angela N.H. Creager and María Jesús Santesmases, "Radiobiology in the Atomic Age: Changing Research Practices and Policies in Comparative Perspective," *Journal of the History of Biology*, vol. 39, no. 4 (2006): 637–647.

56 Failla, Quimby, and Dean, "Problems of Radiation Therapy," 479.

mid-1920s. Clinical practice differed little from laboratory experimentation in the opportunity it presented to physicians and scientists to test different aspects of radium therapy and its physiological effects. Both were a means for making radium therapy more scientific. The intersection of radium therapy and research at Memorial Hospital did not just begin in the 1920s when Failla increased the number of scientists employed in his laboratory and laboratory research grew. Janeway's and Barringer's treatment of more than 400 patients with radium therapy between 1915 and 1917 was an ongoing experiment. The duality of their work was evident in the language used throughout their report. On the one hand they wrote of patients, treatment, improvement, and cures. On the other, they wrote of hypotheses, testing, and reproducing results. Notably, though, they never wrote of subjects because they did not regard their patients as subjects.

As physicians like Janeway, Barringer, Dean, and others called for and helped facilitate an expansion of experimental research and technical development related to their clinical practice, they helped unite science and medicine.[57] They created an environment in which physicists, biologists, and chemists stood alongside clinicians to observe patients in the procedure room and physicians joined scientists at the laboratory bench to test instruments and measure radiation doses.

Conclusion

The history of radium therapy at Memorial Hospital during the first decade in which radium was sufficiently accessible reveals a fluid relationship between the practices of medicine and science. Indeed, the extent to which physicians embraced the knowledge and practices derived from scientific experimentation, as well as their attempts to apply the methods of experimentation in their clinical work, resulted in a blurring of clinical therapy and scientific experiment. Physicists and an increasingly diverse mix of biomedical scientists encouraged this trend by committing themselves to the development of radium therapy as a scientific and technologically precise practice. For physicians and scientists both, the scientific investigation of radium therapy and technological refinement of the clinical practice bolstered scientific medicine,

57 Memorial Hospital pathologist, James Ewing, is notable for his commitment to research, especially to the study of cancer in humans. See, for instance, Ewing, "Radiation Upon Carcinoma," 331–336; and Murphy, "James Ewing," 45–60.

contributed to improved clinical practice, and created an avenue for professional advancement.

In the midst of physicians' and scientists' pursuit of radium therapy and their intersecting professionalization, cancer care and the role of the hospital in providing it, evolved. So too, did the role of patients. Cancer patients who received radium therapy were used as experimental subjects. Physicians did not consider their patients to be subjects, or at least not in any way that, to them, seemed ethically questionable. However, patients effectively became subjects due to their central role in the process through which both clinical technique and biomedical knowledge were improved. That cancer patients doubled as experimental subjects was overshadowed by the general faith that existed regarding the role of science in medicine. For physicians, their collaboration with scientists and application of experimental methods to their clinical work provided greater oversight to a new and technologically sophisticated therapeutic practice. Rather than viewing their patients as subjects of experimentation, they continued to see their patients as therapeutic bodies; the subjective experience of experimentation was subsumed in a context that prioritized the rhetoric of scientific care over scientific research. Physicians and scientists did see a shift, though, in the sort of treatment they were able to provide to patients. Increasingly, they were able to provide cancer therapies they deemed to be "the best treatment known to science."[58] The esteemed role of science and technology in medicine is fundamental to understanding the failure of physicians such as Henry Janeway or Benjamin Barringer to fully appreciate that their efforts to improve radium therapy, often in collaboration with physicists, cast patients in a dual role of patient-subjects.

58 Wood, "Fundamental Research in Cancer," 659.

CHAPTER 8

Anthropometry, Race, and Eugenic Research: "Measurements of Growing Negro Children" at the Tuskegee Institute, 1932–1944

Paul A. Lombardo[1]

From 1932 until 1944 researchers from Long Island's Eugenic Record Office conducted detailed annual measurements of the bodies of students at Alabama's Tuskegee Institute. The measurements were done with the tools of anthropometry, and the students participated as research subjects. Their bodies were the focal point of an experiment that was structured to generate scientific knowledge about traits thought to characterize people of their "Negro" race. The study was planned by Charles Davenport, the most prominent leader of the American eugenics movement and carried out primarily by Morris Steggerda, his colleague at the Eugenics Record Office.[2] The Davenport/Steggerda study fit neatly into a long-term experimental program of comparative racial research designed to provide data to support the conclusion that the "races" are separated by hereditary differences. This paper describes the Davenport/Steggerda study, an episode in eugenic history that has drawn almost no attention by scholars, but which generated data that eventually appeared in more than a dozen scientific articles. It summarizes the origins of anthropometry and its use as a technique meant by eugenic researchers to concretize physical differences among different "racial" groups. It also places the Davenport/Steggerda study among the larger body of racial research that typified work at the Eugenics Record Office for several decades, and reflected attitudes of both Davenport and Steggerda about the need for social separation of people from different races.

The Tuskegee Institute is located in Macon County, Alabama, where the infamous forty- year-long "Tuskegee Study of Untreated Syphilis in the Negro Male" occurred. That study involved some 400 men infected with syphilis who

1 Bobby Lee Cook Professor of Law, Georgia State University. The author wishes to acknowledge the helpful comments of the editors as well as the close reading and valuable critique of Gregory M. Dorr.

2 The records of the Tuskegee anthropometric study are contained in the Morris Steggerda Collection, Otis Historical Archives, National Museum of Health and Medicine, Armed Forces Institute of Pathology, Washington, D.C., hereinafter, Steggerda Collection.

 | DOI 10.1163/9789004286719_010

were deceived into thinking they were receiving medical treatment even though the research goal was to understand the natural history of the disease, and not to deliver medical care. It is now widely recognized as a model of biomedical research abuse.[3] The Davenport/Steggerda anthropometric experiment began at the Tuskegee Institute the same year as the infamous syphilis study, and the two projects continued for more than ten years in tandem. The syphilis experiment rested on the premise, as one prominent physician put it, that "Syphilis in the negro is in many respects almost a different disease from syphilis in the white."[4] Charles Davenport wanted to work at Tuskegee in order to probe the relationship between hereditary "racial resistance" among Blacks to diseases such as cancer, yellow fever, and diphtheria. There is no evidence of any formal connection between the members of the Public Health Service Corps who devised the syphilis study and the researchers from the Eugenics Record Office whose experiment employed anthropometric measurements; nevertheless, the ideological backdrop for both studies included a heightened attention to race that mirrored eugenic thought.[5]

Anthropometry and Its Origins

Anthropometry is a set of techniques for measuring the human body using specialized instruments.[6] By the middle of the 20th Century, more than six hundred anthropometric instruments had been developed allowing for gross measurements such as height, span of arms or breadth of shoulders, and more focused metrics such as the thickness of ear lobes or the distance within the opening of a nostril. Fields as disparate as law enforcement, medical education and cultural studies eventually adopted anthropometric techniques;[7]

---

3 See James H. Jones, *Bad Blood* (New York: Free Press, 1981) and Susan Reverby *Examining Tuskegee: The Infamous Syphlis Study and its Legacy* (Chapel Hill: University of North Carolina Press, 2009).

4 Dr. Joseph Moore quoted in Jones, *Bad Blood*, at 106.

5 See Paul A. Lombardo and Gregory Michael Dorr, "Eugenics, Medical Education, and the Public Health Service: Another Perspective on the Tuskegee Syphilis Experiment," *Bulletin of the History of Medicine*, vol. 80 (2006) 291–316.

6 The term "anthropometria" first appeared in 1654 and its original Greek roots simply denoted the "measurement of humans." See Angelo Albrizio, "Biometry and Anthropometry: From Galton to Constitutional Medicine," *Journal of Anthropological Sciences*, vol. 85 (2007) 101–123, at 101.

7 Edward Mussey Hartwell, "A Preliminary Report on Anthropometry in the United States," *Publications of the American Statistical Association*, Vol. 3 (Dec., 1893) 554–568.

they also played an important part in the evolution of the field of physical anthropology.[8]

Two people are credited with popularizing anthropometry. The Belgian mathematician Adolphe Quetelet (1796–1874) was a known for his application of statistics and probability theory to social phenomena. He launched the first International Statistical Congress in 1871, and publishing *Anthropometry, or the Measurement of the Different Faculties of Man* the same year.[9] Quetelet condensed massive amounts of anthropometric data as a means of describing group differences, and claimed that it was possible to calculate an average for any group, thereby identifying *l'homme moyen* – the "typical" or average man.[10]

Alphonse Bertillon (1853–1914) followed Quetelet, and used anthropometry as a tool for criminal identification in the late 19th and early 20th Century. By making detailed measurements of any person's body, such as the length of finger joints or the circumference of eye sockets, Bertillon claimed that a unique data set could be created to set every individual apart from every other.

Race and the Uses of Anthropometry

Anthropometry was used both during the 18th and 19th century in an attempt to quantify regular physical differences and essential features that set the conventionally defined "races" apart.[11] Charles White, a surgeon from Manchester, England, made the first reported use for racial analysis. White's book, *Account of the Regular Gradation of Man,* appeared in 1799 and included the results of measurements made of "fifty African negroes in order to determine certain differences between them and European peoples."[12] The use of anthropological

8 Lucile E. Hoyme, "Physical Anthropology and Its Instruments: An Historical Study," *Southwestern Journal of Anthropology*, vol. 9 (1953) 408–430, at 410.

9 Quetelet, LAJ. Antropométrie ou mesure des différences facultés de l'homme. [Anthropometry, or the measurement of the different faculties of man.] (Brussels, Belgium: Muquardt, 1871).

10 Ibid (p. 16).

11 Hoyme, "Physical Anthropology"at 412. See also, more generally, John S. Haller, Jr., *Outcasts from Evolution: Scientific Attitudes of Racial Inferiority, 1859–1900* (Urbana: University of Illinois Press, 1971) particularly, Chapter 1: "Attitudes of Racial Inferiority in Nineteenth Century Anthropometry," 3–39. Frederick Blumenbach initially recognized four racial groups (Mongolian, American, Caucasian, and African). Blumenbach later added Malayan as a fifth group. Haller, p. 3–5.

12 Charles S. Myers, "The Future of Anthropometry," *Journal of the Anthropological Institute of Great Britain and Ireland,* vol. 33 (1903) 36–40, at 36.

measurements to mark the supposed boundaries between races became commonplace. Benjamin Gould used anthropometry in his investigations for the U.S. Sanitary Commission during the U.S. Civil War.[13] His study broke the larger racial groups into smaller ethnic units, with numerous comparisons of "French," "Irish," or "Teutonic" recruits. Later, life insurance actuary Frederick Hoffman used anthropometry to bolster the argument that the Black race was slowly deteriorating, as a result of emancipation, and would soon be extinct.[14] The measurement technique carried the cachet of precision and scientific objectivity. Military statistics expert J.H. Baxter described the role it had earned by the late 19th Century, saying: "the measurement of the physical and intellectual qualities of man, to which the title of anthropometry has been formally affixed, has now a settled recognition in statistical science."[15]

Eugenics and Anthropometry

Francis Galton, widely recognized as the father of eugenics, established a laboratory at the 1884 International Health Exhibition in London and eventually measured more than 9000 people there. Galton pursued anthropometry according to Bertillion's system, but it served a purpose beyond mere identification. Like the composite photography that Galton employed as a tool to find the "essence" of a certain type of person – such as criminals, or people with tuberculosis, or Jews – anthropometry yielded clues to the common features that could be distilled as the key characteristics of any group.[16] Galton linked anthropometry to a study of race differences in the book where he coined the

13 Benjamin Apthorp Gould, *Investigations in the Military and Anthropological Statistics of American Soldiers* (New York: Hurd and Houghton, 1869).

14 Frederick Hoffman, *Race Traits and Tendencies of the American Negro*, (New York: American Economic Association, 1896) 176. John Haller cites Civil War anthropometry as providing a seemingly "scientific" empirical foundation for the institutionalized racism of the late 19th Century, see "Civil War Anthropometry: the Making of a Racial Ideology," *Civil War History*, vol. 16 (December 1970) 309–324.

15 *Statistics, Medical and Anthropological, of the Provost-Marshal-General's Bureau, Derived from Records of the examination for Military Service in the Armies of the United States during the Late War of the Rebellion, of over a Million Recruits, Drafted Men, Substitutes and Enrolled Men*, vol. I (Washington: GPO, 1875) lxxxii.

16 Francis Galton, *Inquiries into Human Faculty and its Development* (London: J.M. Dent & Sons, 1883) 6–9. On composite portraits of Jews, see Karl Pearson, *The Life, Letters and Labours of Francis Galton* , Vol 2 (Cambridge University Press, 1924) 293–294.

word *eugenics*.[17] Always an inventor and inveterate quantifier, Galton also produced many of his own anthropometric instruments.[18] In a memorial comment at his death, the *New York Times* described Galton as the man who "made anthropometry a science."[19]

Ales Hrdlicka, a eugenic enthusiast and curator of the division of anthropology at the U.S. National Museum (Smithsonian), said that the goal of anthropometry was the "complete elimination of personal bias" in measurement. He would also declare that eugenical analysis was one key purpose of performing anthropometry.[20]

Planning Research at Tuskegee

For almost a generation, the Eugenics Record Office (ERO) at Cold Spring Harbor, New York provided a focal point for eugenic research in the United States. A brief contact in 1912 between ERO founder and director Charles Davenport and Booker T. Washington, founder and principal at Alabama's Tuskegee Institution, eventually led to the eugenic anthropometric study

17 Id., 17.

18 Albrizio, p. 108–109.

19 "Topics of the Times: Another Giant Departs," *New York Times*, Jan. 20, 1911, p. 10.

20 Alex Hrdlicka, *Anthropometry* (Philadelphia: Wistar Institute, 1920) 8. The many uses of anthropometry were commonly referenced in the literature on eugenics. The logo of the American Eugenics Society shows anthropometry alongside disciplines such as mental testing, ethnology and anatomy as the knowledge base – metaphorical roots of the tree – that formed the foundation for eugenic science, Harry H. Laughlin, *The Second International Exhibition of Eugenics held September 22 to October 22, 1921, in connection with the Second International Congress of Eugenics in the American Museum of Natural History, New York* (Baltimore: William & Wilkins Co., 1923). Dr. J.H. Kellogg, of the Race Betterment Association in Battle Creek, Michigan, wrote a pamphlet on anthropometry, Kellogg, J.H. *Outline Studies of the Human Figure* (Chicago, Battle Creek, Mich.: Mod. Med. Pub. Co., 1893). Kellogg also produced a new "Anthropometrical Dynamometer" to measure muscle strength. "New Dynamometer for Use in Anthropometry," *Scientific American*, vol. LXXI (November 24, 1894) 329. Anthropometry was part of the Better Baby and Fitter Family contests of the early 20th century, see photo of medical team collecting Anthropometric Data, Better Babies Contest, 1913 Louisiana State Fair, in Steven Selden, "Transforming Better Babies into Fitter Families: Archival Resources and the History of the American Eugenics Movement, 1908–1930," *Proceedings of the American Philosophical Society*, vol. 149, (June, 2005), 199–225, at 207.

there.[21] William Stokes, New York real estate mogul, part-time horse breeder and author of a book on eugenics referred Washington to Davenport[22] so that Washington could get copy of Davenport's highly successful text, *Heredity in Relation to Eugenics* (1911).[23] Davenport, always alert for research opportunities, sent the book as requested. He then took the occasion to pose a series of questions to Washington about the "hereditary racial characteristics of the full blooded negro" and other topics, such as racial resistance to yellow fever, malaria, cancer, scarlet fever, diphtheria and suicide.[24] Washington admitted that he had "made no special study" of the issues Davenport raised,[25] but he knew he had captured the attention of an influential scientist, and their correspondence continued. Within a year Washington invited Davenport to take advantage of the "full opportunity for study among our teachers and students" and visit Tuskegee.[26]

Washington was the first African-American to receive an honorary doctorate from Harvard in 1896, only four years after Davenport took his own Ph.D. there. Washington's lunch at the White House with Teddy Roosevelt – another first for an African American – was legendary. To Davenport, having access to the pupils of Tuskegee was an opportunity to work in a setting where he could structure an experiment to explore some of his most important ideas about race, and he could hardly have ignored an invitation to be the guest of one of

21 Washington's connections to the leaders of the American eugenics movement began when he met J.H. Kellogg during a visit to the Battle Creek Sanitarium in 1911, and subsequently consulted him on matters of diet and medicine. Booker T. Washington to John Harvey Kellogg, April 3, 1912, Louis R. Harlan and Raymond W. Smock, eds., *Booker T. Washington Papers, Vol. 11 (1911–1912)* (Urbana: University of Illinois Press, 1981) 510. Washington was also in attendance at the 1914 Race Betterment Conference, over which Kellogg presided. Charles Davenport was a member of the Central Committee for the meeting and gave an address on "The Importance of State Eugenic Investigation," while Washington spoke on "The Negro Race." See *Proceedings of the first National Conference on Race Betterment,* January 8, 9, 10, 11, 12, 1914 (Battle Creek, Mich.: National Conference on Race Betterment, 1914).

22 Washington to Davenport, Jan. 31, 1912, Davenport Collection, APS. Stokes was the author of *The Right to Be Well Born: Horse Breeding and its Relation to Eugenics* (N.Y.: C.J. Obrien, 1917) and he was a correspondent of Washington's.

23 Davenport to Washington, February 6, 1912, Davenport Collection, American Philosophical Society, hereafter, APS. In his thank you note to Davenport, Washington promised to place the book in the Tuskegee library "where students and teachers alike may have access to it." Washington to Davenport, February 20, 1912, Davenport Collection, APS.

24 Davenport to Washington, October 3, 1912, APS.

25 Washington to Davenport, October 11, 1912, APS.

26 Washington to Davenport, January 16, 1913, APS.

Mr. Charles B. Davenport,
Gold Spring Harbor,
Long Island, New York.

My dear Sir:-

I have your kind letter of January 12th.

Our mutual friend, Mr. W. E. D. Stokes, of New York City, has written me from time to time of your studies in Eugenics. We shall be very glad to have you come to Tuskegee at any time that best suits your convenience, and shall afford you full opportunity for study among our teachers and students.

Please be good enough to let us know a few days in advance of your coming.

I am sending you an announcement of our Conference meeting to be held here at Tuskegee, January 22nd and 23rd.

Yours very truly,

Booker T. Washington

E

FIGURE 8.1 *Booker T. Washington to Charles B. Davenport, January 16, 1913.* COURTESY AMERICAN PHILOSOPHICAL SOCIETY.

the most important Black men in America. Fairly jumping at the opportunity for fieldwork at Tuskegee, Davenport told Washington that he would be arriving within the week,[27] but other obligations interceded, and the trip had to be cancelled.[28]

Research at Tuskegee would have complemented Davenport's ongoing interests,[29] but Washington died two years later and Davenport's hopes to visit Tuskegee seemed to be lost. In the subsequent decades, Davenport would become the most prominent leader of the U.S. eugenics movement. He also would establish a reputation as a scholar of racial difference, earned in no

27 Davenport to Washington, January 20, 1913, APS.

28 Davenport to Washington, February 8, 1913, APS.

29 Gregory Dorr describes discussions between Davenport and Virginia eugenicist Harvey Jordan on the potential for racial research, occurring at the same time of the Davenport/Washington correspondence, see Gregory Michael Dorr, *Segregation's Science: Eugenics and Society in Virginia* (Charlottesville: University of Virginia Press, 2008) 55–59.

small part by an experiment that required minute anthropometric measurements of large groups of subjects.

Davenport completed two anthropometric studies as an officer in the Sanitary Corps following America's entry into World War I. The first was a massive compilation of charts showing the results of physical examinations on drafted men.[30] The second tabulated measurements taken of recruits during the War (1917–1918) and at its end (1919).[31] Both were consciously designed to show physical differences – via anthropometric measurements – among different "racial" groups.[32]

Following the War, Davenport attempted to demonstrate the susceptibility of different races to disease and anatomic anomalies. He asked dentists to cooperate with the Eugenics Record Office and collect family histories, then incorporated their responses an article in the first edition of the *Journal of Dental Research,* emphasizing the importance of studying teeth in order to understand the danger of racial mixing. Defective dentition, Davenport suggested, may be due to the mixture of incompatible breeding stocks, and the "hybridization of races" leading to an increasing need for orthodontia and other corrective dentistry. He also emphasized the potential high resistance to dental decay as a likely hereditary, "racial" factor, peculiar to the "full-blooded negro."[33]

Race Crossing in Jamaica

After his Army studies, Davenport's most important anthropometric analysis occurred in *Race Crossing in Jamaica*, which he completed with the help of coauthor Morris Steggerda.[34] Steggerda was born in Michigan in 1900, and took

30 Albert G. Love and Charles B. Davenport, *Defects Found in Drafted Men: Statistical Information Compiled from the Draft Records Showing the Physical Condition of the Men Registered and Examined in Pursuance of the Requirements of the Selective Service Act* (Washington, D.C.: Government Printing Office, 1920) 1663 pages.

31 Charles B. Davenport and Albert G. Love, *The Medical Department of the United States Army in the World War, vol. xv, Statistics, Part one: Army Anthropology, Based on Observations Made on Draft Recruits, 1917–1918, and on Veterans at Demobilization, 1919* (Washington: Government Printing Office, 1921).

32 See Jonathan Peter Spiro, *Defending the Master Race: Conservation, Eugenics and the Legacy of Madison Grant* (Burlington: University of Vermont Press, 2009) 308–313.

33 C.B. Davenport, "The Genetical Factor in Dental Research," *Journal of Dental Research*. vol. I (March, 1919) 9–11.

34 C.B. Davenport & M. Steggerda, *Race Crossing in Jamaica* (Washington: Carnegie Institution, 1929).

FIGURE 8.2
Booker T. Washington, circa 1908.
COURTESY LIBRARY OF CONGRESS.

his M.A. and Ph.D. in zoology at the University of Illinois.[35] Davenport and Steggerda shared an interest in poultry breeding,[36] and Davenport initially arranged summer work for Steggerda in 1925.[37] In 1926, Davenport appointed Steggerda to a full-time position as a field worker.[38] He would eventually replace Arthur Estabrook, another researcher whose career focused on the supposedly dysgenic effects of interracial marriage. Estabrook was the author of *Mongrel Virginians* (1926), at the time the most important study of racial

35 Paul S. Sledzik, "The Morris Steggerda Human Biology Collection," in B. Frohlich, A.B. Harper, and R. Gilberg, eds. *An Aleutian journey*: A Collection of Essays in Honor of *William S. Laughlin*, National Museum of Denmark, Ethnographic Series, vol. 20 (2002) 281–286, at 281.

36 H.H. Laughlin, Memorandum for Dr. H.G. Harris, February 20, 1925, ERO; C.B. Davenport, "Inheritance in Pedigree Breeding of Poultry," *Journal of Heredity, vol. 3* (1907) 26–33.

37 Davenport to Prof. Charles Zeleny, March 20, 1925, ERO.

38 Steggerda's first assignment involved data collection on "negro-Indian crosses on Long Island." Davenport to Steggerda, November 27, 1929, ERO. The Carnegie Department of Genetics where Steggerda held his appointment, worked in cooperation with the ERO in several race related studies, see Oscar Riddle, "Charles Benedict Davenport: 1866–1944," *National Academy Biographical Memoirs*, vol. 25 (1947) 75–110.

FIGURE 8.3
Charles B. Davenport, circa 1929.
COURTESY COLD SPRING HARBOR LABORATORY EUGENICS ARCHIVE.

mixing done by the Eugenics Record Office. Estabrook examined the WIN tribe – a clan of mixed White, Negro, Indian background. His book generated support for revising laws against miscegenation in Virginia.[39]

Steggerda spent two years collecting data in Jamaica that would form the basis for his own doctoral dissertation.[40] Other material from Jamaica became the subject of *Race Crossing in Jamaica*, an investigation involving more than 300 subjects that began as a study of "the adaptability of negroes and mulattoes to civilization."[41] Steggerda measured his subjects' bodies with anthropometric tools and he gave them tests to assess their intellectual and musical abilities and their temperament. He then photographed

39 See Paul A. Lombardo, "Miscegenation, Eugenics and Racism: Historical Footnotes to Loving v. Virginia,"21 *University of California at Davis Law Review* 421 (1988).

40 Steggerda's doctoral dissertation focused on racial hybrids from the Jamaica study. Earl W. Count, "Morris Steggerda, 1900–1950," *American Journal of Physical Anthropology* vol. 9 (March, 1951) 97–106.

41 See "Studies in Jamaica," *Eugenical News,* vol. xi (Oct., 1926) 154; and "Preparations for Negro-White Studies in Jamaica," *Eugenical News,* vol. xi (December, 1926) 188.

each subject.[42] Regular reports on the study's progress were printed in the *Eugenical News*.[43] Some of those reports focused on the significance of specific measurements, such as "Nasal Breadth in Negro x White Crossings."[44] Davenport interpreted the data collected by Steggerda to conclude that racial mixing held great peril, and highlighted the "excessively large number of intellectually incompetent persons" who are born to parents of differing races.[45]

Davenport's conclusions attracted criticism, even from within the eugenics movement. Karl Pearson questioned the size of samples from which Davenport had derived expansive conclusions about Black inferiority, as well as other features of the study methodology.[46] W.E. Castle, a Harvard biologist who had been a student of Davenport's criticized the "broad sweeping terms" which Davenport employed to summarize the dangers of racial disharmony in children of mixed marriages.[47]

An obviously defensive Davenport attempted to rebut his colleagues' critiques,[48] and continued to explore research that might confirm his racial speculations. His later statements about anthropometric measurement and genetic essence were consistent with his earliest speculations about racial difference. "Fundamentally, then, the racial differences are gene differences and

---

42 Some of the controversy surrounding these assessments, and the role of prominent psychologist Arnold Gesell in the work is described by Fredric Weizmann, "From the 'Village of a Thousand Souls' to 'Race Crossing in Jamaica'," *Journal of the History of Behavioral Sciences*, vol. 46 (Summer, 2010) 263–275.

43 *"Negro-white Hybrids in Jamaica – Investigation made under the W.P. Draper Fund,"* Eugenical News *vol. xiii (January., 1928) 21.* On the role of Draper in funding the Jamaica study and other eugenical philanthropy and propaganda, see William H. Tucker, *The Funding of Scientific Racism: Wickliffe Draper and the Pioneer Fund.* Urbana, University of Illinois Press (*2002*).

44 "Nasal Breadth in Negro x White Crossing," *Eugenical News* vol. xiii (March, 1928) 36–37.

45 C.B. Davenport, "Race Crossing in Jamaica," *Scientific Monthly*, vol. 27 (September, 1928) 225–238.

46 K. Pearson, "Review of *Race Crossing in Jamaica*," *Nature*, vol. 126 (1930) 427.

47 William E. Castle, "Race Mixture and. Physical Disharmonies," *Science* vol. 71 (June 13, 1930) 603–606.

48 C.B. Davenport, "Some Criticisms of 'Race Crossing in Jamaica,'" *Science* 72 (November 14, 1930) 501–502. According to some commentary, *Race Crossing in Jamaica* was the end of attempts by geneticists "to emphasize obvious physical disharmonies in race crossing." William B. Provine, "Geneticists and the Biology of Race Crossing," *Science* vol. 182 (November 23, 1973) 790–796, at 793.

when we describe the morphological differences between adults of different races we are describing the end result of the genic differences."[49] *Race Crossing*, like Davenport's other major studies, used anthropometric techniques in an attempt to develop racial typologies and describe ideal racial types. The book became required reading for those who would argue for racial separation and anti-miscegenation law.[50]

In 1931 Steggerda began a longitudinal study of Mayans in the Yucatan. The goal of the visit was "to study the effects of race crossing" and to compare the "minds of men of different races."[51] Steggerda's disappointment showed in his first report to Davenport that Mexico was not a good place to study "race crossing" since mating between indigenous people and the Spanish had happened so many years earlier. Davenport encouraged him to try to identify examples of the "pure Mayan type" and "pure Spanish type" because in the long run, the goal was "to compare the growth of the Indians with the growth of the negro and of the European stock."[52] Pure examples of racial "types" provided more useful research material, and Davenport knew that one of the most prominent locales in the country to study "pure" examples of "negro stock" would be at the Tuskegee Institute.

Steggerda's interest in racial research and his availability to manage field work gave Davenport an opportunity to renew his contact with Tuskegee. Davenport repeated his request to "make some studies upon the children" to Dr. Robert R. Moton, who assumed the presidency of Tuskegee after Washington's death.[53] Moton, reviving the earlier invitation, asked for a plan of study.[54] Davenport's research proposal listed techniques honed over

---

49 Charles B. Davenport, "How Early in Ontogeny Do Human Characters Show Themselves?" Eugen Fischer-Festband, *Zeitschrift. f. Morph. u. Anthrop.*, 34 (1934) 76–78. The manuscript of this "contribution to Souvenier Manuscripts in connection to the 60th birthday of Professor Eugen Fischer, sent 12/21/33" is in the Davenport collection at the American Philosophical Society.

50 The notoriously racist Mississippi Senator Theodore Bilbo used *Race Crossing in Jamaica* as an authoritative source for his own book on the need for racial segregation, Theodore G. Bilbo, *Take Your Choice: Separation or Mongrelization*? (Poplarville, Mississippi: Dream Publishing Company, 1947).

51 "Goes to Mexico to Study Differences in Races," *Science News-Letter*, Vol. 19 (Mar. 28, 1931) 205.

52 Davenport to Steggerda, March 25, 1931, ERO.

53 Davenport to President, Tuskegee Institute, February 10, 1932, APS. ("Many years ago Major Moton invited me to make studies on the children at Tuskegee. [I]...would be able to stop for two or three days at Tuskegee if you could arrange to have me stay there.")

54 Telegram, Moton to Davenport, February 13, 1932.APS.

FIGURE 8.4 *Morris Steggerda, Christine Petty, Cleve Abbott, circa 1940.*
COURTESY OTIS HISTORICAL ARCHIVES, STEGGERDA COLLECTION, NATIONAL MUSEUM OF HEALTH AND MEDICINE.

the years in previous work, such as body measurements and "correlation of characters in negroes and negro-white crosses," as well as "changes in proportions of children from birth to maturity."[55] En route to Yucatan to review Steggerda's field work, Davenport stopped at Tuskegee in 1932 to begin data collection. The next year, Steggerda took on the responsibility for measuring the children of Tuskegee. That work continued under his direct purview for more than 12 years.[56]

Conducting the Study

Davenport planned the Tuskegee project as a longitudinal survey of a single school class through twelve years, from first grade through high school. Study sites included not only the Tuskegee Institute but also several Macon County, Alabama grade schools in the area, as well as the local Children's Home. Tuskegee staff members assisted Steggerda, completing measurements and collecting other information, acting as liaisons to the families of the study subjects, and maintaining records between the times of Steggerda's annual visits. Christine Evans Petty, the women's track and field coach at Tuskegee, functioned as the study coordinator and the main link between Steggerda and the students in the study. She was later listed as co-author on several Steggerda publications.[57] Cleve Abbott, the head football coach at Tuskegee, also assisted Steggerda.[58]

Many features of Steggerda's earlier work in Jamaica reappeared in the Tuskegee project. He completed lengthy data forms for over 200 subjects. He performed sixty-five different measurements of various body dimensions annually. They ranged from the prosaic (e.g. weight and height) to the esoteric (e.g. the depth of ear pits). Sixty-six additional observations were charted, such as the slope of the forehead and shape of the nose. Steggerda counted and

55 Davenport to Moton, February 15, 1932.APS.

56 Davenport to Moton, April 3, 1933, APS. (Alerting Moton that he was sending Morris Steggerda to follow up on work of the previous year, and requested the continuing assistance of staff members from the Physical Training Department per the previous visit.)

57 For example, Morris Steggerda and Christine Evans Petty, "An Anthropometric Study of Negro and White College Women," *Research Quarterly* vol. 11(October, 1940) 110–118. Petty died unexpectedly in 1942, the last year data was collected for the Davenport/Steggerda study.

58 Cleve Abbott was head football coach at the Tuskegee Institute from 1923–1955. He was hired by Booker T. Washington in 1916 as an agricultural chemist and coach. The Tuskegee stadium was named after him in 1996.

evaluated student teeth,[59] and gave tests for perception, taste sensitivity and color blindness. He also made finger and palm prints. Every child was photographed, and each gave a hair sample for analysis. Steggerda incorporated his observations into charts of eye, skin and hair color.[60] Students wrote narrative family histories and eventually displayed their family backgrounds graphically as formal pedigree charts. Researchers noted scars and other bodily anomalies and a tally was kept of the number of children who wore glasses; pedigree notes identified unusually large families.

In 1941, as data collection at Tuskegee was nearing an end, Steggerda prepared a mimeographed manual of fifty-five pages entitled "Anthropometry of the Living." It explained his research procedures and described the techniques he used to analyze some 6500 people he had measured over the course of fifteen years, including Tuskgee students.[61] His field work spanned the "racial groups" that were the focus of his many publications: Mayan Indians in the Yucatan, Zuni Indians in Arizona and New Mexico, Whites from Michigan and the Netherlands, and Blacks from Jamaica and Tuskegee. He credited the three masters of anthropometry and leaders in eugenical research from whom he had learned anthropometric technique: Charles Davenport of the ERO, Earnest Hooten of Harvard and Ales Hrdlicka of the National Museum (Smithsonian).

Steggerda showed sensitivity to the issue of consent for research, an idea that within a decade would become a primary consideration in the ethics of experimentation.[62] He considered it "essential" to receive permission for data collection from government officials, representatives of the community, "including

59 Steggerda received a grant from the American College of Dentists in 1942 to study "the relation between growth and the eruption of the teeth in whites, Negroes and Indians," see "Grants-in-Aid of the American College of Dentists," *Science*, vol. 95 (March 13, 1942) 268, and the data from Tuskegee generated several dental publications, see Morris Steggerda, "Eruption Time of Teeth among Whites, Negroes and Indians," *American Journal of Orthodontics and Oral Surgery*, vol. 28 (June, 1942) 361–370, and Morris Steggerda, "Anthropometry and the Eruption Time of Teeth," *Journal of the American Dental Association*, vol. 32 (March, 1945) 339–342.

60 *Science* reported that the variation of hair sizes among people of different races was "found to be very great," and that Steggerda had, using a new device, "set tentative standards of hair sizes for races." "Hair Sizes in Races and Individuals," *Science*, Vol. 93 (Feb. 28, 1941) 6–8, at 7.

61 Morris Steggerda, "Anthropometry of the Living," (1941) Steggerda Papers. Steggerda also wrote an article on new "Anthropometric Instruments," see M. Steggerda, *American Journal of Physical Anthropology*, vol. 7 (September, 1949) 473–474, and announced a "New Weighing Scale," see M. Steggerda, *American Journal of Physical Anthropology*, vol. 7 (December, 1949) 600–602.

62 See *"Ethically Impossible" STD Research in Guatemala from 1946–1948* (Washington, D.C.: Presidential Commission for the Study of Bioethical Issues, 2011)15–18. Government

the most influential citizens," heads of families and "of course, the subjects themselves." In the U.S. no compensation was given to "whites and negroes" but for Indians in both the U.S. and Mexico "a bar of candy or a rubber balloon" was given to children and "small gifts" to adults. Steggerda discouraged the payment of cash, if possible, but allowed that money not to exceed 25¢ could be provided to adults who had to leave their work to be measured. Subjects were also photographed, and each received a copy of the photo as an additional reward.[63]

The bulk of the manual contained detailed explanations on how to perform anthropometric measurements, with pictures to illustrate the process, sketches of instruments used and data collection blanks. A brief primer on statistical analysis was also included.[64] Steggerda appended actual data charts from studies carried out at Smith College, Holland, Michigan and the Tuskegee Institute, as well as materials from Charles Davenport's own manual on anthropometry[65] and model pedigree charts developed at the Eugenics Record Office.[66]

The pamphlet provides a key to many of the goals of the study, which were focused on determining which characteristics could be identified as part of the essence of a race. It explains why at Tuskegee, as at other study sites, "racial characteristics" were charted for every subject and additional data was collected for children of mixed race. At Tuskegee, for example, particular attention was paid to children designated as "Indians." Notations like "semitic features – omit" were common. They indicated that anyone not of the "pure type" for a race must represent some deviation from the traits thought to be common to the group, and should be removed from calculations meant to distill those most typical "racial characteristics" into mathematical formula. Children considered defective or those with physical anomalies were also thought to improperly skew the data and were thus omitted from average computation for type.[67] When an unusual trait reappeared in a family, it was also noted, such as "has web toes – Father has this trait also."[68] Similarly,

---

officials were attuned to the possibility that incentives for experimental subjects could unduly influence their consent at least as early as 1943.

63 Steggerda, *Anthopometry of the Living*, 9.

64 Steggerda, *Anthopometry of the Living*, 43–45.

65 Chas. B. Davenport, *Guide to Anthropometry and Anthroposcopy* (Cold Spring Harbor, New York: Eugenics Research Association, 1927).

66 Steggerda, *Anthropometry of the Living*, 31. See also Charles B. Davenport, Harry H. Laughlin and A.J. Rosanoff. "How to Make a Eugenical Family Study" *Eugenics Record Office, Bulletin No. 13* (1915). Steggerda left a sketch of instrumentation used in his studies. It contains eye and hair color charts, skin color standards , and a dynamometer – an instrument for measuring hand strength.

67 "Cretin – omit" box 30. Steggerda Collection.

68 See, e.g. box 31, M. Family file, Steggerda Collection.

twins gain additional research attention, with extra questioning in more extensive interviews.

All of Steggerda's studies grouped subjects by racial type in an attempt to create a standard for racial typology. Tuskegee officials sent follow-up letters to families to confirm the racial background of grandparents and other collateral relatives.[69] In some accounts by students, racial mixture was denied, though a researcher's notation: "plenty of light people in the family" might suggest otherwise.[70] Longitudinal pedigree charts were made for all mixed race families, consistent with earlier Eugenics Record Office racial research, such as *Mongrel Virginians*. Steggerda identified 100 students thought to typify "the Negro" for the publication that would focus on racial essence.

Students who left Tuskegee were followed for additional survey details, and Petty traveled to New Orleans, Louisiana, Chattanooga, Tennessee, and Pine Bluff, Arkansas, as well as towns in Connecticut, Wisconsin, Texas and Oklahoma to secure complete data.[71]

Steggerda wrote updates on his work at Tuskegee for the *Eugenical News*. In one report, he described his return to Tuskegee following a winter visit to Yucatan and Guatemala to conduct a longitudinal anthropometric study of the Maya that paralleled his work in Alabama. Then he traveled to the Southwest U.S. to study Navajos and Zunis. Year's end was spent in his home town of Holland Michigan, studying Dutch children whose families came from the northern provinces of the Netherlands. The plan Steggerda outlined involved measuring each group of children annually as they grew in order to compile "standards…for each of these races with which children of families may be compared for genetic studies."[72]

Black Anthropometry

Why were Washington, Moton and staff members at Tuskegee so happy to open the school to Davenport, a man who had been publicly criticized, even

69 "Tuskegee Institute is cooperating with Dr. Davenport of Carnegie Institution of Washington in the attempt to obtain the following information. Please answer all the questions as accurately as possible." Signed: Dr. R.R. Moton, Principal. Feb. 29, 1932 (box 29, item 64a, Steggerda Collection).

70 See box 32, "Negro pedigrees," Steggerda Collection.

71 Morris Steggerda and Christine Evans Petty, "Body Measurements on 100 Negro Males from Tuskegee Institute," *Research Quarterly*, vol. 13 (October, 1942) 275–279. at 275.

72 Morris Steggerda, "Notes on Field Activities," *Eugenical News*, vol. xix (January–February, 1934) 32. Steggerda also wrote about "racial and environmental growth patterns" from data derived from seven-years of study of Dutch (Mich.), Negro (Ala.), Navajo (Ariz.) and Maya (Yucatan), see "Section on Anthropology," *Science*, vol. 89 (February 3, 1939) 89–112, at 107.

by his colleagues in the eugenics movement, for the way his attitudes on race had spilled over into his scientific work? First, Davenport was a respected scientist. He had worked at Harvard and the University of Chicago, was funded by the philanthropy of the Harrimans, Rockefellers and Carnegies, and was a leading researcher in the American Association for the Advancement of Science, the National Academy of Sciences, and the U.S. Army. Second, there was nothing inherently suspect about anthropometry, and it was accepted as part of scientific study in the U.S. for more than a generation. It had been used by racists and anti-racists alike. Early in the 20th Century, Franz Boas did his own measurements to disprove the assertions of the immigration restrictionists who claimed that the head shape of immigrants signaled a decline in the American citizenry.[73]

Anthropometry had been used as a tool of government population assessment for decades. Davenport's own studies of soldiers were well known, and prominent anthropologists like Alex Hrdlicka were allowed to measure immigrants to Ellis Island before World War I, and he reported his preliminary data in an anthropological journal.[74] Anthropometry had also been employed to assess the best white families in America. For example, during his first years working on Long Island, Morris Steggerda measured the grandchildren of Theodore Roosevelt and performed anthropometric surveys at elite white colleges.[75]

In 1906, several years before Davenport approached Booker T. Washington with a plan to measure the children of Tuskegee, W.E.B. Dubois published *The Health and Physique of the Negro American*,[76] a study specifically designed to counter claims of Black inferiority, containing the results of extensive anthropometric measurements on college students. Dubois spent a year pouring over studies by the Army, the Census Bureau and others. He

73 Franz Boas, "Changes in Bodily Form of Descendants of Immigrants," *American Anthropologist*, vol. 14(1912) 530–562. Boas argued that in later generations, immigrants changed in bodily forms and thus such measurements were not a static representation of 'race'.

74 Hrdlicka was invited to Ellis Island by Public Health Service Surgeon General Rupert Blue, see Ales Hrdlicka, "Anthropology of the Old Americans," *American Journal of Physical Anthropology*, vol. v, (July-Sept., 1922) 209–235, at 234.

75 Surveys of his own child, and the children of Kermit Roosevelt survive in the Steggerda collection. See also, Morris Steggerda, Jocelyn Crane, and Mary D. Steele, "One Hundred Measurements and Observations on One Hundred Smith College Students," *American Journal of Physical Anthropology*, vol. xiii (July-September 1929) 189–254. In the appendix to this article, Steggerda notes similar studies at Hollins College (Virginia) and Vassar.

76 W. E. Burghardt Dubois, *The Health and Physique of the Negro American*, (Atlanta: Atlanta University Press, 1906).

then surveyed approximately 700 Black physicians and collected data on 1000 African American students as part of his own "scientific study of the Negro body."[77] DuBois assembled anthropometric measurements of Black students from Virginia's Hampton College. He included an extensive bibliography that listed the work of "racial scientists" of his and earlier eras, such as Robert Bennett Bean. Claiming the supposed precision of anthropometry for his own uses, he placed a picture of his Atlanta laboratory early in the book to show where "exact measurement" could take place in a non-white facility.[78]

Caroline Bond Day was another Black researcher who completed important anthropometric studies. Day studied under Earnest Hooton at Harvard and produced a photo study of mixed race, "Negro-White" Families as her 1929 master's thesis.[79] She measured them with the tools of anthropometry.[80] Hooton and Day shared the then conventional view that racial characteristics were fixed and essential; racial categories clear and distinct.[81] Day knew DuBois and was a graduate of Tuskegee.[82] She listed "blood quantum" numbers in her analysis of families, such as "7/16 white."[83] The NAACP magazine, *The Crisis* printed a review of Day's book, praising it as "one of the most intelligent scientific studies of the Negro yet made."[84]

Though the Davenport/Steggerda study at Tuskegee was the first work that was carried out there with by researchers employed by a preeminent institution of eugenics, at least two other researchers had measured the children at Tuskegee and other Black schools earlier. Northwestern University anthropologist Melville Herskovits did a similar study at Howard University, published in

---

77 Dubois to Franz Boas, quoted in Maria Farland, "W.E.B. Dubois, Anthropometric Science, and the Limits of Racial Uplift," *American Quarterly*, vol. 58 (December 2006) 1017–1045, at 1022.

78 Farland, "W.E.B. Dubois," 1034.

79 Caroline Bond Day, A *Study of Negro-White Families in the United States*, with a forward and notes on the anthropometric data by Earnest A. Hooton (Cambridge, MA: Peabody Museum of Harvard University, 1930; Reprinted, New York: Negro Universities Press, 1970).

80 Heidi Ardizzone, "Such Fine Families": Photography and Race in the Work of Caroline Bond Day, *Visual Studies*, vol. 21 (October, 2006) 106–132, at 106, 111. Hooton supplied a sixty page commentary on anthropometry in Day's book.

81 Ardizzone, p. 107. Ardizzone asserts that "Day's main aim in her study was to provide information about the inheritance of racial traits in mixed race families." at 109.

82 Ardizzone, p. 111.

83 Ardizzone, Shown on fig. 6b, p.112. This estimate of percentage of "blood" from different races was a staple not only of race study but of standard genealogy as well.

84 Ardizzone, p. 128.

1930.[85] According to Herskovits, at least three different studies were done by him and two other investigators from 1915 until 1930 at Howard, Tuskegee, Fisk (Nashville), and Hampton Virginia, all on Black school or college students. Herskovits also studied Black school children in Harlem.[86]

Nor was it surprising that Cleve Abbott and the Tuskegee coaching staff would cooperate with Davenport and Steggerda; body typing had long been popular among physical education teachers.[87]

Just as the Davenport/Steggerda study was beginning, Harvard University planned an exhibit featuring an Anthropometric Laboratory at the Chicago World's Fair that would start in 1933. According to Harvard anthropologist Earnest Hooten, the goal of the exhibition was to show visitors to the Fair the newest methods of anthropometric technique and statistical analysis, while simultaneously compiling a sample of the physical characteristics of fair-goers, with the objective "to establish standards of the American Type."[88] Eventually some 6000 people were measured,[89] photographed and analyzed according to family background and "racial type." Hooten was surprised that, with past associations of anthropometry with criminal identification, such a large number of people would present themselves voluntarily for analysis, but the result suggests that in the year after Davenport and Steggerda began their work at Tuskegee, visitors to the Chicago Fair felt no particular discomfort at being research subjects in an anthropometric study.[90]

85 Melville J. Herskovits, *The Anthropometry of the American Negro* (N.Y. Haskell House, 1930). Herskovitz studied under both E.L. Thorndike and Franz Boaz.

86 Ibid, p. 2–6 . Herskovits credits scholar Beatrice Blackwood for work at Fisk University and Tuskegee. Blackwood visited Tuskegee to record the skin color of various students in an attempt to understand the inheritance of racial characteristics, see Herskovits, p. 5; and Beatrice Blackwood, "Racial Differences in Skin Colour as Recorded by the Colour Top," *Journal of the Royal Anthropological Institute of Great Britain and Ireland*, vol. 60 (January–June, 1930) 137–168.

87 "Patricia Vertinsky, "Embodying Normalcy: Anthropometry and the Long Arm of William H. Sheldon's Somatotyping Project," *Journal of Sport History*, vol. 29 (Spring, 2002) 95–133, at 114.

88 *Official Guidebook of the World's Fair of 1934* (Chicago: Cuneo Press, 1933) 94.

89 "Earnest A. Hooton, "Development and Correlation of Research in Physical Anthropology at Harvard University," *Proceedings of the American Philosophical Society*, vol. 75 (1935) 499–516, at 511.

90 Details of Harvard's Chicago World's Fair exhibit are taken from Earnest Hooten, "The Harvard Anthropometric Laboratory at A Century of Progress Exposition," E.A. Hooten Papers, Peabody Museum, Harvard University.

Davenport Retires

Davenport monitored Steggerda's activities throughout the twelve years of research at Tuskegee. He described the work as part of the larger group of studies that looked at "the comparative anthropology of the Navajo Indians, the Mayan Indians in Yucatan, the negroes in Tuskegee and a certain group of whites in Holland, Michigan,"[91] and regularly apprised colleagues about the "repeated measurements of growing negro children at Tuskegee, Alabama."[92]

But the security of Davenport's protégés became increasingly fragile after he stepped away from daily ERO business in 1934. Harry Laughlin, perhaps the most public face of eugenics after Davenport, found his own research under attack following a 1935 audit by the Carnegie Institution. The work in "racial cleansing" for which he had won an honorary doctorate from the Nazis was dismissed as "a vast and inert accumulation...unsatisfactory for the scientific study of human genetics." It had also simply become too controversial. When Vannevar Bush became President of the Carnegie Institution in 1939, further review of Laughlin's work led to his forced retirement in 1940.[93] Racial researcher Arthur Estabrook, author of *Mongrel Virgin*ians, had been dismissed years earlier.[94] Steggerda was among the last people remaining at the ERO who owed his job to Davenport and had continued to follow his mentor's path in the study of race.[95]

By 1940, Bush was determined that the Carnegie Institution make a clean break from its early identification with eugenics. So when Steggerda solicited his mentor's counsel on future career prospects at Cold Spring Harbor, Davenport steered him away from further study in areas where "race crossing" was prevalent, closing the door on what had been a major interest of his own career as well as Steggerda's. He suggested that if Steggerda was "less interested in human genetics or in the inheritance of particular traits than in the study of race crossing, or of primitive peoples," he needed to make his preferences clear to Bush.[96] But Bush had already set his agenda, and no new position

91 Davenport to Raymond R. Willoughby (Clark University), October 16, 1934, ERO.

92 Davenport to Alfonso Caso (Mexico), November 17, 1937, ERO.

93 Garland E. Allen, "The Eugenics Record Office at Cold Spring Harbor, 1910–1940: An Essay in Institutional History," Osiris, 2nd Series, Vol. 2 (1986), pp. 225–264, at 254.

94 For the details of Estabrook's dismissal from the ERO in 1929, see Lombardo, *Three Generations, No Imbeciles: Eugenics, the Supreme Court and Buck v. Bell* (Baltimore: Johns Hopkins University Press, 2008), 184.

95 When Davenport's announced his own retirement, Steggerda wrote him a "letter of appreciation." In the intimate tribute, he called Davenport "a noble character...a second Darwin in the field of biology." Steggerda to Davenport, January 10, 1940, ERO.

96 Davenport optimistically concluded "I certainly hope that you will be able to present a case that will result in Dr. Bush wishing you to remain with the Department." Davenport

materialized. When Steggerda was offered a part-time teaching position, Bush eased his exit by promising to subsidize his lower pay at the new job.[97]

Davenport's Final Thoughts on Race

Some commentators have suggested that both Davenport and Steggerda had begun to distance themselves from a belief in genetically fixed mental traits,[98] but Steggerda's final comments on Davenport require a different conclusion. Steggerda said that Davenport's insistence on the "constitutional, hereditary, genetical basis" of difference, in mental and physical traits, particularly between races, was a "dominant idea." He also pointed out the apparently unchanging assertion of his mentor that racial mixture – miscegenation – led to physical, mental and temperamental disharmonies in offspring, and thus should be avoided.[99] Davenport, like Francis Galton before him, used composite photographs, then anthropometry, aiming to identify a "pure type" from various population groups. "It is the essential notion of a race," said Galton, "that there should be some ideal form from which the individuals may deviate in all directions . . . and toward which their descendants will continue to cluster."[100] Davenport apparently clung to this formulation most of his career and it was a motive in his design of the Tuskegee anthropometric research.

Davenport's long-standing racial biases were in evidence less than a year before he died, when an article in *Time* Magazine brought his true feelings about race to the surface. *Time* argued for the repeal of the Chinese Exclusion Act, first enacted in 1882, but still in place and causing some embarrassment in light of the alliance between the U.S. and Nationalist China during the 2nd World War. In its place, a bill was pending that would allow Chinese to migrate to the U.S.

---

to Steggerda, May 25, 1940, ERO. Steggerda's response was polite but he made it clear that his revised proposal had emphasized his many publications, and he noted, defensively, that during his time at the ERO he had done "much more than merely gather data." Steggerda to Davenport, May 29, 1940, ERO.

97 Letter from Vannevar Bush to Dr. Robbins Wolcott Barstow, November 30, 1943, ERO.

98 Wilton Marion Krogman, "Fifty Years of Physical Anthropology: The Men, the Materials, the Concepts, the Methods," *Annual Review of Anthropology*, vol. 5 (1976) 1–14, at 9.

99 Morris Steggerda, "Charles Benedict Davenport (1866–1944) the Man and His Contributions to Physical Anthropology," *American Journal of Physical Anthropology*, vol. 2 (June, 1944) 167–185, at 175.

100 Francis Galton, *Inquiries into Human Faculty and its Development* (London: MacMillan, 1883) 10.

under a quota system that would allow approximately 105 immigrants a year.[101] Davenport was so incensed by the article that he wrote a letter to the editor.

In it, he spun a parable about a bee hive, operating in "perfect harmony" because of the "homogeneity of instincts" of its inhabitants. But, when outsider, interloper bees from other hives intruded, particularly when their motive was "sharing the bounty" of a strange hive, they were "killed forthwith." So too with human society, Davenport suggested. People of "dissimilar instincts" should be kept apart, he said, and when "large hearted and short sighted sentimentalists" label such a conclusion "race prejudice," they fail to understand that foreigners are "better fitted by instinct and behavior to their own land." Even the critics "would not like to have injected into our body politics ten million Japanese, fifty million Bantu Negroes of African origin, fifty thousand of the pigmies of Africa or ten thousand of Australian 'black fellows'."[102] This comment echoed Davenport's conclusions from years earlier, following his anthropometric study in Jamaica. He believed that studying the physical form of different kinds of people could yield insights that could inform public policy on topics such as marriage laws and immigration restriction.

The Jamaica experience had taught him that "there exists in mankind a strong instinct for homogeneity. Even children tend to mock at the cripple or deformed person. A homogeneous group of white people will always be led by its instincts to segregate itself from Negroes, Chinese and other groups that are morphologically dissimilar from themselves. We should consider the psychological, instinctive basis of this feeling. It is not sufficient merely to denounce it. It probably has a deep biological meaning. ..."[103] Davenport monitored the end of the Tuskegee data collection, which occurred just two years before his death in 1944.

Steggerda Leaves the Eugenics Record Office

In June, 1944, *Science* magazine announced that Steggerda's fifteen years at the Carnegie Institution at Cold Spring Harbor would come to an end with his appointment as professor of anthropology at the Kennedy School of Missions, a division of the Hartford Seminary Foundation.[104] Only two years earlier,

---

101 "The Congress: 105 Chinese," *Time* (June 14, 1943) 1.

102 Davenport to Editor of *Time*, June 11, 1943, ERO.

103 Dr. C.B. Davenport, "Race Crossing in Jamaica," *Scientific Monthly*, vol. 27 (September, 1928) 225–238, at 238.

104 "Scientific Notes and News," *Science*, Vol. 99, (June 16, 1944) 488–489. That same year, Steggerda's sometimes fawning obituary of his mentor described Davenport as "...a zoologist, biologist, geneticist, eugenicist and anthropologist." Morris Steggerda, "Charles

Steggerda had suffered what he characterized as a "serious illness" and "collapse." He subsequently had a religious awakening and for the final six years of his life taught the techniques of anthropological study to missionaries in training. During that time he wrote *Meditations*, a book in which he mused about Bible verses. He discussed the story of Esau, whose marriage of two foreign women caused "bitter disappointment" to his parents. Steggerda drew this lesson from the story: ". . . people ought to marry those of their own kind, and of their own upbringing, their ways of thinking, their same economic level, and certainly as much as possible of their own religion."[105] He included thoughts entrusted to his diary years earlier about his time at Tuskegee in *Meditations*.

Steggerda reflected on the time he had spent "on the campus of one of our great Negro schools," collecting data that would eventually be analyzed in dozens of scientific reports and papers. Despite the professional success he had experienced, his assessment of the work at Tuskegee was marked by disappointment. He described the behavior of his hosts at Tuskegee, who exhibited "a unique politeness which is almost to the point of uneasiness." Though he praised the amenities he was afforded at the school, in his attempts to make new friends, he complained that he had "not gotten very far."

In fact, Steggerda had experienced ". . . more loneliness on this American campus than I have in the mountains of Jamaica, the jungles of Mexico or the heart of the Navajo desert." In eleven years of visits, he was invited to a total of two homes, even though he had invited fifty of his "Negro friends" to his own. None ever came. He concluded that he had failed in "breaking down the barrier between me, a sympathetic anthropologist, and Negroes whom I like to think of as my friends."[106]

At least when it was first conceptualized, the work of Davenport and Steggerda at Tuskegee fit within the goals of colleagues like Ales Hrdlicka and Harry Laughlin who attempted to develop legislation that would allow for categorization of "True Americans" and a legal definition of "The American Race" for purposes of immigration restriction and marriage law – all based on a metric that would rely on racial differences that were quantifiable and considered by Davenport "quite certainly genetic."[107] But Steggerda was of a different generation, and looked at race with a newer lens that older members of the eugenic

---

Benedict Davenport (1866–1944) the Man and His Contributions to Physical Anthropology," *American Journal of Physical Anthropology*, vol. 2 (June, 1944) 167–185, at 170.

105 Morris Steggerda, *Meditations: Fifty-Two Religious Essays* (Holland, Michigan: Holland Letter Service, 1947).

106 Steggerda, *Meditations*, 93.

107 Paul A. Lombardo, "The American Breed: Nazi Eugenics and the Origins of the Pioneer Fund,"*Albany Law Review*, vol. 65 (2001–2002) 743–830.

corps. The coarse race-baiting that characterized eugenic stalwarts like Madison Grant was not his style, nor was the more circumspect but no less noxious "scientific" posturing of his mentor Davenport. Nevertheless, he shared many of the attitudes of Davenport about the need for racial separation and the distinctions he believed inherent to the different "races." Yet he was unable to discern how spending a lifetime collecting data meant to bolster a system of racial separation might be met with suspicion among his subjects. Steggerda died in 1950.

At this remove, the Davenport/Steggerda anthropometric study at Tuskegee – via the instruments used, the ideological proclivities of the eugenicists who designed and carried it out and the racial typologies it was meant to yield – not to speak of the assistance of members of the Tuskegee faculty & staff – represents a unique window into racial research in the U.S. It generated data that was used in over two dozen publications.[108]

108 Steggerda's periodic accounts of research at Tuskegee were published by the Carnegie Institution of Washington in its Annual Reports; other publications from the Tuskegee data included: Morris Steggerda and P. Densen, "Height, weight, and age tables for homogeneous groups, with particular reference to Navajo Indians and Dutch Whites," *Child Development*, vol. 7 (1936) 115–120; Steggerda, Morris, "Testing Races for the Threshold of Taste, with PTC," *Journal of Heredity*, vol. 28 (1937) 309–310; Morris Steggerda and Christine Evans Petty, "An Anthropometric Study of Negro and White College Women," *Research Quarterly*, vol. 11 (October, 1942) 110–118; Morris Steggerda and Christine Evans Petty, "Body Measurements on 100 Negro Males from Tuskegee Institute," *Research Quarterly*, vol. 13 (October, 1940) [275–278]; Morris Steggerda, "Cross Sections of Human Hair from Four Racial Groups," *Journal of Heredity*, vol. 31 (November, 1940) 474–476; Morris Steggerda, "Physical Measurements on Negro, Navajo, and White Girls of College Age," *American Journal of Physical Anthropology*, vol. 26 (March, 1940) 417–431; Barbara S. Burks and Morris Steggerda, "Potential marital selection in Negro college students," *Sociology and Social Research*, vol. 24 (May–June, 1940) 433–441; Morris Steggerda and Henri Seibert, "Size and Shape of Head Hair from Six Racial Groups," *Journal of Heredity*, vol. 32 (September, 1941) 315–318; Morris Steggerda, "Form discrimination test as given to Navajo, Negro and white school children," *Human Biology*, vol. 13*(1941) 239–246;* Morris Steggerda, "Change in hair color with age," *Journal of Heredity*, vol. 32, (1941) 402–403; Morris Steggerda and Thomas J. Hill, "Eruption time of teeth among Whites, Negroes, and Indians," *American Journal of Orthodontics and Oral Surgery* vol. 28 (June, 1942) 361–370; Morris Steggerda, "Significance of Racial Factors in Physical Measurements of Normal or Defective Children," *American Journal of Mental Deficiency*, vol. xlvi (April, 1942); Morris Steggerda and H.C. Seibert, "The size and shape of human head hair," *Journal of Heredity*, vol. 33, (1942) 302–304; Georg Wolff and Morris Steggerda, "Female-Male Index of Body Build in Negroes and Whites: An Interpretation of Anatomical Sex Differences," *Human Biology*, vol. 15 (May, 1943) 127–152; Morris Steggerda and H.C. Seibert, "Age and hair form," *Journal of Heredity*, vol. 35 (1944) 345–347; and, Morris Steggerda, "Anthropometry and the eruption time of teeth," *Journal of the American Dental Association*, vol. 32 (March, 1945) 339–342.

CHAPTER 9

Nazi Human Experiments: The Victims' Perspective and the Post-Second World War Discourse

Paul Weindling

Whereas much has been written about the perpetrators of Nazi medical atrocities, little is known about the victims either in terms of how many, or who they were, or their experiences of being subjected to experiments. Such a ground up reconstruction in turn informs on how, when and where the experiments occurred, and the victims' narratives can say much about the conduct and scale of the experiments. This chapter develops a balanced perspective, in that it brings victims into the frame. It draws on the survivors' voice which although fragmentary and fragile in terms of injuries was eloquent and informative about what had been experienced. A debate arose in the aftermath of WW2 on the conduct and quality of the Nazi experimental work, and here victims were vocal about their experiences, and in responding to questions of the validity of the research.

It should be observed that the bioethical focus on coercion and the lack of consent has been top down – primarily on the perpetrators. The Nuremberg Medical Trial from December 1946 until August 1947 did give a voice to victims in terms of their testimonies, but even so the main concern was to locate the perpetrators in command hierarchies stretching up to Himmler and Hitler.[1] The issue of whether the perpetrators informed the research subject and obtained consent is one that at best sees the victim as passive, denying agency and thereby a conscious sensibility, reflective capacity and voice. This chapter proposes a wider framework. This is made possible by analysing the identities of the victims, the structure and extent of the human experiments, and giving attention to victim voices and experiences.

1 The Crystallization of a Programme

German medicine is distinguished by a high level of scientific and experimental research with the title of "Herr (and from around WW1, "Frau") Doktor"

1 Paul Weindling, "Victims, Witnesses and the Ethical Legacy of the Nuremberg Medical Trial", in Kim Priemel and Alexa Stiller (eds), *The Nuremberg Trials* (New York: Berghahn Books), 74–103.

 | DOI 10.1163/9789004286719_011

contingent on a research thesis for the MD.[2] Such scientisation contributed to the international renown achieved by German universities, research institutes and clinics until WW2. Such research was inherently problematic when it came to human subjects. Some researchers solved this by an ethic of self-experimentation, or experiments on their own children. This ethic remained throughout the twentieth-century: the Nobel-prize winning research by Werner Forssman on heart catheterization based on self experimentation demonstrates how voluntaristic styles of experiments with the scientist as his own research subject ran parallel to coercion on an unprecedented scale. While there were therapeutic trials on sick persons, increasingly, deliberate infection was used. The coerced Nazi experiments – on a colossal scale (matched only by the Japanese wartime experiments) – occurred within a wider pattern of self-experimentation and clinical research, making the Nazi experiments all the more remarkable.

Prior to 1933 came nearly 50 years of highly charged public discourse on experimental medicine in Germany. There had been a series of protests against scientists exploiting patients, and institutionalised children and adolescents for research in the years from 1899 until 1933.[3] Opposition to the coercive Nazi experiments arose from the point of view of the ethic of self-experimentation. Chief Luftwaffe medical inspector Erich Hippke is a case in point as he invoked self-experiments to disapprove of the notorious low pressure experiments at the concentration camp of Dachau that took research subjects to the point of death.[4] Sporadic medical critique (albeit weak and attenuated) and reference to Reich guidelines for consensual experiments, as well as disengagement and distancing by scientists from the excesses of their colleagues ran parallel to resistance and sabotage by the victims and other prisoners.[5]

Compulsory sterilization measures meant that research was conducted on the illnesses and disabilities associated with the law such as hereditary

---

2 Thomas Neville Bonner, *Becoming a Physician: Medical Education in Britain, France, Germany, and the United States, 1750–1945* (Oxford: Oxford University Press, 1995).

3 Wolfgang U. Eckart, Andreas J. Reuland, 'First Principles: Julius Moses and Medical Experimentation in the late Weimar Republic', Eckart, ed., *Man, Medicine, and the State: The Human Body as an Object of Government Sponsored Medical Research in the 19th Century* (Stuttgart: Franz Steiner Verlag, 2006), 35–38.

4 Staatsarchiv Nürnberg Rep 502 KV-Anklage Interrogations Generalia H 143 Erich Hippke.

5 Paul Weindling, *Victims and Survivors of Nazi Human Experiments: Science and Suffering in the Holocaust* (London: Bloomsbury, 2014). Volker Roelcke, Simon Duckheim. 'Medizinische Dissertationen aus der Zeit des Nationalsozialismus: Potential eines Quellenbestands und erste Ergebnisse zu „Alltag", Ethik und Mentalität der universitären medizinischen Forschung bis (und ab) 1945', *Medizinhistorisches Journal* 49, 2014, 260–271.

deafness. From 11 March 1935 Nazi race hygienists and civil servants decided on the sterilization of the mixed race children who were stigmatized as "Rhineland Bastards", the name reflecting the term for a cross-breed.[6] In all 385 children and adolescents were rounded up and evaluated by scientific commissions, including the anthropologists Wolfgang Abel, Eugen Fischer, and the assistant of the racial hygienist Otmar von Verschuer, Heinrich Schade.[7] The sterilizations established a pattern – first, using administrative machinery to identify a group of racial undesirables; then academic study and evaluation of the adolescents; finally, their sterilization. This process happened time and time again.

The testimony of one of the Rhineland victims shows how the procedures occurred without explanation. Hans Hauck was born in Frankfurt am Main on August 10, 1920 as the son of an Algerian soldier and a German mother. He joined the Hitler Youth in 1933. But he was summoned for sterilization. He recollected when Gestapo officers collected him and fellow adolescents: "we all were too scared to object. I suspected something would happen, but did not know about sterilization and castration." He had to sign that he would not marry or have sexual relations. He was held for 14 days with a group of mixed race adolescents, who were all very scared, while research was conducted on them.[8]

The coerced experiments as a concerted programme emerged only at the time of the Second World War, and ran for a few years making the high number of victims remarkable. Experiments occurred in clinical contexts, notably psychiatric hospitals, ghettoes and in Polish hospitals under German occupation. Concentration camps existed in Nazi Germany since 1933 but they were not at first used for research. The extermination camps, Belczec, Sobibor and Treblinka were not used for research, as the priority was efficient killing of large numbers. By way of contrast, anthropological research and experiments occurred from 1939 in the concentration camps that held prisoners for forced labour, notably Buchenwald, Dachau, Mauthausen, Natzweiler, Neuengamme, Ravensbück and Sachsenhausen. Auschwitz as a dual extermination and forced labour camp (actually a set of camps) offered from 1943 an immense

---

6 Iris Wigger, *Schwarze Schmach am Rhein. Rassistische Diskriminierung zwischen Geschlecht, Klasse, Nation und Rasse* (Münster: Westfälisches Dampfboot, 2006)

7 Hans-Walter Schmuhl, *Grenzüberschreitungen. Das Kaiser-Wilhelm-Institut für Anthropologie, menschliche Erblehre und Eugenik, 1927–1945* (Göttingen: Wallstein, 2005), 291–299.

8 Hans Hauck, Shoah Foundation Interview code 41964. Hauck was born Aug 10, 1920 at Frankfurt am Main. Tina Campt, *Other Germans: Black Germans and the Politics of Race, Gender and memory in the Third Reich* (Ann Arbor: University of Michigan Press, 2005).

human population for research, and holding facilities extensive facilities for forced experimentation.

There were two types of research in the concentration camps: first on physical anthropology by the Reich Health Office's Criminal Biology Departmemt. The other type of research was toxicological or involved experimental physiology. Research at Sachsenhausen was initiated by a camp doctor in 1939. It was thus only with the outbreak of war that concentration camps became used for human experiments. Sachsenhausen was sited close to Berlin at Oranienburg which was both the centre of SS administration and where the Auer Company was developing scientifically based armaments production. Here, in September 1939 experiments on the poison gas "Yellow Cross" or Yperit (a type of mustard gas and lewisite combination) were conducted on 31 prisoners. These research subjects were Germans, although one Czech was among them. The initiative came from the camp doctor and SS officer, Hugo-Heinz Schmick. The experiments ceased when Schmick moved to the SS hospital of Hohenlychen, which was itself set to become a major medical research centre involved in coercive experiments.[9]

Other research was conducted in prisoner of war camps, as with the military doctor and bacteriologist Friedrich Meythaler infecting Allied prisoners with hepatitis on Crete in 1941.[10] Hepatitis research then exemplifies the shift from field research to the concentration camp – in 1943 the bacteriologist Arnold Dohmen went from Berlin to Auschwitz to select eleven Jewish boys, as they arrived in Auschwitz. They were then transferred to Sachsenhausen where the experiments were conducted during 1944.[11]

9 Wolfgang Woelk, 'Der Pharmakologe und Toxikologe Wolfgang Wirth (1898–1996) und die Giftgasforschung im Nationalsozialismus', Woelk, Frank Sparing, Karen Bayer, Michael Esch, eds, *Nach der Diktatur. Die Medizinische Akademie Düsseldorf nach 1945* (Essen: Klartext, 2003), 269–287, 277. Judith Hahn, *Grawitz, Genzken, Gebhardt, drei Karrieren im Sanitätsdienst der SS* (Münster: Klemm & Oelschläger, 2007), 271–272. Alexander Neumann, *"Arzttum ist immer Kämpfertum". Die Heeressanitätsinspektions und das Amt "Chef des Wehrmachtsanitätswesens" im Zweiten Weltkrieg (1939–1945)* (Düsseldorf: Droste, 2005), 290–291. Astrid Ley, Günter Morsch, *Medical Care and Crime. The Infirmary of Sachsenhausen Concentration Camp 1936–1945* (Berlin: Metropol, 2007), 333.

10 F. Meythaler, 'Zur Pathophysiologie des Ikterus', *Klinische Wochenschrift*, 21 no. 32 (8 August 1942) 701–706. B. Leyendecker, F. Klapp, 'Human Hepatitis experiments in the Second World War', *Zeitschrift für die gesamte Hygiene*, 35.12 (1989) 756–760. B. Leyendecker, B.F. Klapp,'Deutsche Hepatitisforschung im Zweiten Weltkrieg', Ärztekammer Berlin in Zusammenarbeit mit der Bundesärztekammer, eds, *Der Wert des Menschen; Medizin in Deutschland 1918–1945* (Berlin: Edition Hentrich, 1989).

11 Ley, Morsch, *Medical Care and Crime.* Saul Oren-Hornfeld, *Wie brennend Feuer. Ein Opfer medizinischer Experimente im Konzentrationslager Sachsenhausen erzählt* (Berlin: Metropol, 2005).

War initially disrupted medical research with physicians being called up for military service – the years 1940–41 saw little in the way of coerced experiments. The neurologist Georg Schaltenbrand analysed cerebrospinal fluid and blood lipoids. He experimented on multiple sclerosis at Schloss Werneck clinic in an attempt to prove that it was an infectious disease. He injected human spinal fluid into apes and then from apes into patients. Schaltenbrand viewed multiple sclerosis as damaging fertility, and its concealment as damaging marriages. He considered multiple sclerosis should be grounds for forced abortion and sterilisation.[12] Schaltenbrand's experiments on multiple sclerosis were disrupted by the requisitioning of hospitals by the military, when patient-research subjects were "transferred", many to be killed.

The sense that Germany was slipping behind the Allies in terms of medical research was well shown by the Allied development of penicillin. The confrontation with an intractable range of war-related medical issues such as wound infection and epidemic control meant that in late 1941 the SS began to support experiments in concentration camps, mainly on military related topics such as survival in extreme conditions or infectious diseases being encountered on the Russian front. By 1942 large scale experimental research in concentration camps as well as opportunistically in psychiatric hospitals was underway.

II Holocaust Historiography

The first person on the Allied side to draw attention to the coerced experiments was the scientific intelligence officer, John W. Thompson. At the time he was seconded from the RCAF to a British disarmament unit, and then from November 1945 to a field intelligence gathering organisation, FIAT. Through encountering survivors of Bergen-Belsen, he became aware of experiments at Auschwitz. He widened this by contacting medical and scientific and war crimes organizations regarding the experiments as a special type of "medical war crimes". He then set about a full documentation of all experiments, irrespective whether the perpetrators were available for trial or known to have

12 Georg Schaltenbrand, *Die Multiple Sklerose beim Menschen* (Leipzig: Thieme, 1943), 252. Thomas Schmelter, Christine Meesmann, Gisela Walter, Herwig Praxl: Heil- und Pflegeanstalt Werneck. In: Michael von Cranach, Hans-Ludwig Siemen, eds, *Psychiatrie im Nationalsozialismus. Die Bayerischen Heil- und Pflegeanstalten zwischen 1933 und 1945* (Munich: Oldenbourg, 1999), 35–54.

died in the war.[13] Thompson's documentation on perpetrators provided the basis for the US decision to hold a special trial at Nuremberg concerned with medical experiments. Here, a delegation of German medical observers published fundamental accounts based on documents presented in court.[14]

Although the human experiment victims are in many ways iconic of the worst atrocities of the Holocaust (one has only to think of the Mengele twins), the Nazi human experiments scarcely figure in the historiography of the "final solution" and Nazi genocide. Indeed, the experiments have only a tenuous historical position, even though they are widely recognized as the very worst of the Nazi atrocities. The doyen of Holocaust historians, Saul Friedländer makes the important point that while research on Jews was intended to generate an ideological stereotype, it impacted on myriad Jewish lives and voices. For the human experiment victims, it is important to recognize their agency and individuality. Friedländer cites just few examples of the very worst atrocities. He shows how Mengele commanded "Zwillinge heraus" (twins to the fore) when selecting arrivals at Auschwitz-Birkenau for the gas chambers. No historian (as opposed to significant documentation by journalists) has engaged to any extent with the experiences of the twins, individually and collectively, or other experiment survivors. Friedländer draws attention to research on living victims who were then killed, as by the anatomist Paul Kremer in Auschwitz, seeking to study the effects of starvation on internal organs.[15] Again, historians neglected both Kremer's rationales (he researched environmental effects in evolution), and the life histories of those whose bodies were used to fuel the gargantuan appetites of German medical scientists for anatomical specimens.[16] Yet figures like Kremer and Mengele were only extreme instances

13 Paul Weindling, *John W. Thompson, Psychiatrist in the Shadow of the Holocaust* (Rochester, NY: Rochester University Press, 2010). My thanks to Larry Stewart for identifying Thompson's medical career.

14 Alexander Mitscherlich, Fred Mielke (eds), *Das Diktat der Menschenverachtung* (Heidelberg: Lambert Schneider, 1947). Alexander Mitscherlich, Fred Mielke (eds), *Medizin ohne Menschlichkeit, Dokumente des Nürnberger Ärzteprozesses* (Frankfurt am Main: S. Fischer, 1960). 1 edn Heidelberg: Lambert Schneider, 1949 as *Wissenschaft ohne Menschlichkeit*. Alexander Mitscherlich Fred Mielke (with contributions by Ivy, Taylor, Alexander and Deutsch), *Doctors of Infamy: The Story of the Nazi Medical Crimes* (New York: Henry Schuman, 1949). Alice Platen-Hallermund, *Die Tötung Geisteskranker in Deutschland: Aus der deutschen Ärzte-Kommission beim amerikanischen Militärgericht* (Frankfurt: Verlag der Frankfurter Hefte, 1948).

15 Saul Friedländer, *The Years of Extermination* (New York: Harper Collins, 2007), 505, 507.

16 Sabine Hildebrandt, 'Anatomy in the Third Reich: an outline, part 2. Bodies for anatomy and related medical disciplines', *Clinical Anatomy*, 22/8 (2009), 894–905.

illustrative of a dynamic of experimentation and murderous research, a dynamic that took a wide range of forms, and included a high proportion of non-Jewish victims – Jewish corpses from concentration camps were considered too emaciated for dissection, and a potential research subject found to be Jewish might be ejected from an experimental block at Buchenwald.[17]

The Coryphaei of historians of Nazi Germany confirm the tenuous position accorded to racial science as a driving force of Nazi policy. Peter Longerich as biographer of Himmler gives scarcely any attention to the Reich leader of the SS's role in medical research. Michael Kater in a pioneering study balanced experimental projects of the SS Ancestral Research organization in medical research with extensive treatment of archaeology and pre-history.[18] Richard Evans' grand narrative of Nazi Germany identifies human experiments in Auschwitz as normal science. Yet his account of Mengele is detached from racial policy as well as racial rationales, and fails to clarify Mengele's sources of support.[19] The depiction is one of Mengele as "engaged in pure research without any obvious application" within parameters of a previously nazified profession.[20] Yet far from detached, Mengele appears to exploiting his position as camp doctor in the "Gypsy Camp", when researching children with Noma (a deficiency disease causing facial disfiguration), or opportunely using his position on the Auschwitz ramp to identify twins and dwarves among the streams of arriving Jews. Mengele was anxiously seeking material for his intended habilitation under Otmar von Verschuer, while applying racial science in his selections at Auschwitz.

The efforts at grand narrative synthesis find only limited place for coerced experimentation. One unfortunate consequence is that this has resulted in a focus on a select number of perpetrators, and too narrow a focus on SS atrocities. A related misconception is that the experiments solely arose from an ideology of racial purity. The idea that this was "pseudo-science" (here an honorable exception may be made for Evans who however leaves wider questions unanswered) leaves the mainstream German science and medicine community wholly detached and uninvolved. One can instead see scientific agendas

17 Saul Friedländer, *The Years of Extermination* (New York: Harper Collins, 2007), 505, 507, 655–656.

18 Peter Longerich, *Heinrich Himmler* (Oxford: Oxford University Press, 2012), 275–277. Michael H. Kater: *Das "Ahnenerbe" der SS 1935–1945* (Munich: Oldenbourg Verlag, 2006), 1 edn 1974.

19 Richard Evans, *The Third Reich at War: How the Nazis Led Germany from Conquest to Disaster* (London: Allen Lane, 2008).

20 http://www.telegraph.co.uk/science/science-news/3540339/How-Hitler-perverted-the-course-of-science.html (accessed 1 March 2012).

TABLE 9.1 *Death and survival*

Circumstances of death	Confirmed victim	Pending	Total
Killed for research	2956	50	3006
Died or killed after the experiment	781	23	804
Died from experimental procedures	383	171	554
Grand total	4120	244	4364

being pursued regardless of the human cost.[21] While ambitious young researchers like the physiologist Sigmund Rascher in Dachau and Mengele in Auschwitz were SS members, the organization of the experiments was contingent on other administrative structures. Despite Rascher's close connections with Himmler, the deadly experiments that took prisoners to the point of death in freezing and low pressure were conducted under the auspices of the Luftwaffe, the German air force. Similarly, Mengele's twin research at Auschwitz had the support of the German Research Fund (*Deutsche Forschungsgemeinschaft*), rather than being in any way centrally directed by the SS. There were over 1000 victims of experiments which were fatal, or who died afterwards (see Table 9.1).

The experimenters were on occasions not SS officers nor even Nazi Party members. Claus Schilling, the malariologist who experimented in Dachau from 1942, was neither; Schilling gained Himmler's support for infecting prisoners and testing dangerous drug treatments from 1942 until US forces liberated the camp in April 1945. After liberation Schilling asked permission to continue the experiments, albeit with volunteers. In the event Schilling was executed after being placed on trial.[22] The chemicals and pharmaceutical corporation, IG Farben sponsored experiments in concentration camps with its products such on treatment of malaria and typhus. Mostly (but not always), such experiments had support from Himmler, and a cluster of senior SS doctors, keen to develop competing research hierarchies.

21 Volker Roelcke, 'Fortschritt ohne Rücksicht. Menschen als Versuchskaninchen bei den Sulfonamid-Experimenten im Konzentrationslager Ravensbrück', Insa Eschebach, Astrid Ley, eds *Geschlecht und Rasse in der NS Medizin* (Berlin: Metropol, 2012), 101–114.

22 Cf Marione Hulverscheidt, 'German Malariology Experiments with Humans supported by the DFG until 1945', *Man, Medicine and the State*, 226–227. Hana Vondra, "Malariaexperimente in Konzentrationslagern und Heilanstalten während der Zeit des Nationalsozialismus". Dissertation an der MH Hannover (Abtlg. Geschichte der Medizin), 1999.

The SS financial administrator Oswald Pohl had to grant resources. He did this in collaboration with the Reich SS Physician, Ernst Grawitz. In open rivalry to Grawitz, Wolfram Sievers developed a special department for military research within the Ancestral Research organization of the SS. Sievers supported the murderous Jewish skeleton collection involving the selection in Auschwitz and gassing of 86 victims at Natzweiler-Struthof in Alsace, and other initiatives of the anatomist August Hirt, as well as those of virologist Eugen Haagen, requiring Roma transports from Auschwitz to Natzweiler in Alsace. Carl Clauberg's research on sterilisation by means of inter-uterine injection of chemical and hormonal substances was supported by Himmler. Clauberg thus obtained financial support deriving from the funding of the General Plan for the East. A further tier of authority came through the Hygiene Institute of the Waffen SS under the epidemiologist Joachim Mrugowsky. He had under his command satellite institutes at Buchenwald (the site of the notorious experiments of the bacteriologist Erwin Ding) and at Raisko near Auschwitz. At the same time, researchers had contacts with outside providers of research resources as Ding with the IG Farben conglomerate, and Mengele with the Berlin-based director of the KWI for Anthropology, Otmar von Verschuer and other scientists from the Kaiser Wilhelm Society. Mengele appears to have improvised resources for his research in Auschwitz from April 1944, a year after arriving at Auschwitz from Berlin. That the majority of his research subjects were Hungarian confirms this, as the deportations from Hungary began only in spring 1944.[23]

In most situations it was the scientists, who took the initiative in proposing experiments to the SS. This is clearly the case with Rascher's deadly low pressure experiments and Schilling's malaria experiments, both at Dachau. It was also the case with the orthopedic surgeon, Karl Gebhardt's disabling (and in several cases deadly) tetanus experiments at the concentration camp of Ravensbrück on 20 men and 74 Polish women, and on his near neighbor at the sanatorium August Heissmeyer's TB experiments on 20 Jewish children, who were tragically killed a few days before the end of the war. Gebhardt was ambitious to expand the SS sanatorium of Hohenlychen as a research centre.[24] The SS Inspector of Nutrition, Ernst Günther Schenck took the initiative in proposing nutritional experiments at Mauthausen. The brain anatomist Julius

23 Weindling, *Victims and Survivors of Nazi Human Experiments: Science and Suffering in the Holocaust* (London: Bloomsbury, 2014).

24 John Silver, 'Ludwig Guttmann (1899–1980), Stoke Mandeville Hospital and the Paralympic Games', *Journal of Medical Biography*, 20 no. 3 (2012) 101–105. Hahn, *Grawitz, Genzken, Gebhardt*, 13, 167–169, 179, 183, 227.

Hallervorden suggested that he be sent brains from psychiatric victims of Nazi euthanasia killings. Rather than an explicit racial agenda, the experiments were on a diverse range of clinical problems. Mengele was interested in processes of heredity and growth, for example in his studies of dwarfism, or cleft palate. The twins, whom he collected so assiduously in their hundreds, were objects for studies of the role of heredity in pathology, physiological development and race on a comparative basis.[25] A belligerent medical establishment adopted experimental methods and techniques in an aggressive and feverish search for rapid scientific triumphs. Mainstream science was overwhelmingly involved at every level.

III Victim Perspectives

The focus on the perpetrators very much derives from concerns with the accused at the Nuremberg Doctors Trial. Their victims appear in the literature as passive and incidental, although the prosecutors were determined to give the victims a voice at the trial. In fact, victim interrogations in preparation for the trial are an overlooked source.[26] On the one hand the type of research subject explains much about the intentions of the experiments. On the other, resistance and sabotage are overlooked in the presumed model of passivity. The one group to have been given considerable historical attention is that of the Polish victims of the sulphonamide leg wounding experiments. They were called by fellow prisoners "Rabbits". The "Rabbits" protested to the camp commandant against the experiments as a denial of their rights as prisoners. Not only did three testify eloquently at the Nuremberg medical trial, and some further victims at the British-administered Ravensbrück Trial in 1946–47, but they also took a key role in the struggle for compensation.[27]

25 Paul Weindling, "Die Opfer von Menschenversuchen und gewaltsamer Forschung im Nationalsozialismus mit Fokus auf Geschlecht und Rasse. Ergebnisse eines Forschungsprojekts", Insa Eschebach und Astrid Ley ed, *Geschlecht und Rasse in der NS-Medizin* (Berlin: Metropol Verlag, 2012), pp. 81–100.

26 See the victim interrogations at Staatsarchiv Nürnberg Rep 502 KV-Anklage interrogations: eg H 36 Haremza, Ignacy b. 7.vii.1920 in Oberhausen. Polish. Rascher experimental subject; H 158 Hoellenreiner, Karl born 9iii1914 Prisoner in Auschwitz, Buchenwald and Dachau. Seawater drinking; H 192 Hornung, Hans, born 4iii1902 in Würzburg. Imprisoned for "Rassenschande". Versuchsperson with Dr Rascher.

27 Aleksandra Loewenau, 'The Impact of Nazi Medical Experiments on Polish Inmates at Dachau, Auschwitz and Ravensbrück', PhD, Oxford Brookes University, 2012.

TABLE 9.2 *Nationality as of March 1938*

Nationality	Confirmed	Pending	Total
Austrian*	782	17	799
Belgian	16	32	48
British	20	4	24
Czechoslovakian	262	1019**	1281
Danish	2	1	3
Dutch	265	26	291
French	156	57	213
German	2252	123	2375
Greek	426	17	443
Hungarian	612	1395	2007
Irish (Republic)	1	–	1
Italian	71	6	77
Latvian	1	1	2
Lithuanian	4	2	6
Luxembourgian	1	–	1
Norwegian	11	1	12
Polish	2727	4167**	6894
Romanian	51	39	90
Soviet	1022	24	1046
Spanish	22	4	26
Stateless*	449	4	453
Swedish	1	–	1
Swiss	3	–	3
Yugoslav	537	3422**	3959
Unknown	6057	1647	7704
Grand total	15,751	12,008	27,759

* "Stateless" includes 440 stateless residents of Vienna in addition, subjected to anthropological investigation, and so can be added to the Austrians.

** These numbers relate to persons compensated but for whom no reliable data has been created to date.

The rationales for the research included the wartime lack of availability of laboratory animals, notably primates and rabbits. As forced labor became ever more widespread, the idea arose that the bodies of prisoners could be exploited for research just as they were in any case being worked to dearth. Ethical and

TABLE 9.3 *Religion of victims*

Religion	Confirmed victims	Pending	Total
Jewish	3076	792	3,868
Other or unknown*	12,678	11,217	23,891
Grand total	15,754	12,008	27,759

* Christians (Catholics, Protestants, and Orthodox), Muslims, Jehovah's Witnesses, Seventh-Day Adventists and atheists.

humane inhibitions fell away. On the non-racial side were experiments on the efficacy of vaccines and disease transmission (e.g., for proving how hepatitis was an infectious disease). At times choices had a racial as well as an immunological rationale. Thus for "Fleckfieber" (typhus – there is endless confusion with the German *Typhus* meaning typhoid) research at Buchenwald, criminal German prisoners rather than Jews were used, as exposure to the disease in areas where it was endemic meant that the person would have acquired a degree of immunity.[28] Yet increasingly groups who were ideologically stigmatized by the Nazis were used. Priests (mainly Polish, but also a few Germans and Czechoslovakian), Sinti and Roma, Polish women, and Jewish children were variously exploited for experimental research.

Religion and ethnicity have been recorded, as known for a proportion of the total numbers of research subjects (see Table 9.3). From 1943 we find a higher proportion of Jewish victims. This would again indicate that there was an intensification of racial research. Statistics on gender indicate a male to female ratio of approximately 2:1 (see Fig. 9.1). One possible reason was the high number of military experiments as related to infectious disease. With the upswing in experimentation in 1941/2 the priority was strategically related research on infectious diseases and wound infections. The ideal subject would equate to the German soldier or – as in the freezing and low pressure experiments – fighter pilot at high altitude, or needing to survive in the freezing waters.

There were four types of coercive research running parallel to "euthanasia" killings of psychiatric patients:

First, to perfect more efficient killing procedures: here there were examples of the testing of poison gas and chemicals, lethal injections with phenol and luminal, and electrocution.

28 Paul Weindling, *Epidemics and Genocide in Eastern Europe* (Oxford: Oxford University Press, 2000).

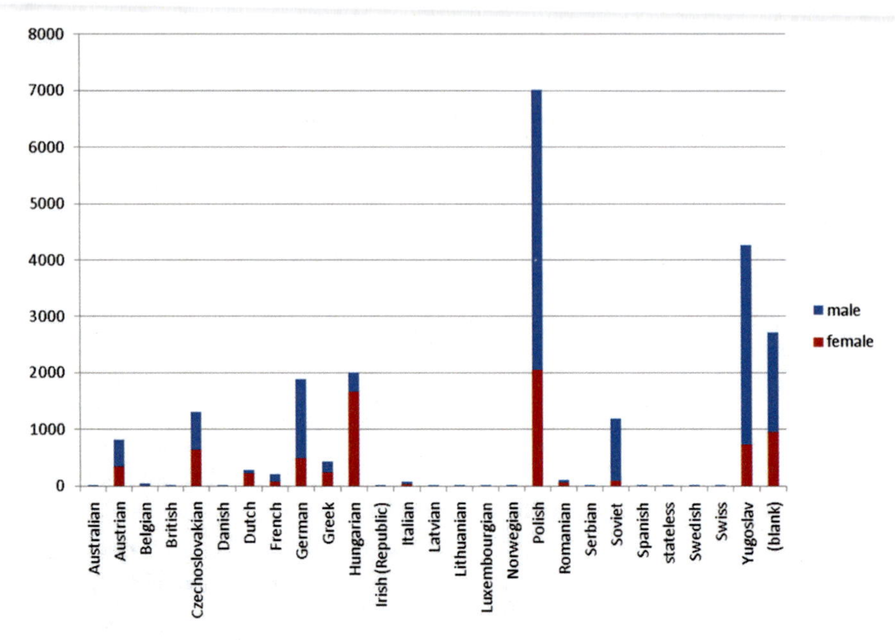

FIGURE 9.1 *Gender distribution*

The second type of experiment was designed so as to understand the causes of psychiatric illness. This had a dual role. One was to contribute to Nazi race policy, by understanding how illness was transmitted on a hereditary basis – one might see this research as a preliminary to policies of eradicating the lives of the sick and disabled. Second, there were intrinsic scientific reasons. We find clusters of these experiments from 1939–1945.

In the third type of experiment, psychiatric hospitals were regarded as reservoirs of human experimental "material" for clinical research. We see this in a series of experiments on tuberculosis immunization on young psychiatric patients at the twin Bavarian hospitals of Bad Irsee/Kaufbeuren. Parallel types of research were carried out in concentration camps.[29] Finally, brain anatomists sought to collect what they perceived as "interesting cases" from the "T4 killing centers". The question is exactly whose brains, were selected by

29 Paul Weindling, "Menschenversuche und „Euthanasie" – das Zitieren von Namen, historische Aufarbeitung und Gedenken", Arbeitskreis zur Erforschung der nationalsozialistischen "Euthanasie" und Zwangssterilisation, ed, *Den Opfern ihre Namen geben. NS-"Euthanasie"-Verbrechen, historisch-politische Verantwortung und Erinnerungskultur* (Bad Irsee: Impulse, 2011), 115–132.

whom, in what numbers, and where these brains and body parts ended up?[30] (see Table 9.1)

The year 1942 was characterised by the start of long-term experiments on malaria at Dachau and typhus (Fleckfieber) at Buchenwald. From 1943 Auschwitz emerged as a major centre of coerced experimentation. This involved a range of experimental research on Jewish men and women, and in Clauberg's infertility experiments wholly on women, with a clear racial motivation. Examples were: sterilization research by Horst Schumann using X-rays, and excising of the victims' gonads for evaluating the effects of radiation. The infertility research by Carl Clauberg offered an alternative approach to mass sterilization, while Clauberg also had interests in how hormones could enhance fertility. These researches preceded the racial and genetic research by Mengele. The victims were Jews, especially Greek Sephardic Jews for Clauberg, Schumann and the Jewish skeleton collection, whereas Mengele started his research in late 1943 or early 1944 on Sinti and Roma, as he could opportunistically exploit his role as physician responsible for the "Gypsy Family Camp" at Auschwitz. Mengele's researches were wide-ranging and complex. The twin researches represent a complex set of agendas on the inheritance of physical and psychological traits, and on growth. Other concerns were with physiology and disease.

Auschwitz became a central dispersal point for Jews selected for racial research. Noted cases were: the intended Jewish skeleton collection at Strasbourg for which SS anthropologists Bruno Beger and Hans Fleischhacker selected a group of 115 Jews for anthropological measurement, from whom 86 were dispatched to be killed at Natzweiler in Alsace. Jewish children from Auschwitz at were transferred at Sachsenhausen for hepatitis research, and for TB research at Neuengamme, ending with their murders at the Bullenhusen Damm school cellar in April 1945. The adolescent boy hepatitis victims at Sachsenhausen were also due to be killed, but prisoners intervened to save their lives.[31]

IV Between Compliance and Resistance

Victims often did not know what was happening to them, as the Hauck testimony has already indicated. Others, were consciously defiant. While the

30 Paul Weindling, '"Cleansing" Anatomical Collections: The Politics of Removing Specimens from German Anatomical and Medical Collections 1988–92', *Annals of Anatomy*, 194 no. 3 (2012), 237–242.

31 Saul Oren-Hornfeld, *Wie brennend Feuer. Ein Opfer medizinischer Experimente im Konzentrationslager Sachsenhausen erzählt*. Berlin: Metropol Verlag, 2005, 107–60.

assistant of Gebhardt, Fritz Fischer developed a biological approach to anthropology based on genetic analysis over generations, anthropologists in Vienna pursued approaches based on study of living persons. In September 1939 a commission of eight anthropologists under Josef Wastl from the Natural History Museum in Vienna selected from the great variety of the internees. As one recollected, "I doubt the researchers found the stereotypical Jew the Nazi press displayed in their perverse cartoons."[32] Others thought the anthropologists were trying to prove the stereotype of a large nose and thick lips. The commission made plaster face masks from 440 "stateless" Eastern Jews (out of about a thousand viciously rounded up by storm troopers and police) in the Prater Sports Stadium – some had lived in Vienna since 1905. The subjects were photographed, measured and hair samples were taken. Having a mask made was not painful, although it involved brief discomfort to ensure the nostrils were clear of plaster for breathing. One research subject, the sixteen-year old Gustav Ziegler (later, Gershon Evan) remembered the anthropologist as polite: "A man in a white coat, the only person in the room, received me in a friendly manner, and throughout the performing of the work tried to set my mind at ease." The experience of having mask made did not hurt amidst the acute distress of the arrest. Gustav perceived the experience as an act of defiance in disproving the Nazi racial stereotypes, as displayed at an exhibition at the Natural History Museum on "The mental and physical appearance of Jews". Yet the research opportunistically exploited Nazi racial measures.[33] From the group of Jews, who were sent to Buchenwald for "protective custody", most were killed, some rapidly, so the face mask of a living person became if not a death mask then a mark of a person experiencing life-threatening persecution.

The Greek young women and girls held in Block Ten did not understand what was happening when a young woman said that she spent time in a barrel and nothing happened – in fact, she was subjected to a deliberately covert X-ray sterilisation. Later on, having learned what had happened she resisted. "My number was called for the operation. I did not answer when they called my number...people thought I was hysterical."[34]

---

32 Gershon Evan, *Winds of Life. The Destinies of a Young Viennese Jew 1938 to 1958* (Riverside, CA: Ariadne Press, 2000).

33 Gershon Evan, *Winds of Life. The Destinies of a Young Viennese Jew 1938 to 1958* (Riverside, CA: Ariadne Press, 2000).

34 Lore Shelley, *Criminal Experiments on Human Beings in Auschwitz and War Research Laboratories: Twenty Women Prisoners' Accounts* (San Francisco: Mellen Research University Press, 1991), 77–86.

The 74 Ravensbrück "Rabbits" (they adopted the name given to them by fellow prisoners), protested that their rights as prisoners were violated, and subsequently refused to undergo further experiments. They were part of a pattern of resistance and sabotage of Nazi experiments and indeed more generally by prisoner researchers. The French prisoner Claudette Bloch sabotaged Auschwitz research on developing a plant for producing rubber by deliberately mixing up seeds.[35] The vaccine research at Buchenwald involved a group of prisoners which included the renowned philosopher-researcher Ludwik Fleck: they delivered fake vaccine to the Germans.[36] Yet they were also not involved in the murderous experimental research in a neighboring block. There was a wider pattern of sabotage and manipulation. The instances include fake twins as Kuhn, Gyorgy (born on 01/23/1932) and his younger brother Kuhn, Istwar (born on 12/17/1932).[37] Vera Kriegel recollected that she was not well behaved and not afraid of Mengele. He obtained medical data, and she in return gained food; in their symbiotic interactions, she gave Mengele trouble any way that she could.[38] Partial operations to remove ovaries were carried out by prisoner surgeons among whom was the tragic Maximilian Samuel at Auschwitz. But as operative removal of both ovaries not fully carried out by Samuel, victims could become pregnant after the war.[39]

At times the prisoner assistants conducted sabotage – for example, it was claimed that the water temperature in the freezing experiments was manipulated. The attenuation of vaccines (in the case of the twenty Jewish children injected with TB bacilli) was carried out to make their effects less severe, when the French prisoner nurses (who were also to be killed) tampered with the cultures that were injected. It was said that during the seawater drinking experiment on Sinti/Roma at Dachau, drinking water was smuggled in. When the Swiss conscientious objector Wolfgang Furrer was experimented on at Dachau as one of Claus Schilling's malaria human guinea pigs, the prisoner assistants told him, "We prisoners do the utmost to

35 Claudette Bloch Kennedy interview with author.

36 Paul Weindling, 'The Fractured Crucible: Images of the Scientific Survival. The Defence of Ludwik Fleck', Johannes Fehr, Nathalie Jas and Ilana Löwy (eds), *Penser avec Ludwik Fleck – Investigating a Life Studying Life Sciences* (Zurich: Ludwik Fleck Centre, Collegium Helveticum, 2009), 47–62.

37 http://www.candlesholocaustmuseum.org/learn/about-survivors.htm (viewed 18 January 2016).

38 Vera Kriegel, 'Endlich den höchsten Berg gefunden…', Sachse, *Auschwitz*, 76–82.

39 Hermann Langbein, *Menschen in Auschwitz* (Frankfurt am Main: Ullstein-Verlag, 1980). Sari J. Siegel, 'Treating an Auschwitz Prisoner-Physician: The Case of Dr. Maximilian Samuel', *Holocaust and Genocide Studies* vol 28 (2014) 450–481.

ensure that no one dies" ("Wir Gefangenen tun das Möglicheste, dass Keiner sterben muss…").[40]

The prisoner assistant Walter Neff first saw a fatal experiment carried out by the physiologist Hans Romberg. The prisoner was taken to a low pressure of what would pertain at 15,000 meters, and the oxygen was switched off; the research subject died. Neff stated that he sabotaged equipment in mid-May 1942 when he filed down the scale of the pressure measurement apparatus – then it broke on next experiment. But Romberg repaired the scale quickly, so that 20 deaths then occurred. The attempt however shows that equipment could be – and was – manipulated.

Individuals attempted subterfuge and evasion. In the case of Zarfati Barouche, her left ovary was irradiated by Schumann; she worried that her right side would be sterilized – "I walked up to Mengele naked, hiding my scar with my skirt. Luckily he did not notice, and once again I was saved from the experiments."[41] Other individual protests were in vain. In Auschwitz and finally at Ravensbrück Jewish women were forced to submit to sterilization injections devised by Clauberg, carried out so they could not see what was happening to their lower body. These injections were given despite protests.[42]

Judith Vr. in Auschwitz in June 1944 was forced to submit to three grueling weeks of inter-uterine injections.[43] While her protests were in vain, she provides a precise narrative of the procedures and examinations to which she was subjected over a three-week period, when she felt increasingly sick and distressed.

There was a clandestine series of photographs of the disfigured legs of three of the "Rabbits", including Bogumila Babinska (Jasiuk) in Ravensbrück concentration camp. The photographs were part of a wider collective solidarity when prisoners' common suffering brought them together.[44]

Gustl Nathan, originally from Dusseldorf, ironically referred to the "experimental club" at Block 10 in Auschwitz, with its daily routine of experiments

---

40 Yad Vashem Archives 0.33/4000 Wolfgang Furrer, Chapter 8, Menschen als Versuchskaninchen.

41 Lore Shelley, *Criminal Experiments on Human Beings in Auschwitz and War Research Laboratories: Twenty Women Prisoners' Accounts* (San Francisco: Mellen Research University Press, 1991). The doctor was likely not to have been Mengele (who was not involved in X-ray sterilization nor chemical sterilization experiments).

42 Bundesarchiv Koblenz (BAK) B 126/27770.

43 BAK B 126/27770.

44 Aleksandra Loewenau, 'Die "Kaninchen" von Ravensbrück: Eine Fotogeschichte', in Insa Eschebach and Astrid Ley, eds, *Geschlecht und Rasse in der NS-Medizin* (Berlin: Metropol Verlag, 2012), 115–139.

and forced labor. She attempted to create a spirit of solidarity to induce a sense of self-worth and self confidence. Clauberg injected painful liquid into her uterus, and then observed the results by X-raying her. The procedure, carried out on multiple occasions, was often accompanied by bleeding and inflammation. Her one thought was surviving this incredible pain. Otherwise there were injections in her breast, and a psychological experiment on ten of the women.[45] Gustl Nathan did not find the doctors brutal in the everyday conduct. Block 10 was also where a group of French prisoners tried to promote greater solidarity, and optimism. Slavka Kleinova saved a group of 17 mostly French prisoners. The greater solidarity changed the atmosphere of the block for the general benefit of prisoners. Keizer founded bridge club and discussion circles for politics and literature. Slavka Kleinova provided a detailed description of the block, smuggled out of Auschwitz by Tadeusz Holuj and the prisoner doctor Stanislaw Klodzinski. The complete summary was forwarded to London, where it was kept at the Polish documentation centre.

v Medical Resistance

Considerable force was used at times. Survivor accounts refer to the Austrian doctor Wilhelm Beiglböck, who conducted the seawater drinking experiments at Dachau in 1944, brandishing a pistol to ensure compliance. There was also resistance – or at least non-compliance – from individual physicians. As with non-participation in other aspects of the Holocaust, not least sterilization and euthanasia, participation in the experiments was a matter for the conscience of individual German physicians. Rascher felt that the doctors Siegfried Ruff and Weltz were against his experiments (although they too countenanced high risk experiments) when he took prisoners to the point of death.[46] The chief air force doctor, Hippke, was similarly opposed. At a meeting of military medical doctors, the bacteriologist, Gerhard Rose was critical of the research on infectious disease transmission and therapy carried out by Ding in Buchenwald. But then Rose became caught up in experiments when he supported the evaluation of a Danish produced vaccine against typhus – earning him a guilty verdict at the Nuremberg Trials. The disruption suggests that human experiments were controversial among the medical research establishment, and at times

45 Center for Jewish History, New York, Gustl Nathan, Memoirs.

46 Staatsarchiv Nürnberg Rep 502A KV Verteidigung, Handakten Sauter 37 Material zu Ruff, Max Matthes 31 October 1946. Rep 502 KV-Anklage Interrogations Generalia H 143 Erich Hippke interrogation by Leo Alexander 31 January 1947.

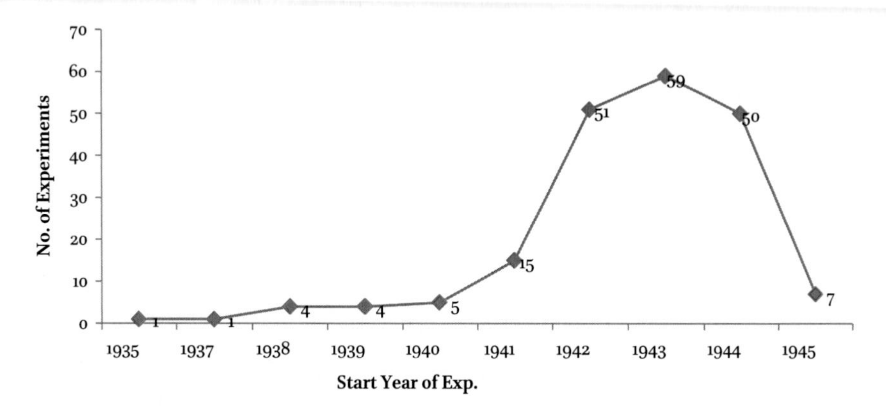

FIGURE 9.2 *Number of experiments*
NOTE: THE YEAR IS WHEN THE EXPERIMENTS STARTED. THE SACHSENHAUSEN SHOE TRACK EXPERIMENT RAN FROM 1940 THROUGH TO 1945

they were sabotaged and undermined. The scope for victims was limited, but evasion, disruption and resistance could be effective.

VI Conclusion

There was no single type of coerced medical or racial experiment, and nor was there a single directing authority. The analysis presented here suggests that multiple types of unethical research occurred under National Socialism. The high point of coerced experimentation in 1942–44 was intensive, and very large numbers of victims were involved (see Figs. 9.2 and 9.3) Here one needs to recognize the multiplicity of victim categories. Jews were at first used sporadically by researchers like Rascher, but then at Auschwitz became the prime victim group. Clauberg and Schumann used Jews exclusively. Mengele initially exploited gypsies and then from May 1944 increasingly turned his attention to Jews. There were massively however also those of other religions (see Table 9.3) There was a wide spectrum of nationalities, mainly Poles, but also fit and healthy Germans (see Table 9.2). Not only were large numbers of victims affected, but also overall numbers of surviving victims was far higher – and far more injured – than postwar compensation authorities have ever recognized. Murderous experiments were a distinctive feature of Nazi medical research. But given that the majority of victims survived, the narratives of survivors allow one to understand more fully the life-shattering and destructive consequences of coerced research.

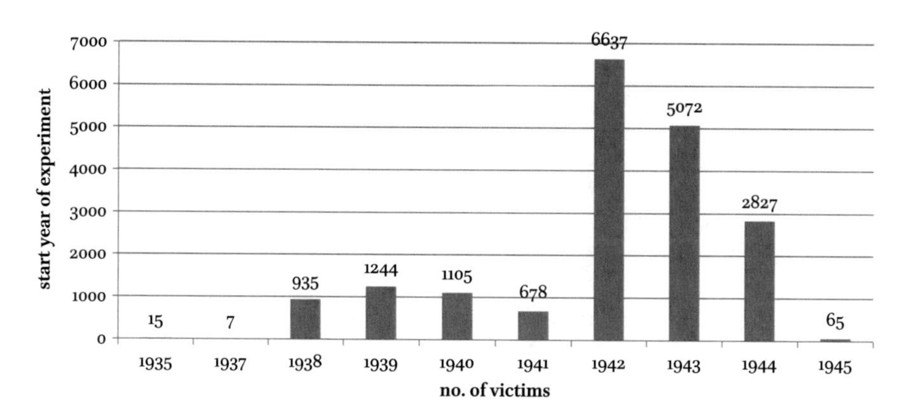

FIGURE 9.3 *Victims by year of the experiment*

Acknowledgments

Wellcome Trust Grant No. 096580/Z/11/A on research subject narratives. AHRC Grant AH/E509398/1 Human Experiments under National Socialism. Conference for Jewish Material Claims Against Germany Application 8229/Fund SO 29.

CHAPTER 10

A Eugenics Experiment: Sterilization, Hyperactivity and Degeneration

Erika Dyck

John Smith first came to the attention of medical authorities at the age of nineteen. He had struggled at school and his teachers claimed that he was dull. His parents were both identified as criminals and John admitted that he had masturbated since he was a boy about the age of twelve. As a young adult he sought medical help for his chronic masturbation. The doctor met with him, assessed his situation, and recommended a vasectomy to relieve his urge to masturbate, which authorities surmised might then alleviate his mental and intellectual problems. Two months after the operation John complained that he had gained weight and that the desire to masturbate had grown even stronger. He pleaded with his doctor to perform a more invasive surgery to remove this mental anguish and physiological burden. At this point his doctor recommended castration and the surgery proceeded apace. A year after the operation John had gained more weight but felt significant relief from his mental pains. He had also stopped masturbating. When the doctor inquired about how he had curbed his masturbation, John responded, "the desire is as great as ever, but I have the will to resist."[1]

John Smith is a pseudonym I have applied to a case that appeared in the *New York Medical Journal* at the turn of the twentieth century. It refers to an unnamed boy in Missouri who sought help at the Indiana Reformatory in the period leading up to the passage of the first eugenics law in North America.[2]

1 H.C. Sharp, "The Severing of the Vasa Deferentia and its Relation to the Neuropsychopathic Constitution," *New York Medical Journal* (1902) vol. LXXV (10), 413.

2 Indiana passed the first involuntary sterilization law in the world in 1907. For a careful study of the American developments, see: *A Century of Eugenics in America: From the Indiana Experiment to the Human Genome Era* (ed) Paul Lombardo, (Bloomington: University of Indiana Press, 2011), p. ix for the Indiana beginning, but see the collection of essays for more on American eugenics. See also: Paul Lombardo, *Three Generations No Imbeciles: Eugenics and Supreme Court, and* Buck v. Bell (Baltimore: Johns Hopkins University Press, 2008); Gregory Michael Dorr, *Segregation's Science: Eugenics & Society in Virginia* (Charlottesville: University of Virginia Press, 2008); Edwin Black, *War Against the Weak: Eugenics and America's Campaign to Create a Master Race* (New York: Four Walls Eight Windows, 2003); Alexandra Minna Stern, *Eugenic Nation: Faults & Frontiers of Better Breeding in Modern America*

 | DOI 10.1163/9789004286719_012

This example illustrates one of the characterizations of deviant or abnormal male behavior and the consequent experiments aimed at boys and men that culminated in sterilization, even including castration. It also indicates the voluntary nature of this exchange, emphasizing that the boy himself sought medical intervention to curb activity that he believed transgressed normalcy. This self-discipline and even self-policing might attract a Foucauldian analysis considering the ways in which medicalized views extended into the private realm, and even invaded the personal, intimate thoughts of this young boy such that he mastered self-control as proof of his restored health.[3] Examining the actions of the medical and psychiatric community, however, suggests that concerns over male sexual habits and actions attracted attention from eugenicists and psychiatrists particularly interested in legitimizing their professional aims through experimental practices.

Historian Angus McLaren has written extensively on mental health and sexuality, and in the context of male masturbation and impotence he argues that medical interventions and individual concerns had a long history of experimentation. For men suffering from impotence, for example, he found that throughout the 19th century doctors searched for reliable aphrodisiacs, which commonly resulted in "external stimulation of the genitals by frictions, flagellations, and galvanism" and even electrical shocks,[4] aligning these

---

(Berkeley: University of California Press, 2005); Mark A. Largent, *Breeding Contempt: the History of Coerced Sterilization in the United States* (New Brunswick: Rutgers University Press, 2008); Ian Robert Dowbiggin, *Keeping America Sane: Psychiatry and Eugenics in the United States and Canada, 1880–1940* (Ithaca: Cornell University Press, 1997); Harry Laughlin, "The Eugenical Sterilization of the Feeble-Minded," in *Mental Retardation in America* (eds) Steven Noll and James W. Trent Jr. (New York: New York University Press, 2004), 225–231; Diane B. Paul, *The Politics of Heredity: Essays on Eugenics, Biomedicine, and the Nature-Nurture Debate* (Albany: State University of New York Press, 1998); Daniel J. Kevles, *In the Name of Eugenics: Genetics and the Uses of Human Heredity* (Cambridge: Harvard University Press, 1985); Wendy Kline, *Building a Better Race: Gender, Sexuality, and Eugenics from the Turn of the Century to the Baby Boom* (Berkely: University of California Press, 2001); and Johanna Schoen, *Choice and Coercion: Birth Control, Sterilization and Abortion in Public Health and Welfare* (Chapel Hill: University of North Carolina Press, 2005).

3 See any range of works by Michel Foucault, including and especially: *Birth of the Clinic: An Archaeology of Medical Perception* (New York: Vintage Books, 1994, first English translation1973), which introduces his concept of the "medical gaze" emanating from clinical institutions and ultimately imposing a discipline and order over bodies in a sophisticated deconstruction of the process of medicalization.

4 Angus McLaren, *Impotence: A Cultural* History (Chicago: The University of Chicago Press, 2007), 138.

historical approaches with the early modern experiments discussed earlier in this volume. Masturbation and impotence were linked, with the medical assumption that the former led to the latter and yet, McLaren argues that patients rarely saw it this way. Vasectomies had been used to control, and in some cases invigorate, male sexuality since at least the 1890s[5] and together with the identification of hormones in the 1920s sterilization and or the manipulation of the endocrine system encouraged a new wave of experiments on males. As fellow authors Paul Lombardo and Paul Weindling have shown in the previous chapters, eugenic theories also permeated this era of experimentation and leading eugenicists, including Charles Davenport, discussed by Lombardo, readily embraced the new science of endocrinology to help justify altering the sex organs. Inferior males were not simply criminal or immoral, but as McLaren shows, Davenport now claimed that they had bad glands that necessitated medical treatment.[6] McLaren further concludes that, "rejuvenation, eugenics, and endocrinology were each in their own ways symptoms of the twentieth-century desire to improve the body. Though their adherents might not all have believed that human behavior could be standardized, they were convinced that sexual behavior could be measured against scientifically established norms."[7] McLaren's work in this area points to the intersections between male sexuality, eugenics and experimentation, and this article explores a set of Canadian studies that sit at that nexus.

In Canada the primary focal point for such experiments was Red Deer, Alberta. The province adopted the first and ultimately the most aggressive eugenics initiative in the country in 1928. This program remained in effect until 1971 and much of its work concentrated on institutionalized children living within the Provincial Training School for Mental Defectives in Red Deer. Over the course of its existence, the Alberta eugenics program recommended sexual sterilization surgeries for 4,725 individuals and ultimately performed operations on 2,822 people, by comparison with the other Canadian eugenics program, in British Columbia, which operated on fewer than 200 individuals. Recommendations for the surgeries in Alberta came from an appointed Eugenics Board and which established five categories: psychotic patients; mental defectives, including individuals with arrested mental development for congenital or acquired reasons before the age of 18; neurosyphilitic patients who did not respond to treatment; patients with epilepsy and psychosis or mental deterioration; and those with Huntington's Chorea disease, following

5 McLaren, 191–193.

6 McLaren, 203.

7 McLaren, 207.

its presumed genetic causes.[8] The Board approved sterilization surgery when its members determined that there was a "danger of the transmission of any mental disability or deficiency to offspring, or a risk of mental injury either to such persons or to the progeny if sterilization were not carried out."[9] British Columbia and Alberta were the only two provinces that passed eugenics laws in Canada, which placed them in the midst of thirty states in the United States, and dozens of other countries worldwide, including India, Germany, Sweden, and Japan.[10] Several other Canadian provinces considered implementing similar programs, but none succeeded in approving legislation.

Sexual sterilizations for boys and men within the purview of Alberta's eugenics program operated on some of the same principles that had been applied to institutionalized women. The fundamental presumption was that these people would pass on genetic defects exhibited in mental deficiency. For sterilized men, their discharge from the institution initially raised alarms that they may engage in more risky sexual activities, or that deviant sexual proclivities were proof of their mental disorders in the first place. After the Second World War, however, those attitudes gradually shifted away from connecting eugenic vasectomies with male sexual acts and toward a better understanding of the relationship between male hormones and masculinity or socially acceptable male behavior. The studies emerging in this vein increasingly focused attention on male children and adolescents in the Provincial Training School for Mental Defectives in Red Deer, and brought hyperactivity and learning abilities under psychiatric surveillance. This renewed attention to hyperactive disorders made boys and young men the targets of psychiatric and horomone-related experiments alongside their sustained role in the sterilization program in ways that have often been overlooked by medical historians.[11]

In the broad context of eugenics, masculinity and appropriate boyish behavior attracted attention from medical professionals interested in policing the boundaries between what was deemed normal and abnormal. In Alberta, the main proponent of that research was also the superintendent of the Provincial Training School for Mentally Defective Children who had unprecedented access to child subjects. Dr. Leonard Le Vann engaged in a number of experimental studies in an attempt to demonstrate a relationship between mental deficiency and schizophrenia. Based on the gendered interpretations

---

8 Law Reform Commission of Canada, Working Paper 24 "Sterilization," 27.

9 Ibid.

10 Paul Lombardo (ed) *A Century of Eugenics in America: From the Indiana Experiment to the Human Genome Era* (Bloomington: Indiana University Press, 2011), ix.

11 McLaren's work is an obvious exception.

of behavior and the pre-existing assumptions about hyperactivity in boys, Le Vann's studies relied on testing theories of degeneracy and inheritability among institutionalized males.

In 1937 the Sexual Sterilization Act was amended in an effort to speed up the process of approving surgeries. The amendment removed the need for informed consent from patients whose intelligence quotients fell below 80. These patients were considered incapable of providing consent or even of appreciating the nature of the program and its consequences. As a result, patients who fell into this category were routinely sterilized without their knowledge or consent. The change in law also had implications for the research carried out at other psychiatric institutions in the province, and in particular at the Provincial Training School for Mentally Defective Children. In an effort to prove the relationship between intelligence, behavior and heredity, psychiatrists engaged in a series of studies that suggested that so-called defective children belonged to a different racial strain, one in which social or cultural interventions had limited or no impact. In other words, children who fell into this category were subjected to surgical sterilizations on the presumption that they were both genetically "defective" and also culturally or racially incapable of contributing to society in meaningful ways.

Sterilization Operations

Men facing sterilization under the eugenics program received one of two orders: vasectomy or orchidectomy. A vasectomy involved severing the vas and usually removing a section of it. There are two vas or tubes running from the testicles, behind and under the bladder and entering the penis. The vasectomy operation requires severing these two tubes, or vas, to ensure that they no longer deliver sperm during ejaculation.[12] The operation does not produce immediate effects, but after a few months sperm are cleared from the system and the man should be rendered sterile. Vasectomies represented the vast majority of the operations recommended for males by the Alberta Eugenics Board. Castration or orchidectomy involved the full removal of the testicles, and thus the body's ability to produce sperm and regulate hormones in the testes. This operation was considerably more invasive and had the combined effect of altering sexual function and desire by involving the endocrine system. Its

12 Donald J. Dodds, *Voluntary Male Sterilization* (Toronto: The Damian Press, 1970), 15, 16, 20–21.

application was reserved for special cases where an adjustment in hormones was considered advantageous.

By the time Alberta inaugurated its Sexual Sterilization Act in 1929, American programs had already been in effect for several years. That very year, the *Journal of the American Medical Association* published a study based on the voluminous data collected in California, which had boasted that it was "where more elective sterilization has been done than in all the rest of the world together."[13] The California program began in 1909, and by the end of 1928 had sterilized 3,232 men and 2,588 women, more in those twenty years than Alberta would sterilize through the duration of its program. In these cases most patients were put under a general anaesthetic, according to the California author, "because they are dealing with the insane and feebleminded."[14] Based on experiences in both California and Indiana, states with long-standing eugenics policies, researchers found no change in sexual desire for males who underwent vasectomies. Physicians were primarily concerned with the risk of increased sexual desire from younger sterilized boys, but discovered that even among adolescent boys whose hormones were still changing, the vasectomy operation did not arouse additional aggressive or excitable activity.[15] Based on his experiences in California, Dr. Dickinson recommended that "sterilization, not castration, should be done." He went on to explain that "sterilization does not cause castration, nor does it unsex either sex. The operation of sterilization is not dangerous to health of life."[16]

The Eugenics Board minutes in Alberta recorded the first scheduled orchidectomy in early 1937.[17] The case notes indicated that a full castration should proceed, but in the event that the parents will not consent to that more invasive operation, a vasectomy would suffice.[18] The following year only a few orchidectomies warranted noting in the meeting minutes, including one for a patient diagnosed with psychopathic personality disorder, though not mentally deficient, and another who was simply described as "Eskimo."[19] A case in

13 Robert L. Dickinson, "Sterilization without unsexing: 1. Surgical review with especial reference to 5,820 operations on insane and feeble-minded in California," *Journal of the American Medical Association* (1929) 92(5), p. 373 [373–379].

14 Ibid., 373.

15 Ibid., 374.

16 Ibid., 379.

17 Eugenics Board Minutes, Meeting 84, July 15, 1937.

18 Ibid., 2.

19 Eugenics Board Minutes, Meeting 91, July 14, 1938, p. 2 and Meeting 95, April 26, 27, 28, 1939.

1944 was downgraded to vasectomy, although Board members felt that due to the patient's "history of exhibitionism" an orchidectomy was preferable. This patient, however, was illiterate and an accurate intelligence quotient rating proved difficult to obtain; in the absence of the legal authority to proceed without consent, the Board accepted the vasectomy as a viable alternative for which parents usually consented.[20]

The absence of a full set of patient records now prohibits a retrospective analysis of male patients and the corresponding recommendations, but contemporary reports and publications offer further insights into the reasons behind some of the male operations. For example, in a medical article based on experiences in the Provincial Training School in Red Deer, physician James Russell Grant reported that the nature of the patient population in that institution, as elsewhere, reflected important gender differences in the reasons for committal. Of those patients who fell into the "moron" category, and were the likely subjects of eugenics efforts, the males "appear as psychopathic personalities and criminal offenders, ..." whereas the females labeled in this category "are more likely to seek help and advice from welfare workers and guidance clinics, or by reapplication to the Training School, chiefly on account of personality disorders or neurotic symptoms."[21] Grant hinted at the degree to which male disorders and abnormal conduct often related to expressions of aggression, whereas disordered females tended to be more withdrawn.

Prior to arriving in Alberta Grant had worked as the Registrar of the Maudsley Hospital in London, England, which was at that time being reestablished with more emphasis on children's mental health after serving during the First World War as a military institution. Maudsley Hospital later gained a particular reputation for shifting towards mental health services for children and youth and developing better criteria for understanding the particular behavioral and psychological features of adolescents.[22] Staff at the Provincial Training School in Alberta soon adopted a similar focus on adolescents and consequently, in the post-war period, the Eugenics Board increasingly began approving sterilization surgeries for patients in this category.[23]

20 Eugenics Board Minutes, Meeting 121, June 15, 1944.

21 James Russell Grant, "Results of Institutional Treatment of Juvenile Mental Defectives over a 30-Year Period," *Canadian Medical Association Journal* (1956) 75(Dec 1), 919–920 [918–921].

22 Edgar Jones and Shahina Rahman, "Framing Mental Illness, 1923–1939: The Maudsley Hospital and its Patients," *Social History of Medicine* (2008) 21(1): 107–125.

23 Jana Grekul, "Sterilization in Alberta," 254. This shift towards adolescents was also reflected in a growing emphasis on the use of Guidance Clinics and school programs to detect mental deficiencies or mental health problems. For more on this topic, see Amy

Grant further described the patient population within the Provincial Training School as an institution best suited for unruly children or children who had become a burden on their families due to low levels of intelligence or physical and mental impairments. Within the "moron" category, he insisted that "low-grade morons are patently aware of their defect," suggesting that these individuals were easier to manage within the system given the widespread acknowledgement of severe disability.[24] By contrast, individuals in the high-grade moron category he felt were more apt to deny their "disability and are never made fully aware of their limitations, perhaps through the constant encouragement they are given in institutional life...thereby they come to meet the world on equal terms, which they cannot in fact do."[25] This category of patients, according to Grant, represented a serious challenge to administrators seeking to deinstitutionalize children or send them home to families, because upon discharge they were liable to being taken advantage of, both socially and sexually. For boys and young men, this challenge might manifest in their aggressive sexual behavior, whereas for girls and young women, they were more likely to become victims of sexual assaults. In either scenario, disabling their reproductive functions at least provided a prophylactic measure against pregnancies, and associated expenses.

More general character disorders represented another category within the Provincial Training School population. As Grant explained: "character disorder has been taken to mean persistent misconduct trying the patience of society, or such that the parents themselves have declared the child beyond their control." He went on to specify "it does not include masturbation, isolated minor sexual offences, or isolated instances of theft or aggressive behavior. It does, however, include extreme single instances, such as murder."[26] Within this category the gendered dimensions became even more evident, as patients were committed and examined on the basis of small crimes and legal transgressions including theft, assault, promiscuity, prolonged unruly conduct and rowdyism.[27] Grant went on to claim that prior to admission to the institution none of the male patients had been "charged with rape or attempted rape, or been involved in a heterosexual offence," whereas "a large proportion of females have been concerned in illegitimate births and promiscuity." Here too, his

Samson, "Surveillance and Coercion," (unpublished PhD dissertation, University of Saskatchewan, in progress).

24 Grant, 919.

25 Ibid.

26 Ibid., 220.

27 Ibid., 920.

interpretation of this difference lay in the way in which other members of society interacted with these youths with presumed character disorders. The young women, he believed, were at a greater risk of becoming "easy prey for the wiles and depredations of 'normal' society, at a time when their character has been insufficiently fashioned to compensate the limitations of their intellect."[28] Young men, on the contrary, showed no previous predilection toward sexual offences prior to committal. Fearing that they may develop such characteristics, it became imperative to examine boys at a young age before they developed a sexual appetite. Curbing their sexual activity was critical for their future release into the community. The eugenics program had been premised on this ideal.

American medical literature similarly revealed a growing concern about the relationship between sterilization and sexual activity. In a 1950 study of vasectomized patients from mental health and penal institutions in the United States, researchers found that patients reported a great deal of "satisfaction". Few claimed to experience an increase in their libido, and the numbers of institutionalized men now requesting vasectomies seemed to be on the rise. Indeed the author noted that "many boys, learning that the procedure was often a step on the road toward parole, had stopped him [Dr. Butler at the Sonoma State Home in Eldridge California] as he passed to ask, 'Doctor, isn't it about time for my operation?'"[29]

Discharge and Deviance

The superintendent of the Provincial Mental Hospital in Ponoka, Alberta, Randall MacLean, periodically pleaded with his colleagues in medicine to pay closer attention to the realities and shortcomings of the mental health system. In 1942 he wrote: "Mental Hospitals are very expensive of construction and primarily are not designed nor staffed nor in a position to house and care for seniles, epileptics, mental defectives or incurable, physically incapacitated individuals. These types actually militate against obtaining best results in the handling of the mentally ill. They constitute very undesirable influences in any active treatment Mental Hospital." Although admissions in these categories represented a minority of cases, MacLean repeatedly lamented that these categories of incurable patients should be cared for at

28 Ibid.

29 Paul L. Garrison and Clarence J. Gamble, "Sexual effects of vasectomy," *Journal of the American Medical Association* (1950) 144(4), p. 293. [293–295].

home or in designated institutions, but should not be mixed with patients suffering from acute psychoses who, according to MacLean, could be rehabilitated: "Mental Defectives likewise are likely to be upsetting to the mentally ill, as they are frequently unstable, mischievous, cruel, scheming and crafty in their behavior."[30]

MacLean's comments tapped into a number of cross currents within contemporary psychiatry. The mental hospitals were grossly overcrowded, and despite efforts to organize patients according to disease, rather than class and gender, the complexity of co-morbid conditions combined with a lack of understanding of disease etiology forced asylum staff to rely on crude observations. In part it is due to these kinds of conditions that, as historians such as Edward Shorter, Joel Braslow and Jack Pressman have pointed out, the period between the 1930s and into the 1950s represented an important turning point in psychiatry.[31] Shorter describes it this way: "In the first half of the twentieth century, psychiatry was caught in a dilemma. On the one hand, psychiatrists could warehouse their patients in vast bins in the hopes that they might recover spontaneously. On the other, they had psychoanalysis, a therapy suitable for the needs of wealthy people desiring self-insight, but not for real psychiatric illness. Caught between these unappealing choices, psychiatrists sought alternatives…."[32] During this period psychiatry embraced a number of radical therapies, bodily interventions such as insulin shock therapy, malaria therapy, electro-shock or electroconvulsive therapy, lobotomies, and by the 1950s a host of pharmacological interventions. While these therapeutic innovations have been described as barbaric in hindsight, malaria therapy and lobotomies nevertheless earned their "discoverers" Nobel Prizes for path-breaking research.[33] Teaming up with neurologists and emphasizing a physical and increasingly scientific approach to mental health, psychiatry, at least

30 Randall MacLean, "Procedures to be Followed in Obtaining the Admission of a Patient to a Provincial Mental Hospital" *Alberta Medical Bulletin* (1942) 7 (4) p. 23.

31 Edward Shorter, *A History of Psychiatry: From the Era of the Asylum to the Age of Prozac* (New York: John Wiley & Sons Inc., 1997); Joel Braslow, *Mental Ills and Bodily Cures: Psychiatric Treatment in the First Half of the Twentieth Century* (Berkeley: University of California Press, 1997); and Jack D. Pressman, *Last Resort: Psychosurgery and the Limits of Medicine* (Cambridge: Cambridge University Press, 1998).

32 Shorter, 190.

33 For example, Julius Wagner-Jauregg introduced the therapeutic application of malaria among patients with dementia paralytica and for that received the Nobel Prize in Physiology or Medicine in 1927; and, Egaz Moniz who developed the "burr-hole" technique for lobotomies received the Nobel Prize in Physiology and Medicine in 1949.

according to Shorter, realigned itself with its biological roots after a brief "hiatus" with Freudian psychoanalysis.[34]

The development of these kinds of interventions, however, grew out of a culture of experimentation in psychiatry fuelled by a sense of desperation, overcrowded institutions, and a growing feeling of disillusionment among psychiatrists with their inability to cure and rehabilitate patients deemed mentally ill. This desolate situation affected patients facing a lifetime in an institution, or alienation in a community, and it also affected psychiatrists whose reputation as a medical sub-specialty remained in jeopardy as the discipline languished separate from the triumphs and progress felt in other fields of medicine. These features of psychiatry helped to pave the way for experimentation within mental hospitals in search of rehabilitation as well as more sophisticated explanations for disorders, drawing especially upon recent developments in biochemistry, neurology and endocrinology.

Within this context, psychiatrists and staff working with patients confined under the Sexual Sterilization law felt relatively free to extend this approach as they explored therapeutic and experimental objectives within institutions. Combining elements of fiscal conservatism, psychiatric experimentation and eugenics, these three imperatives came together in post-World War Two Alberta in a manner that justified the continuation of the eugenics program as part of the broadening scope of psychiatric research, treatment and social policy.[35]

Indeed, under the dual influences from the provincial government to reduce the numbers of people living in the institution and from the medical community to devise more sophisticated etiological and prognostic studies, the Albertan psychiatrists were under pressure to experiment. By the 1950s the testing of psychopharmacological agents amplified this pressure and drug trials increased in the institutions under the mandate of the Eugenics Board. For example, the superintendent of the Provincial Training School, Leonard Le Vann, gauged the reactions of children in the institution to various treatments for schizophrenia and developed a hypothesis that low-grade defectives were in fact exhibiting a kind of proto-schizophrenia. Combining elements from psychodynamic or Freudian theories and older notions of schizophrenia as a disease of adolescence, Le Vann suggested that mental deficiency was actually a remnant of an earlier evolutionary species. The implications of this research were significant for the eugenics program if he could establish a

34 Shorter, 145.

35 This idea played out in California too, see: Braslow, *Mental Ills and Bodily Cures*, Chapter 3.

clinical relationship between psychotic disorders and mental deficiency, especially since a diagnosis of mental deficiency meant that a patient's consent was no longer necessary before proceeding with a sterilization operation.

Dr. Leonard Le Vann had arrived in Alberta in 1949 after graduating from medical school in Edinburgh, Scotland in 1943. He had had a short stint as the staff psychiatrist in Brandon at the Manitoba Mental Hospital before moving to Alberta and working at the Provincial Training School. At that time, the University of Alberta also appointed him as an adjunct lecturer in the Department of Psychiatry.[36] The College of Physicians and Surgeons of Alberta recognized his expertise in psychiatry, granting him a license to practice and ultimately appointing him superintendent of the Provincial Training School, from which position he oversaw all operations and activities related to the institutionalized children at the Training School.

During court proceedings in the 1990s when patients later attempted to sue the provincial government for sterilization abuses, it became clear that Le Vann did not indeed have specialist training in psychiatry, though he did have considerable experience in this field. Correspondence from the 1970s with officials in Alberta's Department of Health indicated that they understood this gap in his training and paid him as a specialist as an acknowledgment of his expertise, but did not extend the title of specialist to him because he had not completed the requisite academic training.[37] Upon further investigation of his qualifications it appeared that he had obtained the British equivalent of a Licentiate of the Medical Council of Canada, or a medical degree, but had also often signed his letters using the initials LRFPS, which was neither a recognized qualification nor could the subsequent legal committee even discern its meaning. He had also claimed to be a fellow of the Royal Society and the London College of Medicine, a designation that had no correlation in Alberta and did not convey any additional specialization in psychiatry. In spite of his lack of qualifications as a psychiatrist, he nonetheless conducted experiments on children at the Training School and even signed his letters for a time as "the clinical psychiatrist," during the early 1950s when he worked at the Red Deer Guidance Clinic. In 1960 he became the superintendent of the Provincial Training School which gave him unprecedented access to the entire institutional population, although he had still failed to meet the formal requirements for practicing psychiatry in the province.[38]

---

36 Robert Lampard papers, personal donation, "Medical History of Dr. L.J. le Vann," curriculum vitae.

37 University of Alberta Archives, Leilani Muir files, Undertaking #11.

38 Ibid., Undertaking #11, p. 8.

Regardless of the observation in hindsight that Le Vann did not have the required prerequisites to practice psychiatry, during the 1950s and 1960s he gained distinction for his research and leadership within the field of children's mental health in the province. He was one of the few staff members who regularly conducted research and published his results in medical journals. His publications reveal Le Vann's interest in somatic or bodily therapies, physical interventions as keys to curing mental ailments. His support for sexual sterilizations stemmed from this broader fascination with physical therapies.[39]

In one of his first articles based on his work in Alberta, Le Vann considered the presence of hallucinations within patients diagnosed with manic depression, and recommended electro-convulsive shock therapies for controlling these symptoms. Although this practice was gaining currency within psychiatry, his case analysis revealed his detection of co-mingling symptoms. He described the example of Anne P., a 29-year old, single girl who had religious hallucinations involving the presence of the Virgin Mary and Jesus Christ. Her parents did not speak English, and the "girl" had masturbated since the age of 13. Le Vann recommended two shock treatments every two days and found the girl's mood changed positively and that she recognized the "silliness" of listening to these voices, though she still had auditory hallucinations. The second case involved Colin I., who was a 55-year old married man who also had auditory hallucinations, though Colin's were more paranoid in nature and revolved around officials conspiring to harm him. Colin had admitted to "severe drinking bouts," which Le Vann felt exacerbated the presence of voices in his head. After two rounds of electric shocks, the intruding voices disappeared and Colin claimed that he wanted to return to work as a bricklayer.[40] Le Vann's clinical case notes in this article reveal his interest in the social and moral or environmental triggers underlying severe psychiatric problems. In the case of Anne, her masturbation and possibly her unmarried and non-English status raised the likelihood of her obtaining a poor intelligence quotient and consequently coming to the attention of Le Vann. In both cases, electric shocks appeared to provide relief.[41] These two cases helped to set in motion a number of studies

---

39 Joel Braslow found the same thing in California, and it is quite likely the Albertan psychiatrists borrowed directly from these examples. See in particular, Braslow, *Mental Ills and Bodily Cures*, chapter four, which discusses sterilizations at Stockton Hospital.

40 L.J. Le Vann, "The Evaluation of Auditory Hallucinations," *Alberta Medical Bulletin* (1950) 15(1), pp. 25–26.

41 For more on the history of Electro-convulsive therapies, see: Edward Shorter and David Healy, *Shock Therapy: A History of Electroconvulsive Treatment in Mental Illness* (New Brunswick: Rutgers University Press, 2007).

that Le Vann later elaborated on with his firm belief that genetics and biology played fundamental roles in dictating human behavior. While mental health researchers continually emphasized the role of the environment in shaping responses, Le Vann connected biology and conduct in a manner that justified the surgical interventions. He developed this theory primarily by studying children in the Provincial Training School in Red Deer in the post-war period.[42]

During his early studies in Alberta, Le Vann claimed that: "indeed the picture of comparison between the normal child and the idiot might almost be a comparison between two separate species. On the one hand the graceful, intelligently curious, active young homo sapiens, and on the other the gross, retarded, animalistic, early primate type individual."[43] On this basis he did not agree with his contemporaries that schizophrenia represented a deterioration of the mind, but instead represented a "prototypical homo sapiens," which Le Vann called a "congenital schizophrenic." The implications of his theory bolstered the genetic argument that mental illness was in fact inheritable, and therefore individuals with mental illnesses, even of the psychotic type, were candidates for sterilization.[44]

Some of his colleagues supported this assertion that schizophrenia may indeed be related to mental deficiency, and thus present in children before the typical onset of psychotic symptoms characteristic of schizophrenia in the adolescent years. James Russell Grant, for instance, further explored the dual existence of symptoms of deficiency and schizophrenia as a syndrome called "propfschizophrenia," a term coined by American colleagues Bromberg and Schilder.[45] In the Albertan study, Grant relied on a preliminary hypothesis to

---

42 Some of his early work focused on alcoholics, but he later developed a greater focus on childhood mental illnesses. See: L.J. le Vann, "A Clinical Survey of Alcoholics," *Canadian Medical Association Journal* (1953) 69, pp. 584–588.

43 This conceptualization appears consistent with theories of recapitulation introduced by Ernst Heinrich Phillipp August Haeckel, a German biologist who built upon Charles Darwin's work in describing evolutionary development. Since Haeckel's "biogenetic law" states that "ontogeny recapitulates phylogeny" it is possible that Le Vann inferred from this that the "idiot" child genetically "acts out" his/her phylogenetically-acquired traits. Thus, seen from a pseudo-scientific standpoint, they (and their ontogeny) could be that of "less-evolved" children and they could be more easily viewed as the "early primate type individual" which he appears to have thought of them as. I am grateful to Adam Montgomery for drawing this connection to my attention.

44 L.J. Le Vann, "A Concept of Schizophrenia in the Lower Grade Mental Defective," *Canadian Medical Association Journal*, 1950, 469–472 (esp. 469, 472).

45 As cited in James Russell Grant, "Propfschizophrenia: A Clinical and Physiological Essay," *Alberta Medical Bulletin* (1956) 21(3), p. 6 [6–9]; W. Bromberg, "Schizophrenic-like

prove the existence of schizophrenia among the population of low-grade mental defectives.[46] He tested this theory on 100 children and young adults who had been classified with high to low-grade IQs. Grant received permission from the Director of Mental Health, Randall McLean, and support from superintendent and colleague Le Vann to conduct the study. The local ward nurses administered the drugs to the institutionalized patients. The youngest children, ages 5 and 6, became flushed and tended to fall asleep within a few hours. Some of the older children and young adults complained of visual disturbances and dry mouth from the drugs.[47] Grant concluded that an overwhelming majority of the cases tested showed signs of underlying schizophrenia and recommended further studies on children to better understand the hereditary underpinnings of degenerative mental deficiencies.[48]

In an attempt to elaborate on this concept, by the mid-1950s as anti-psychotic drugs became increasingly available for controlling symptoms associated with schizophrenia, Le Vann began administering anti-psychotics, including trifluoperazine and dihydrochloride to the children living at the Provincial Training School, effectively treating them for what he believed would develop into schizophrenia. He maintained that this approach was critical because most of the children had been admitted to the hospital with very limited communication and social skills and often came to the residential school on account of "abnormal or hyperactive behavior" at home.[49] The newly available tranquilizing drugs allowed for substantial changes within the institutional setting, including the freedom to "abolish restraints, to open ward doors and to institute

---

Psychosis in Defective Children," *Proceedings and Addresses of the Annual Session* American Association for Mental Deficiency (1934) 39, pp. 226–257; and P. Schilder, "Reaction Types Resembling Functional Psychoses in Childhood on the Basis of an Organic Inferiority of the Brain," *Mental Hygiene* (1935) 19, 439–446.

46 The blood test checked for levels of atropine in the blood, as hypothesized by Abram Hoffer in A. Hoffer. "Effect of Atropine on Blood Pressure of Patients with Mental and Emotional Disease," A.M.A., *Archives of Neurology and Psychiatry* (1954) 71, pp. 80–86; and A. Hoffer, "Objective criteria for the diagnosis of schizophrenia exemplified by the atropine test." Confin. *Neurological* (1954) 14, (6) pp. 385–390. Hoffer's articles discuss the existence of psychosis but do not make any links to mental deficiency or degeneration, as is suggested by the Albertan study.

47 Grant, 8.

48 For more on theories of degeneracy, see: Daniel Pick, *Faces of Degeneration: A European Disorder, c. 1848–1918* (Cambridge: Cambridge University Press, 1989).

49 L.J. Le Vann, "Short Communications: Trifluoperazine, Dihydrochloride: An Effective Tranquilizing Agent for Behavioral Abnormalities in Defective Children," *Canadian Medical Association Journal* (1959) 80, p. 123 [123–124].

educational and training procedures," in all but the "hard core of mentally defective children who have not responded successfully to therapy with chlorpromazine" or other tranquilizing agents.[50] In a study of thirty-three patients, who ranged in age from five to forty years, Le Vann concluded that the use of tranquilizing anti-psychotic drugs "does make considerably easier the care of the hyperactive mentally deficient children."[51] Where previously these cases had been difficult to manage, and posed administrative challenges, the application of tranquilizers improved the capacity for nursing care. In addition to the improved conditions for staff, the results also further encouraged Le Vann to continue exploring the relationship between mental deficiency and schizophrenia.

Le Vann believed that he was on the cusp of making a significant contribution to psychiatry through his focus on childhood disorders. By 1960 he published an article where he boldly placed his own work alongside that of Eugene Bleuler, who had first coined the term schizophrenia in 1908,[52] and the work of the Maudsley Hospital, which had begun focusing on childhood mental health after the Second World War. Le Vann felt that the Training School at Red Deer offered a suitable location for contributing to this growing field of study with its rather closed staff and patient population alongside extensive resources for vocational, occupation and recreational training.[53]

His later studies attempted to demonstrate a link between hyperactivity, learning disabilities, schizophrenia and mental deficiency. In an effort to show the relationships among these categories, he administered a number of new anti-psychotic and tranquilizing substances to his young patients, surmising that a typical response corresponded to a specific disease category. Given the range of disorders that Le Vann identified within this spectrum of mental deterioration, his subjects more often tended to be boys. The ratio of girls to boys in the initial assessment revealed an overwhelming preponderance of boys, with twelve boys and only four girls in his first study. The diagnoses of the children included: juvenile schizophrenia (one girl and two boys); two autistic boys; one boy with epilepsy and obsessive-compulsive disorder; one impulsive and emotionally unstable girl; one anxious and withdrawn girl; one boy with behavioral explosivity; and one abnormally small boy being treated as an outpatient for anxiety related concerns.[54]

50 Ibid., 123–124.

51 Ibid., 124.

52 Edward Shorter, *A History of Psychiatry*, 104.

53 L.J. Le Vann, "A Pilot Project for Emotionally Disturbed Children in Alberta," *Canadian Medical Association Journal* (1960) 83(10), 525.

54 Ibid.

In Le Vann's research he examined the children's physiological responses and suggested that changes due to the use of psycho-sedatives helped to prove his claim that the etiology of mental retardation and deficiency was in fact related to schizophrenia. Under the Alberta laws, Le Vann did not require additional consent from the students or their guardians to conduct these studies. In 1961 he proudly claimed, in "the mental hospitals and residential schools for retarded children these drugs have revolutionized psychiatric care by the realization of the 'open door' policy and also by diminishing the need for the more hazardous type of physical therapies such as electro-shock and insulin-shock."[55] He similarly began administering anti-convulsant drugs to epileptic children, claiming here too that the drugs helped to prove the existence of an underlying mental disease, and that with the use of these drugs that controlled outbursts patients could be returned to the community. By the mid 1960s, Red Deer housed over 800 children classified within the category of mental retardation, and maintained a waiting list of another one thousand children whose families sought accommodation.[56] In one retrospective review of the sexual sterilization program in Alberta, a former Board member recalled the "justification for sterilization was sometimes offered to ease off grounds supervision and off-grounds visits."[57] The conditions for mental health care in Alberta, as elsewhere, remained overcrowded, under-resourced and neglected. Le Vann's experimental theory that tranquilizing drugs had the potential to reduce the number of patients requiring custodial care generated widespread interest from political authorities.

The overall goal from Le Vann and the Provincial Training School staff, in combination with the Ministry of Health, remained fixed upon returning the children to the community. Most of the children would not necessarily return to their families, but instead were released as adults with some modest training and skills intended to help them find employment. To that end, Le Vann

55 L.J. LeVann, "Thioridazine (Mellaril) a Psycho-sedative virtually free of side-effects," in *Alberta Medical Bulletin* (1961) 26 (4), p. 144. His thesis was later criticized by American researchers who found that these drugs had serious side effects, especially for children with mental retardation. See: Alan A. Baumeister, Jay A. Sevin and Bryan H. King, "Neuroleptics," in *Psychotropic Medication and Developmental Disabilities: The International Consensus Handbook* (eds) S. Reiss & M.G. Aman, (Columbus, OH: Ohio State University, 1998), 133–150.

56 Marjorie Montgomery Bowker, "Brief on Mental Retardation for a Special Legislative and Lay Committee appointed by the Legislative Assembly of Albert for the Study of Preventative Health Services in Alberta," December 1965, 3.

57 D. Gibson "Involuntary sterilization of the mentally retarded: a western Canadian phenomenon." *Canadian Psychiatric Association Journal* (1974)19(1), p. 61.

held steadfastly to his belief that tranquilizing medications aided in the management of mentally deficient children and was critical for their social training. He suggested that "the cardinal symptoms in the children must be relieved and, because their education must continue, the children must remain alert."[58] To that end, Le Vann agreed to test a variety of newly available, experimental psychopharmacological drugs on the children in his care.[59] For example, in 1969 he published the results of his research with haloperidol, an experimental drug, to determine whether it may improve mental alertness in children with behavioral disorders. Here again, there was an emphasis on boys, using a study group of 61 boys and 39 girls, approximately half of whom were under twelve years of age and the other half were described as adolescents.[60] In particular he found that haloperidol helped to reduce the levels of hyperactivity, assaults and self-injury. Like other major tranquilizers, however, haloperidol produced side effects that created tremors and muscular rigidity. Le Vann nonetheless felt that the improved capacity for calm conduct and mental alertness overwhelmed these relatively minor concerns.[61]

In a follow-up study, Le Vann reported that by comparing reactions to haloperidol and chlorpromazine, the latter of which had already enjoyed commercial and therapeutic success, haloperidol performed more successfully when addressing problems of aggression, hostility, and impulsivity.[62] Although he did not specify the gender break-down used in that study, his increasing focus on aggression and destructive activity continually drew on the descriptions applied to boys within the Training School, and placed the male residents at a much higher rate of experimentation as a result.[63] Another contemporary study using anabolic compounds relied on a patient group of 74 boys and

58 L.J. Le Vann, "A New Butyrophenone: Trifluperidol: A Psychaitric Evaluation in a Pediatric Setting," *Canadian Psychiatric Association Journal* (1968) 13 (3) p. 271.

59 Triperidol, for example, was a product of McNeil Laboratories, and released to Le Vann only for investigational purposes.

60 L.J. Le Vann, "Clinical Note: Haloperidol in the Treatment of Behavioral Disorders in Children and Adolescents," *Canadian Psychiatric Association Journal* (1969) 14 (2), p. 217.

61 Ibid., 219.

62 L.J. Le Vann, "Clinical Comparison of Haloperidol with Chlorpromazine in Mentally Retarded Children," *American Journal of Mental Deficiency* (1971) 75(6), p. 719.

63 For more on this way in which gendered behaviors in children were understood during this period, see: Matthew Smith, "Putting Hyperactivity in its Place: Cold War Politics, the Brain Race and the Origins of Hyperactivity in the United States, 1957–1968," in *Locating Health: Historical and Anthropological Investigations of Health and Place* (eds) Erika Dyck and Christopher Fletcher (London: Pickering & Chatto Ltd., 2011), 57–70.

26 girls ranging from four to six years of age.[64] A study of Neuleptil, described as "a new phenothiazine" examined the drug's effect on sociopathic conduct, including aggression and self-destructive activity; Le Vann used 19 boys and 11 girls to investigate the influence of this drug on controlling behavior.[65]

Le Vann was not alone in recognizing the relationship between boys and aggression. Ilina Singh has argued that during the 1950s and 1960s drug advertisements for hyperactivity concentrated exclusively on boys. In her view "between 1955 and 1998 all children pictured in advertising campaigns for minor tranquilizers and stimulant drugs are boys."[66] Singh further explained that in the post-war period "a new generation of experts in boys' psychology have argued that we are 'medicalizing boyhood' with stimulant drugs in a contemporary setting which is so fast paced and competitive that boys are forced to live up to a 'culture of masculinity' much before their time."[67] Although Singh's study concentrated on otherwise "normal" boys, she helped to demonstrate the contemporary relationship between medicalized hyperactivity and boys. While Le Vann's children subjects were not expected to become middle-class professionals, he nonetheless worked within this cultural context in assessing the children deemed mentally deficient in Alberta. His concentration on hyperactivity and learning abilities shifted his focus towards boys within the institution and made them more likely subjects of his psychiatric experiments, while the boys' bodies functioned as the instruments for the pharmaceutical trials.

Le Vann kept in close contact with the Eugenics Board. In 1949, when he joined the staff at the Provincial Training School, he was welcomed by the Board Chairman and encouraged to take note of the procedures so that he could assist them in detecting potential candidates for sterilization from his new post.[68] From that point onward, he became a regular attendee of the eugenics board meetings, although he was never formally a member of it. Two years later, Le Vann's activities appeared within the Board's records, first for

64 L.J. Le Vann and R.E. Cohn, "Clinical Evaluation of Norbolethone Therapy in Stunted Growth and Poorly Thriving Children," *International Journal of Clinical Pharmacology, Therapy and Toxicology* ((1972) 6(1), p. 55.

65 L.J. Le Vann, "Neuleptil – A new phenothiazine," *Modern Medicine of Canada* (1972) 27(9), p. 1.

66 Ilina Singh, "Not Just Naughty: 50 Years of Stimulant Drug Advertising," in *Medicating Modern America: Prescription Drugs in History* (eds) Andrea Tone and Elizabeth Siegel Watkins (New York: New York University Press, 2007), p. 137.

67 Ibid., 138.

68 Eugenics Board minutes, Meeting 157, November 17, 1949, p. 2.

helping to arrange for consent from a parent, and next to urge the Board to hold off on sterilizing children until they had reached adolescence. He "felt that the sexual tendencies of those [children] presented may be better evaluated at that time."[69] If a trainee was up for parole, however, the Board intervened and reviewed the case regardless of the age.

The confluence of Le Vann's research and interest in sexual sterilization came together in at least one detailed case within the eugenics program. In 1963 the Board evaluated a young man described as "borderline intelligence quotient – schizoid personality." Because his diagnosis did not include mental deficiency, his sterilization could not be supported unconditionally. The Board indicated that "sterilization cannot be recommended on the presently given diagnosis, but, if epilepsy or simple schizophrenia can be entertained, then the operation subject to the consent of parent would be possible."[70] As in other cases, the Board recommended re-evaluating the patient to see whether or not the diagnosis or intelligence quotient might change to accommodate a waiver of consent. Le Vann's studies on the relationship between childhood mental deficiency and adult mental disorder or degenerative and hereditary disorders promised to provide clinical evidence to prove this link. Retesting intelligence might then ensure that one of the largest groups of patients left in the institutions, the psychotics, could be handled and ultimately discharged under the terms of the Sexual Sterilization Act.

Conclusion

In 1998 Ken Nelson gathered alongside his former trainees from Red Deer at the Legislative Assembly of Alberta to watch Premier Ralph Klein respond to the public demands for compensation for the men and women who had suffered under the eugenics program. Ken had been sterilized in 1960 when he was a boy living at the Provincial Training School in Red Deer.[71] "'I stood in the legislature gallery…and watched the premier of Alberta take the rights away from 700 people,' said Nelson, his voice straining with emotion, after the

69 Eugenics Board Minutes, Meeting 168, February 9, 1951.

70 Eugenics Board Minutes, Letter from Secretary to the Board, Mrs. E.S. James to Dr AD MacPherson, Medical Superintendent, Provincial Mental Institute, Edmonton, June 24 1963.

71 Larry Johnsrude, "Province revokes rights; Government opts out of Charter, limits sterilization victims' right to sue for compensation; Alberta's Sterilization Solution," *Edmonton Journal* Mar 11, 1998, A1.

Conservatives tabled legislation to limit the rights of people sterilized in government-run institutions from 1928 to 1972 to sue for compensation."[72] Alberta's premier, Ralph Klein attempted to invoke the "not-with-standing" clause in the Canadian Charter, which permitted for an assertion of provincial autonomy, after "he noted the sterilizations, part of a government-run eugenics program, occurred before many of today's taxpayers were born. 'We think it will be in the best interests of the people to limit the liability for something we were not responsible for many, many years ago,'" Klein told the *Edmonton Journal*.[73]

The historical pathologization of boys and men as potentially more aggressive and sexual had made them especially vulnerable to psychiatric and genetic experiments aimed at confirming links between sexual behavior, low intelligence and schizophrenic disorders. The Eugenics Program in Alberta created an opportunity for eugenicists and psychiatrists to work together in a culture of medical experimentation explicitly aimed at reducing the numbers of institutionalized individuals, while at the same time allegedly controlling the spread of mental diseases and so-called defective families. The financial and professional justifications for these studies ensured that a number of institutionalized children became the subjects of experiments under a banner of progress and under the auspices of the eugenics movement. Those same children as adults were then denied compensation on the basis that those experiments were part of the past.

Alberta's unique situation regarding a lack of consent provisions for individuals whose intelligence quotients fell below 80 thereby created specific and fertile conditions for experimentation. Individuals who fell into this category were amongst the most vulnerable and had few if any resources at their disposal to resist either sterilization or experimentation. The subsequent statement from the premier added another layer to the characterization of institutionalized individuals, especially those considered mentally deficient, as suitable subjects for medical experimentation.

---

72 Arnold T. "Wave of anger, outrage greets proposed law: A Time Line," *Edmonton Journal*, Mar 11, 1998, A1.

73 Ibid.

Index

Printed in the United States
By Bookmasters